Environmental Impacts of Sugar Production

The Cultivation and Processing of Sugarcane and Sugar Beet

Environmental Impacts of Sugar Production

The Cultivation and Processing of Sugarcane and Sugar Beet

Oliver D. Cheesman

CABI Bioscience UK Centre
Surrey
UK

CABI Publishing

CABI Publishing is a division of CAB International

CABI Publishing
CAB International
Wallingford
Oxfordshire OX10 8DE
UK

Tel: +44 (0)1491 832111
Fax: +44 (0)1491 833508
E-mail: cabi@cabi.org
Web site: www.cabi-publishing.org

CABI Publishing
875 Massachusetts Avenue
7th Floor
Cambridge, MA 02139
USA

Tel: +1 617 395 4056
Fax: +1 617 354 6875
E-mail: cabi-nao@cabi.org

WWF International (Freshwater Programme)
Avenue du Mont-Blanc 27
1196 Gland, Switzerland
Tel: +41 22 364 9027
Fax: +41 22 364 0526
Website: www.panda.org/freshwater

A catalogue record for this book is available from the British Library, London, UK.

Library of Congress Cataloging-in-Publication Data

Cheesman Oliver.
The environmental impacts of sugar production / by Oliver Cheesman.
p. cm.
Includes bibliographical references (p.).
ISBN 0-85199-981-6 (alk. paper)
1. Sugar crops--Environmental aspects. I. Title.

SB217.C55 2005
636.6--dc22

2004014573

ISBN 0 85199 981 6

Typeset by AMA DataSet Ltd, UK.
Printed and bound in the UK by Cromwell Press, Trowbridge.

Contents

Preface

This is a CABI-WWF co-publication prepared as part of WWF's sustainable sugar initiative. For the purposes of this review, the term 'sugar production' refers to the combined activities of the sugar industry: cultivation of sugar crops (cane and beet) and the primary processing of raw materials derived from these crops. Secondary processing and utilization of primary processing wastes as by-products are also considered. Environmental impacts include impacts on human health, but not (for the purposes of this review) any related to sugar consumption. It is noted that patterns of trade influence elements of sugar production that have an impact on the environment, and trade patterns certainly affect the livelihoods of many connected with the sugar industry. However, trade and livelihood issues are not covered in detail in this review. It is not the main intention to compare cane and beet systems, although comparisons are sometimes made for contextual purposes.

The environmental impacts of the processing (but not cultivation) of sugar crops have been summarized previously by UNEP (1982), and other texts on aspects of sugar production often include some coverage of environmental issues (e.g. Wilson *et al.*, 1996; Keating and Wilson, 1997). However, this appears to be the first attempt to collate and review information on the environmental impacts of sugar production as a whole. It was originally commissioned specifically as a literature review, and its structure and style should be seen in that context. The emphasis is on an environmental perspective, although agronomic priorities are generally acknowledged where appropriate. The one area where a consistent difference in viewpoints has become apparent is in relation to soil quality. From an environmental perspective, any perturbation of (say) soil nutrient balance is seen as degradation; this only tends to be the case from an agronomic perspective when the effect is sufficient to reduce yields.

The relevant literature is extensive, and there are undoubtedly gaps in this review. Many of the specific studies cited here were identified through CAB Direct. CAB Direct is a database of nearly 5 million abstracts from the world's literature on agriculture and applied life sciences, based on material going back to 1973, and managed by CAB International (CABI). CABI also produces searchable databases of abstracts relating to specific aspects of agriculture and particular commodities, including *Sugar Industry Abstracts*. The CAB Direct database contains nearly 125,000 abstracts referring to sugarcane or sugar beet. Whilst a relatively small proportion of these are devoted exclusively to environmental issues, many contain relevant information. Clearly, it has not been possible to read every one of these, much less assimilate them into this review. Some areas are poorly served by the literature. In particular, there appears to be a lack of data on air pollution (and human health impacts) arising from poorly managed aerial application of agrochemicals. Similarly, short-term water pollution events arising directly from an application of fertilizer or pesticide appear not to be reported.

A range of general texts are available covering sugarcane, and this review draws particularly on Bakker (1999) for contextual information on this crop. Unfortunately, the timing of the review has meant that it was prepared without James's updated and revised version of Blackburn (1984) being available as a published source. In relation to sugar beet, the main contextual source used here is Cooke and Scott (1993). In addition to these (and other) general sources, volumes of technical papers on cane cultivation and processing are published regularly, arising from meetings of the International Society of Sugar Cane Technologists (ISSCT), and local associations like the South African Sugar Technologists' Association (SASTA), the Australian Society of Sugar Cane Technologists (ASSCT), the American Society of Sugar Cane Technologists (also ASSCT) and the West Indies Sugar Technologists (WIST). Similar volumes for beet are produced by the Institut International de Recherches Betteravières (IIRB) in Europe, and the American Society of Sugar Beet Technologists (ASSBT). Of course, the extent to which environmental issues are covered by general texts and volumes of technical papers varies considerably.

In broad terms, the literature on environmental aspects of cane sugar production is dominated by contributions from Australia, South Africa and to a lesser extent India and Mauritius. In relation to beet sugar production, most contributions to the literature come from Europe (published in a range of languages) and the USA. This should not be taken to suggest that environmental impacts, or measures to reduce them, are necessarily of greatest significance here, merely that relevant studies from these countries dominate the literature. There is also a danger in extrapolating general trends from the available literature, given its considerable geographical bias.

Oliver D. Cheesman

Acknowledgements

This book is a CABI-WWF co-publication. WWF is grateful to HSBC for its support in the production of this book through the Investing in Nature Programme. For more information please visit www.hsbc.com/hsbc/society.

Despite being unable to fully synthesize the overwhelming amount of material available, or overcome the geographical bias of studies reported, I hope that the broad conclusions of this review are reliable, and the examples cited representative. Whilst I accept full responsibilty for the inadequacies of this review, I am grateful for the assistance and guidance of a number of colleagues at CABI, notably Bibi Ali, Mark Palmer, Nicola Ward, Rebecca Stubbs, Rachel Robinson, Janny Vos, Gretel White and Jim Waller. I am also grateful to Rachel Wiseman, Richard Perkins, Richard Holland and Nasir Mahmood Nasir at the World Wide Fund for Nature (WWF) for their contributions to this review. In addition to some of those already named, a number of people kindly provided comments on an early draft of the manuscript. In this respect I would like to acknowledge David Eastwood, Harold Davis, Ahnand Rajkumar, Mike May, Keith Jaggard, Rene Ng Kee Kwong and Francis Lopez. Whilst this book would have been much weaker without their contributions, their involvement should not be seen as an endorsement of the final product. I am also grateful to Linda Fenwick for her patience and support throughout the protracted preparation of the manuscript.

Foreword

Sugar is a product that most of us consume on a daily basis; indeed it is so common that most people do not give a second thought to where or how it is produced and with what consequences.

Although sugar production contributes to development in many poor countries, by producing employment and reliable incomes for many, there is a range of negative issues associated with its production. These include poor working conditions and child labour, as were highlighted in a recent report by Human Rights Watch. While these are problems typical of farming and manufacturing in many developing countries, they also pose potential risks to multinational companies, such as Coca-Cola or Cargill, which buy a significant part of the approximately 145 million tonnes of sugar produced world-wide each year.

Sugar has hit the headlines in recent years because of health fears, with accusations and counter-claims that too much sugar and fat in people's diets contribute to a growing 'obesity epidemic'. However, despite these warnings, the demand for sugar has continued to rise steadily, increasing by about 70% in total since 1980.

This important commodity also has taken centre stage in international trade arena. In a case filed by Brazil, Thailand and Australia, a World Trade Organization (WTO) panel recently ruled that the EU is illegally dumping millions of tonnes of subsidized sugar on world markets. This ruling on sugar is important, since it will require the EU to take full account of the impacts of its agricultural and trade rules on the rest of the world.

Meanwhile, the environmental impacts of sugar production have been largely ignored. Sugarcane plantations in many tropical and sub-tropical countries have led to perhaps the largest losses of biodiversity of any single agricultural product. Although much of this habitat and species loss is historic, sugar production today – whether from cane or beet – has a wide range of negative impacts on soil, water and air in parts of the world that environmental organizations, such as WWF, have identified as globally important. The Great Barrier Reef off Australia's coast, which suffers from effluents and sediment from sugar farms, is one such case; the Konya basin in central Turkey, a vital nature conservation area, is another where more than 300,000 hectares of sugar beet require between 50 and 80% of the usable water in the basin. This is why WWF has identified sugar as a commodity for special attention.

Sugar is a much more complex matter than the simple spoonful that we add to our coffee or tea might indicate. I welcome Oliver Cheesman's thoughtful analysis, as well as the invaluable collaboration between CAB International and WWF from which this book stems. I believe that this publication will contribute strongly to the emerging body of knowledge on the sustainability of sugar, by laying out clearly the main issues to be tackled, as well as identifying a range of solutions that are already in use. Better understanding of the environmental issues around sugar, and their root causes, is the first step to ensuring greater sustainability for this ubiquitous commodity.

Dr Claude Martin,
Director General, WWF International

August 2004

1

Background

The following information is derived from a range of sources, principally (where not otherwise indicated): Bakker (1999) for sugarcane; Cooke and Scott (1993) for sugar beet; and SKIL (2003) for general information, particularly with respect to processing.

Sugar

Sugars (saccharides) are a family of naturally occurring carbohydrate compounds. They are produced by plants through the process of photosynthesis, which combines carbon dioxide and water to generate oxygen and glucose (one of the simplest forms of sugar, a monosaccharide):

$$6CO_2 + 6H_2O \rightarrow C_6H_{12}O_6 + 6O_2$$

carbon dioxide + water → glucose + oxygen

This reaction is driven by the energy of sunlight, and energy can later be recovered from glucose, in the process of respiration. In order to store glucose, and to provide the physical materials for growth, plants build more complex sugars by chemically linking glucose molecules together. Hence, they form compounds like sucrose, a disaccharide (two glucose molecules joined together), and starch and cellulose, polysaccharides (many glucose molecules joined together). Sucrose may subsequently be broken down, in the process of inversion, into two monosaccharide molecules (one of glucose and one of fructose).

The requirement of photosynthesis for water and sunlight provides two of the main limits on plant growth, and hence on the cultivation of agricultural crops. Crop plants can only be grown effectively in areas where sunlight and water are freely available (and where soils are sufficiently fertile to provide nutrients). Whilst sunlight cannot realistically be provided by artificial means, water availability can be artificially enhanced, through irrigation, and soil fertility can be manipulated with amendments such as fertilizers. Such activities influence some of the most important environmental impacts of agriculture, along with the effects of clearing natural habitats for cultivation in the first place, and practices such as the control of weeds, pests and diseases.

As well as providing dietary sources of energy, sugars have been widely exploited by humans as sweeteners and preservatives. From very early times, fruit (rich in fructose) was prized for its sweetness, as was honey (rich in fructose and glucose). However, it proved difficult to extract sucrose from plants in significant quantities, and for many years this form of sugar was an expensive commodity, particularly in Europe. However, historical developments in the exploitation of sugarcane and sugar beet substantially increased access to refined sugar, and demand continues to grow. Demand from developed countries may slow down in the long term, however, as it is already high, health concerns have been expressed over consumption of refined sugar, and alternative sugars (e.g. isoglucose from cereals or high fructose corn syrup from maize) and artificial sweeteners (e.g. saccharin, aspartame and cyclamate) are increasingly investigated.

History of sugar production

Studies suggest that sugarcane was first domesticated in Papua New Guinea, from where its cultivation spread throughout the Pacific and to India, where the first crude forms of sugar were produced around 2000 years ago. Western Europeans probably encountered such sugar during the crusades, and a lucrative trade developed as a consequence. Sugar was an expensive commodity, and Europe sought to obtain its own supplies. However, the plant only thrives in tropical areas (within about 35° of the equator). This led Portuguese settlers in Brazil to initiate cane cultivation in the 16th century (probably the first example of plantation agriculture – Courtenay, 1980). By the 17th and 18th centuries, when European colonial powers had established huge slave-based sugarcane plantations in the Caribbean, sugar held a central position in the world economy. It is difficult to transport refined sugar over long distances, and cane sugar was (and still is) imported to Europe in raw form and refined locally. Political upheaval in the Napoleonic period led to French blockades of sugar imports into Europe. These, along with slave revolts on Caribbean cane plantations, drove a search for alternative sources.

Sugar beet was first identified as an alternative source of sugar, and one that could be grown in temperate regions, in the mid-18th century. However, processing methods were not developed, and beet was not grown widely in Europe until the 19th century and the blockades of cane sugar imports. Plant breeding, mechanized cultivation, the use of fertilizers, technical improvements in beet processing and trade barriers to cane all developed very quickly, and by the beginning of the 20th century most of the world's sugar was produced from beet. An international conference led to the relaxation of trade restrictions, and a recovery of the predominance of cane as a sugar source.

By the end of the latter part of the 20th century, worldwide sugar consumption had increased over 100-fold in 150 years. World production has increased steadily in recent decades (see Fig. 1.1). Between the mid-1960s and 1990s, the largest expansion of sugar production occurred in India (from 3 to 15 Mt) and Brazil (from 5 to 10 Mt), but other countries have seen a decline in production, including Barbados and Cuba (Hartemink, 2003). Sugar is produced in around 120 countries, with cane sugar production (about 70% of the global total – see Fig. 1.1) concentrated in tropical areas, and beet in temperate

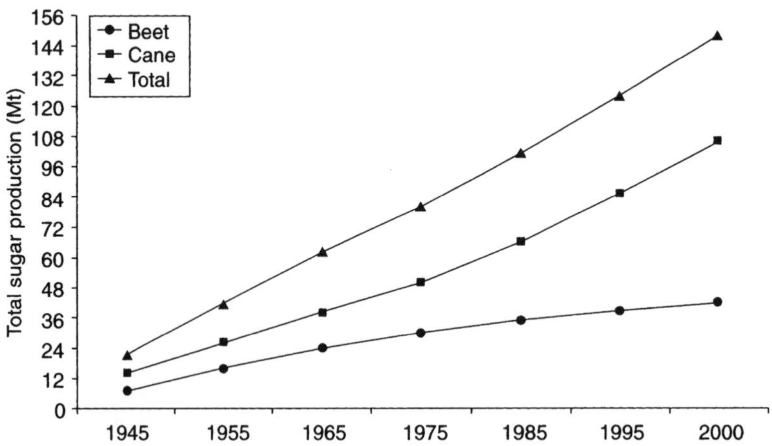

Fig. 1.1. World sugar production from cane and beet 1940–2000 (after Draycott and Christenson, 2003).

regions; in some areas, including parts of Spain, North Africa, Iran and Pakistan, both crops may be grown. In addition, a small quantity of sugar is derived from the sap of maple trees (Canadian Forest Service, 1995), this industry being more or less confined to parts of the north-eastern USA and eastern Canada.

Increasing global sugar production reflects changes in cultivation systems (including those associated with mechanization, use of agrochemicals and plant breeding), increased areas under sugar crops in some countries and technical advances in the processing of sugar crops. The influence of such developments on environmental aspects of sugar production are complex; while individual activities may have become less polluting, the intensification of production may have amplified negative effects. The newest technological advances, notably in the development of transgenic crops, raise important questions in relation to environmental protection. Systems of cultivation vary considerably between localities, with smallholder versus plantation production of cane, for example. Cultivation systems and associated socio-economic factors influence patterns of productivity, uptake of technologies and likely environmental impacts. Sugar production and its environmental context in two very different situations (Australia and Papua New Guinea, respectively) are outlined in Boxes 1.1 and 1.2.

Box 1.1. Sugar production in Australia.

> The rewards have undoubtedly been large . . . but 'success' in the context of the Australian sugar industry has been achieved with substantial cost to the natural environment (Johnson *et al.*, 1997)

Australia is currently amongst the largest sugar producers in the world. After some 70 years of increasing productivity, a yield plateau affected the industry between about 1970 and 1990, but productivity began to increase again shortly thereafter (Garside *et al.*, 1997b). In 1997, the country produced 5.6 Mt of raw sugar from 40 Mt of cane, an increase of around 70% on a decade earlier (Ballantyne, 1998). Average annual cane yields are around 75 t/ha (Hartemink, 2003). Each 100 t of cane processed is estimated to generate 14.3 t raw sugar, 27.2 t bagasse, 5.2 t filter cake, 2.6 t molasses and 50.7 t waste water. Bagasse is generally used to fuel low efficiency boilers in cane mills (and, in some cases, to generate surplus electricity for export), molasses tends to be sold at low value for animal feed or as distillery feedstock, and filter cake and boiler ash are returned to cane fields as soil amendments. However, there is an increasing interest in more effective utilization of by-products (Allen *et al.*, 1997).

Much of Australia's cane cultivation is centred on family-owned farms, and sociological aspects of the industry are considered by Lawrence and Gray (1997). In 1998, Ballantyne (1998) estimated that the average Australian cane farm was around 70 ha in area, producing over 6000 t of cane per year, although Johnson *et al.* (1997) suggest that the average farm size per grower is increasing. Unlike many sugar producing countries, Australia exports a substantial majority of its output, making sensitivity to the global market a key economic characteristic of the industry here. However, this is just one of many factors influencing profitability of operations for Australian cane growers, which include: farm size, soil type, climate and weather, availability of water for irrigation, farming practices, pests and diseases, level of debt, currency fluctuations, government policies and charges, costs of fuel, labour and agrochemicals (Chapman and Milford, 1997; Ballantyne, 1998). Cane growing in Australia is highly regulated; government and industry regulations in Queensland effectively control the locations of farms, numbers of growers, areas cultivated and quantities of sugar produced (Johnson *et al.*, 1997).

The distribution of cane growing in Australia is largely determined by climatic limitations on production (which are examined by Muchow *et al.*, 1997). The Queensland industry accounts for 95% of Australia's raw sugar output (and almost all of its exports), while a separate industry in the northern part of adjoining New South Wales accounts for most of the rest. A new, small centre of sugar production was developed in the 1990s in the Ord River region of Western Australia. The location of the main areas of cultivation brings the industry into potential conflict with sensitive environments, being concentrated along 2000 km of the subtropical east coast, mostly in high rainfall areas with important river systems, often adjacent to World Heritage-listed rainforests or close to the Great Barrier Reef (Johnson *et al.*, 1997; Ballantyne, 1998).

continued

Box 1.1. *Continued.*

Arthington *et al.* (1997) note that data on environmental impacts are more readily available in Australia than in most cane sugar producing countries. A number of studies have highlighted impacts of cane cultivation and the link (in a number of cases) between these, agricultural practices and yield declines. For example, Wood (1985) found a decline in soil quality under cane in an area (Herbert Valley) where yield was declining, and identified excessive cultivation, insufficient fallowing, the burning of crop residues and application of large quantities of N fertilizers as likely contributory factors. Such studies have led to widespread changes in cane cultivation methods in Australia, including reduced tillage, and a shift from preharvest cane burning to green cane harvesting/trash blanketing (Dick and Hurney, 1986). More recently, there has also been investigation of the potential impacts of sugar cultivation on downstream environments. Arthington *et al.* (1997) note that three consequences of sugarcane production have been particularly detrimental: (i) extensive clearance of riparian and flood-plain habitats; (ii) soil erosion and stream sedimentation; and (iii) pollution of water bodies with agrochemicals. Rising water-tables and salinization, and acid sulphate soils are also acknowledged environmental problems, and Meyer (1997) and Arthington *et al.* (1997) also draw attention to hydrological impacts related to irrigation, leading to over-commitment and degradation of many river systems. Whilst there is often little solid evidence to connect cane growing with downstream impacts (Johnson *et al.*, 1997), considerable concern remains over potential effects on sensitive ecosystems like the Great Barrier Reef (Thorburn *et al.*, 2003a). Studies of sugar industry impacts in Australia have also focused recently on effects in relation to the dynamics of greenhouse gases (Weier, 1998).

Sustainable development has been an increasing concern in Australia (as elswhere), as competition for natural resources among conflicting uses becomes more intense and environmental concerns are more widely expressed, and the local sugar industry has been identified as one focus for this (e.g. Johnson *et al.*, 1997; Syme *et al.*, 1997). In recent years, the Australian sugar industry has shown an increasing interest in developing programmes for reducing environmental impacts, for example with the development of the CANEGROWERS' *Code of Practice for Sustainable Cane Growing in Queensland* (CANEGROWERS, 1998; see also Christiansen, 1999), and the launch of COMPASS (Combining Profitability and Sustainability in Sugar) workshops for cane growers (Azzopardi *et al.*, 2002). Other programmes include research into Sustaining Soil and Water Resources (Kingston and Lawn, 2003) and Protecting the Environment (Rayment, 2003), under the auspices of the Cooperative Research Centre for Sustainable Sugar Production (CRC Sugar).

Box 1.2. Sugar production in Papua New Guinea.

Papua New Guinea is considered to be the centre of origin of sugarcane, but until the early 1980s it relied on imports to meet its domestic sugar demand. Although sugarcane is grown by farmers on a small scale, and wild species (*Saccharum robustum* and *Saccharum spontaneum*) occur along river banks and on fallow land, it is only recently that attempts have been made to establish a commercial sugar industry here (Hartemink and Kuniata, 1996). An area of natural grassland, some forest and swamp vegetation, on the alluvial soils of the Ramu Valley, was selected for cane cultivation (Chartres, 1981; Hartemink, 2003). An initial planting of just 3 ha on the Ramu Sugar Estate in 1979 was increased to 1592 ha in 1981, 5011 ha in 1983 and had expanded to 7546 ha by 1997 (Kuniata, 2000). Up to four ratoons can be obtained following harvest of the planted cane, and cane yields have varied from 27 to 106 t/ha/year, stabilizing during the 1990s at around 60 t/ha/year (Hartemink, 2003).

Although overhead irrigation is used in the nurseries, cane cultivation in Papua New Guinea is an exclusively rain-fed operation (Kuniata, 2000). Erosion is a threat in some areas, and terraces have been installed to control surface water flow (Hartemink and Kuniata, 1996; Hartemink, 2003). As the crop plant is indigenous, the industry has faced particular challenges from pests, diseases and weeds, most of which are also native and may have co-evolved with the ancestors of the crop plant, and these have been the main factors in determining yields (Hartemink and Kuniata, 1996; Kuniata *et al.*, 2001; Magarey *et al.*, 2002). The most important pest is the stem borer *Sesamia grisescens*, which can cause sugar production losses of up to 18% (Kuniata and Sweet, 1994), and against which an integrated system of control has been adopted (Kuniata, 2000; Lloyd and Kuniata, 2000; Box 2.7). There is evidence that soil fertility is declining on the estate (see Box 6.2), and may become an increasing constraint on yield (Hartemink and Kuniata, 1996; Hartemink, 2003).

In the context of this review, it is worth noting that costs of compliance with environmental regulations are amongst the factors that affect the economics of sugar production, and hence patterns of trade. This is a relatively recent feature of this (as other) industry, and is only part of a complex of factors that influence the market, including sugar quotas, price regulations, subsidies, quality standards, etc. Most sugar is consumed within the country of production and only around 25% of world production is traded internationally. None the less, global sugar trade issues are contentious. As with other commodities, manipulation of markets has long been a feature of the trade in sugar. For example, when competition with other crops depressed cane planting in the early 1600s, statutes were simply introduced restricting the cultivation of ginger, tobacco and other rival commodities (Alexander, 1985). Currently, sugar prices are heavily subsidized in the European Union (EU). Consequently, over 5 Mt/year of refined beet sugar is exported annually, while 1 Mt of raw cane sugar is imported. This latter activity is a form of overseas aid, which is also practised by the USA. The EU's perceived overproduction has been subjected to General Agreement on Tariffs and Trade (GATT) requirements, which will probably see a substantial cut-back in production over the next few years.

Sugar products

Numerous types of sugar and syrup are available to domestic and industrial users. White (refined) sugar is essentially pure sucrose, although different manufacturers produce crystals of different sizes, and this leads to some apparent differences in characteristics. For example, smaller crystals dissolve more readily, and may therefore seem to be sweeter. Brown sugars come in many different styles but are essentially of two types: sticky or free-flowing. Historically, sticky brown sugar was essentially raw cane sugar, and this type of product is still available locally based on traditional or small-scale manufacturing systems (for example, 'jaggery' or 'gur' from India – see below). However, most brown sugar of either type is now made by mixing a refined (or at least purified) sugar with a suitable syrup. The colour of the sugar and the syrup determines the colour of the final product, and the ratio of syrup to sugar (and the effect of drying treatments) determines whether the product is sticky or free-flowing. Syrups range from pure sucrose solutions to heavily treated products incorporating flavours and colours. Refiners' or 'golden' syrup is a sugar solution containing inverted sucrose, the glucose and fructose content helping to ensure that crystallization does not occur during storage. Treacle is a similar product, made from molasses rather than a pure sugar solution.

Sugar consumption and human health

There is increasing concern over the impact of increased consumption of refined sugar on human health. This issue is outside the remit of this study, but it is worth noting that some authors have identified some irony in the idea that agricultural practices (such as cane cultivation), which may have serious environmental impacts, underpin the supply of food products which may impair consumer's health (e.g. Sager, 1995 – with particular reference to the USA and Australia). In the interests of balance, it should also be noted that some recent analyses have challenged many of the perceived health risks associated with sugar consumption (e.g. Clay, 1998).

Cultivation and Processing of Sugar Crops

In terms of cultivation, it is worth noting that sugar crops differ from many others in that the economic product is not derived from the reproductive portion of the plant (e.g. as for cereals harvested for grain), but from the vegetative structures (the stalk of cane and root of beet). Differences between cane and beet cultivation practices reflect differences in the anatomy and physiology of the plants, as well as the geographical and other circumstances under which they are typically grown. There are, however, many similarities in the ways in which the respective crops are processed to recover sucrose. One of the main differences, in this repect, is that cane is generally processed to raw sugar (and then transported, often over transcontinental distances, for refining), whereas beet is typically processed to refined sugar on one site.

Sugarcane

Sugarcane is a tropical grass, a group of *Saccharum* species somewhat resembling bamboo, which stores sucrose in its stem. There may be considered to be two confirmed wild species of *Saccharum* and four domesticated species. The wild forms are *S. spontaneum* (found throughout tropical Oceania, Asia and Africa) and *S. robustum* (restricted to Papua New Guinea and neighbouring islands). The domesticated forms are *S. officinarum* (noble cane, amongst the first forms of 'chewing' cane to be cultivated, but now grown in very few localities), *S. edule* (apparently restricted to Melanesia and Indonesia), *S. barberi* (from which sugar was first manufactured) and *S. sinense*. Virtually all commercial varieties of sugarcane are hybrids obtained by selective breeding based on the above species. Such breeding programmes assisted the sugar industry in increasing cane sugar yields from around 1–1.5 t/ha to around 8–17 t/ha in the course of the 20th century.

Cultivation of sugarcane

As with most commercial crops, many different varieties (cultivars) are available, some growing up to 5 m tall. Planting material is typically in the form of cuttings (setts, seed-pieces) from nursery stock. Optimum periods for growth and harvesting vary between localities, according to climatic seasonality. Depending on local conditions, the cane growing period varies between about 10 and 22 months, and cane yields range from around 50 to 120 t/ha/year (Ruschel and Vose, 1982). Once harvested, the roots of the old crop may be ploughed out, and the field replanted. However, under the right conditions, cane will re-grow from the old root stock. This repeated cropping is known as ratooning, and may be continued for 6 or 7 years, providing for some economic advantage over annual crops (although yields typically decline somewhat with each ratoon). The number of ratoons varies with the vigour of the cane and with local conditions. In Australia, for example, four to five ratoons are typical (Garside *et al.*, 1997b). At least two ratoon crops would normally be expected, resulting in the cane remaining in the field for at least 3 years. Ratooning results in cane cultivation sharing some characteristics with the repeated cropping of perennials, rather than annuals, and the typically short fallow period between eventual ploughing out and replanting (rather than rotation of crops) exacerbates the effects of intensive monoculture of this plant.

Sugarcane is renowned for its efficiency in converting solar energy to organic material (see Box 1.3). In order to fulfil this considerable growth potential, the cane plant requires strong sunlight and abundant water.

Harvesting of sugarcane

Cane typically ripens in the cooler and drier part of the year, and water stress enhances sucrose accumulation. Although some leaves remain green, much of the foliage is dead once the cane is mature. In many systems of cultivation, controlled burning before harvesting is used to clear dead leaves ('cane straw' or 'trash') and some of the waxy coating of the cane. In some areas of cultivation, cane is cut by hand, providing significant employment for local communities, but incurring substantial labour costs to the producer. Mechanized harvesting, however, can also be expensive, in terms of the capital and running costs of equipment, and is only practical where land conditions are suitable (for example, on relatively flat ground). Mechanical harvesting also increases the proportion of extraneous material removed from the field with the cane. At harvest, the cane is cut at approximately ground level, the top green leaves are cropped off and cut stalks are bundled together and transported to the mill. Discarded tops and trash, where

Box 1.3. Sugarcane as a fixer of solar energy.

Alongside the extensive utilization of waste materials as by-products, the exceptional efficiency of cane in the conversion of solar energy to biomass is amongst the most widely cited examples of positive environmental features of the sugar industry. Payne (1991) asserts that sugarcane is a crop unparalleled in its capacity to trap sunlight energy and has the highest harvest index (ratio of material utilized to material grown in the field) of all crops. Except where otherwise indicated, the following account is based on Alexander (1985), who considers that perennial tropical grasses like cane, grown in the tropics as total biomass commodities, are potentially the most efficient farm crops on earth in their usage of expensive agricultural inputs.

The great efficiency of cane as a fixer of solar energy arises from a combination of features. First, in evolutionary terms, *Saccharum* species tend to interbreed readily, and there is even substantial interbreeding potential between *Saccharum* and related genera. As well as providing a wide range of options for modern plant breeders (and making hard work for taxonomists), this characteristic has assisted the natural spread of traits conferring adaptive advantage between these plants.

In terms of its anatomy and growth form, the plant also has valuable characteristics that enhance growth potential. *Saccharum* leaves are as effective a receptive surface to 'capture' sunlight as those of any extant higher plant. The perennial crown and expanding stool cluster, from which new buds, shoots and stalks develop, provide for continuous growth, while organic matter is continuously deposited above and below ground as redundant parts of the structure die and decay. Its vigorous growth allows sugarcane to outcompete other plants, and thereby to monopolize available resources (contributing to its ability to suppress weeds in a cultivated situation).

Perhaps the key characteristics, however, are physiological. Sugarcane and related species use the C4 photosynthetic pathway, as distinct from the C3 (or Calvin cycle) pathway used by most other plants. In terms of efficiency of photosynthesis, the C4 pathway allows a plant to exploit lower concentrations of carbon dioxide and a greater range of light intensities, while eliminating photorespiration. Sugarcane has also been shown to utilize a wider range of wavelengths of solar radiation within the visible spectrum than most plants. Physiological adaptations also include the use by these plants of sucrose (as opposed to, for example, starch) as a principal photosynthate, facilitating easy translocation of carbohydrate for growth.

As a consequence of such adaptations, sugarcane fixes around four times as much solar energy as most temperate crops, and can consequently yield around 50 t dry matter/ha/year (Paturau, 1989). Another illustration of the yield efficiency of sugarcane is given by UNEP (1982), which estimates that 1 million kcal of energy in the form of sugar requires (on average) 0.07 ha of sugarcane for its production, whilst the equivalent amount of energy in the form of beef requires 7.70 ha to produce.

this has not been burnt, can provide a valuable mulch (see Box 2.4). If the whole plant is to be harvested (e.g. for biomass – see Box 8.1), preharvest burning is inappropriate, and tops and trash will be collected as well as stalks. As a source of sugar, cut cane can deteriorate rapidly (e.g. Larrahondo *et al.*, 2002), for example through the action of microorganisms, and mills are generally sited close to cane fields to minimize transport time. Typically, cane is processed within 24 h of cutting. Refined sugar is obtained from cane in two stages, the first set of processes yielding raw cane sugar, which is subsequently refined (often some distance from the site of cane cultivation and processing).

Sugar beet

Sugar beet is a temperate, biennial root vegetable, the wild ancestors of which are thought to have evolved in coastal areas and first been domesticated as garden vegetables on the shores of the Mediterranean in ancient times. Although the taxonomy is confused, all cultivated types can be considered to be forms of *Beta vulgaris*, falling into one of four categories: leaf beets, garden beets, fodder beets and sugar beet (which is sometimes referred to as *Beta maritima*).

During its first year of growth, the beet stores sucrose in a bulbous root, similar in appearance to a fat parsnip. Marggraf and

Achard (in the 17th and 18th centuries) were the first to develop means by which beet could be used as an alternative source of sucrose to sugarcane and to selectively breed varieties of the plant for sugar production. Such breeding programmes are complicated by physiological constraints on the development of the beet plant, that typically result in a negative correlation between root yield and sugar content (e.g. Hoffman and Marlander, 2001; Jansen and Burba, 2001). Most commercial sugar beet varieties currently grown in Europe are triploid hybrid forms.

Cultivation of sugar beet

As with sugarcane, many different varieties are available. Beet is typically grown from seed. Timing of operations varies according to local conditions, but much beet is sown in the spring (after ploughing in autumn), and then harvested in the first autumn/early winter before it has a chance to flower and set seed. However, the crop may be sown in the autumn in areas with a Mediterranean climate. Sugar beet is not generally grown in the same fields year after year, but as part of a rotation with cereals and other crops (often providing the most important cash crop in the rotation).

Sugar content for mature beet is typically around 17% by weight, but varies considerably according to variety and growing conditions. This is substantially more than for sugarcane, but the yield of beet per hectare is much lower, such that sugar production is often only 7 t/ha (although it is possible to obtain yields up to 15 t/ha).

Harvesting of sugar beet

The beets are typically dug out of the ground in autumn/early winter, using mechanized harvesters. The green leaves are cut from the tops of the plants, and may be left in the field. Beet is usually transported to the sugar factory in large trucks, because the distances involved are typically greater than in the cane industry. This is partly a result of sugar beet being a rotational crop which requires nearly four times the land area of the equivalent cane monoculture, and partly due to the slower rate of sugar loss in postharvest beet.

Processing of sugar crops

The fundamental steps in the processing of sugarcane and sugar beet are sufficiently similar to be considered together here. Both crops require cleaning on arrival at the mill/factory (which results in some loss of sugar). As a root crop, beet requires the greater amount of cleaning. It is often fed down flumes on delivery to the factory, and further washed to remove mud, stones, leaves and other waste material. A great deal of waste water is generated by this process, from which solid wastes (notably mud) are settled out.

Juice extraction

Sucrose extraction from beets is easier than extraction from cane for a number of reasons. First, beet can be stored for several weeks after harvesting without substantial loss of sugar. Secondly, sucrose is readily diffused out of whole cells in beet, whereas the cells of cane need to be broken open prior to extraction – beet can therefore yield a higher purity juice, without much of the cell material and other impurities found in cane juice.

Once cleaned, beet is cut into 4–5 mm thick slices ('cossettes'), which are fed into a tower or rotating drum diffuser. Sugar is removed in solution by a counter current diffusion process. Typical raw juice from diffusion contains around 14% sugar. Spent cossettes are fed into large screw presses, which extract residual juice. Pressed pulp is recovered at around 70% moisture content, and fed into driers, large rotating drums, using air at 600–900°C. This drying process can account for about a third of the total factory energy consumption. Newer, more efficient, drier designs use steam.

Cleaned cane is typically shredded with rotating knives or crushed using hammer mills. Sugar is removed by washing the shredded cane with hot water and running it through rollers (milling), or by diffusion in a counter current process. Extracted cane juice typically contains around 15% sugar. Residual cane material ('bagasse') is recovered at about 50% moisture content. Bagasse is often used as the main fuel source in the mill, firing boilers for co-generation (production of thermal energy, as steam, and the generation of electricity, see Box 8.2).

Water consumption in juice extraction can be substantial. More water removes more sugar by diffusion, but also makes the extracted juice more dilute, increasing the energy required in the next stages of processing. Extracted juice is cleaned (by clarification) and concentrated (by evaporation) before the sugar it contains is crystallized.

Clarification and evaporation

Clarification involves removal of impurities from the juice, and adjustment of its chemical properties to enhance this process and subsequent concentration of juice by evaporation. Lime (calcium hydroxide) is used to treat both beet and cane juice in the clarification process, although beet factories tend to use more lime than cane mills. Magnesium oxide has been used as a partial substitute for lime in some cases. In addition, carbon dioxide may be bubbled through juice in the clarification process. Sulphitation (the addition of sulphur dioxide or derivatives) is also common in cane and beet sugar processing at this stage and the next, to prevent impurities reacting to cause coloration in the juice. Phosphatation (addition of phosphoric acid) may also be carried out to assist clarification. Treated juice is typically pumped slowly through a tank (clarifier), where waste solids settle out. These still contain valuable sugar, which is recovered by slurrying (mixing solid wastes with water) and/or passage through vacuum filters or filter presses, leaving a lime-rich waste material (filter press mud, filter cake, lime mud).

After clarification, the clear juice is still relatively dilute, and must be concentrated prior to crystallization. This is typically achieved by passage through multiple-effect evaporators. It is not unusual to see six (sometimes seven) effects in a beet factory evaporator, although many cane mills use only three or four. In either situation, the effective management of the evaporators is very important in determining the energy efficiency of the whole processing operation.

Boiling

Crystallization is carried out by 'boiling' the mother liquor (concentrated juice) in vacuum pans. Crystals tend to form as pure sucrose, so impurities (e.g. invert sugars – fructose and glucose) become concentrated in the mother liquor. Boiling is repeated (typically three times) in order to recover as much sucrose as possible, but the impurities increasingly inhibit crystallization. The mixture ('massecuite') of sugar crystals and mother liquor from each boiling is centrifuged to separate the main components. As well as sugar, a sweet, viscous waste material (molasses – essentially, the exhausted mother liquor) is ultimately generated by this process.

Such processing of cane juice yields raw sugar, the crystals of which are brown and sticky and have a distinctive taste. Generally, this is stored at the mill, for transport to refineries closer to the point of consumption of the ultimate (white sugar) product. Refining involves affination (partial dissolution of raw sugar, followed by centrifuging), chemical treatments similar to those described above for clarification (and possibly decoloration with activated carbon), followed by boiling to recover white sugar crystals. Because of the different nature of colour chemicals in beet juice, it is possible to produce white sugar as part of a single process at the beet factory. However, some factories produce a form of raw beet sugar, which is subsequently refined elsewhere.

Traditional and alternative systems of cane sugar processing

As noted above, the growing of cane for sugar is an ancient practice in many parts of the tropics. Old systems of sugar production persist, producing traditional forms of sugar, some of which operate on a semi-industrialized scale (UNEP, 1982). For example, 'gur' or 'jaggery' ('juggeri') is a form of raw sugar manufactured by traditional systems in India. It is generally used for sweetening in cooking, or for fermentation, and has poor storage qualities. A similar product called 'panella' is made in parts of South America. These forms of raw sugar are made by a small-scale process of milling (traditionally, with a pestle and mortar), evaporation (in an open, bagasse-fired pan) and solidification (further heating in an open pan, producing a syrup thick enough to solidify on cooling). 'Khandsari' sugar is also produced in India, based on a development of gur manufacture which involves a partial industrialization of the process, including some clarification/decolouring and centrifuging to yield a dry crystal sugar (UNEP, 1982). Whilst traditional sugar processing of this kind can contribute to local economies, its products are unlikely to be widely traded, and it can have negative environmental impacts. The processes are energy inefficient, and may (for example) lead to over-exploitation of fuel wood supplies, local deforestation and soil erosion (UNEP, 1982). However, such systems of sugar production are much smaller scale, and less intensive, than fully commercialized cultivation and processing, and their environmental impacts should be seen in that context.

2

Overview

Environmental considerations are increasingly important for the sugar industry, as a consequence of pressure from conservationists and local communities (Johnson *et al.*, 1997; Ballantyne, 1998; Mallawaarachchi and Quiggin, 2001; Mallawaarachchi *et al.*, 2001), increasing regulation (ISO, 2001; SASA, 2002) and pressures applied through markets (SASA, 2002). Consequently, the reduction of environmental impacts is an increasing focus for the industry both for cane (e.g. Singh, 1994) and for beet (e.g. Cooke and Scott, 1993). Whilst this can be seen as a potential source of conflict between producers and other stakeholders, it can also be interpreted as a opportunity to enhance productivity and the long-term viability of the industry. In some areas, increasing numbers of cane farmers have become enthusiastic about the social and economic benefits of adopting ecologically sound practices, and there is evidence that some measures that reduce environmental impacts also contribute to increased efficiency of production (e.g. Ballantyne, 1998). Clearly, the development of new equipment and technologies can be important factors in the reduction of environmental impacts, as for beet cultivation (e.g. Hruska, 1991) and processing (e.g. Goncharov, 1994; Grabka, 1995). In some respects, it may be possible to transfer technology between the beet and cane sugar industries (Avram-Waganoff, 1990). However, in predicting the actual uptake of novel methods, it is important to consider the full range of constraints that affect how farmers adopt new technologies, including those with a particular environmental emphasis (e.g. see Lawrence and Gray, 1997).

Like many industries, particularly those based on intensive agriculture, the sugar industry must address a complex range of environmental concerns. Kropff *et al.* (1997) discuss intensive agriculture and opportunities for sustainable production, with particular emphasis on the sugar industry. Principal challenges include the need to improve production systems to maximize water and nutrient use efficiency, conserve soils and better control weeds, pests and diseases while reducing other impacts of pesticides. The relative importance of particular concerns may differ from country to country, as will the local structures and regulations that exist to manage them, but a similar range of issues pertains across the sugar industry as a whole.

Although the cultivation of sugar crops and the processing of the raw materials that they yield can be seen as separate, they are interrelated. For example, because the sucrose content of cane declines rapidly following harvest, active liaison between growers and processors is important (e.g. Bakker, 1999). Stockfisch and Marlander (2002) consider that sugar beet is particularly well suited to the development of sustainable production concepts, specifically because cultivation and processing are closely linked. Such linkages can contribute to the limitation of environmental impacts, for example, through the use of processing wastes (by-products) as soil amendments, reducing the need for applications of other types of fertilizers (Lescure and Affret, 1997). From an environmental perspective, the extent of by-product utilization is an impressive aspect of the (particularly cane) sugar industry, although maximum benefits rely on responsible handling of potentially polluting materials.

In any given situation, there may be difficulties in unambiguously assigning environmental impacts to the sugar industry. At a landscape scale, cane growing is likely to be just one (although possibly a dominant) land use contributing to environmental degradation. Similarly, beet growing is just part of a mosaic of activities at a landscape scale. However, with this crop there are additional difficulties in distinguishing its environmental impacts from those of other land uses, even at a field or farm scale, and certainly over medium to long time scales, because beet is typically grown as part of a rotation. In some areas, sugar processing may be just one of a range of activities undertaken by a group of factories or other industrial units, making impacts (say, of effluent discharges) difficult to assign to any individual set of operations.

Although trade issues are not specifically examined in this review, it is also important to acknowledge that patterns of trade affect patterns of production, and therefore environmental impacts. Any substantial shift in global sugar trading conditions could significantly alter the environmental impacts of sugar production, for example, by shifting the balance between cane and beet, rendering significant areas of sugar cultivation uneconomic, or creating opportunities for economically viable sugar production in new localities. Whilst some changes in trade patterns may be somewhat unpredictable, others are contingent on known prospective political and economic changes, such as enlargement of the European Union (EU). The latter stands to affect patterns and methods of beet sugar production and trade in eastern Europe and the countries of the former Soviet Union, where methods have often been relatively inefficient and reliant on relatively old technologies. Characteristics of the sugar industry (including farm sizes, factory capacities and expected yields) in eastern European countries are summarized by Urbaniec (1996) and Treadgold (1998).

The following sections summarize the environmental impacts of the cultivation and processing of sugarcane and sugar beet, and measures that can be taken to reduce these impacts. Subsequent sections present information on a number of general agricultural practices (tillage, agrochemical usage, etc.) that contribute to multiple environmental impacts, and general measures (effective planning, alternative systems of agriculture, etc.) that can reduce multiple environmental impacts. Later chapters go on to examine impacts of sugar production on particular aspects of the environment (water resources, soils, etc.), and additional measures that have been suggested to reduce these impacts. The final chapter examines the uses and impacts of by-products of sugar production.

A Summary of the Environmental Impacts of Cultivation of Sugarcane and Sugar Beet

The environmental impacts of the cultivation of sugar crops are summarized below, and their sources relative to key processes and inputs are illustrated in Figs 2.1 (for cane) and 2.2 (for beet). A summary of the findings of a recent review of the environmental aspects of sugar beet cultivation in the UK is given in Box 2.1.

Impacts on biodiversity

Loss of natural habitats

Historically, substantial areas have been cleared for cane cultivation, leading to the loss of habitats including rainforest, tropical seasonal forest, thorn forest, semi-desert scrub and savannah. In some places, including South America, South-east Asia and Australia, the area under cultivation has continued to expand in recent years. Land clearance not only results in the direct loss of species and habitats, but underlies a range of wider impacts on ecosystem function,

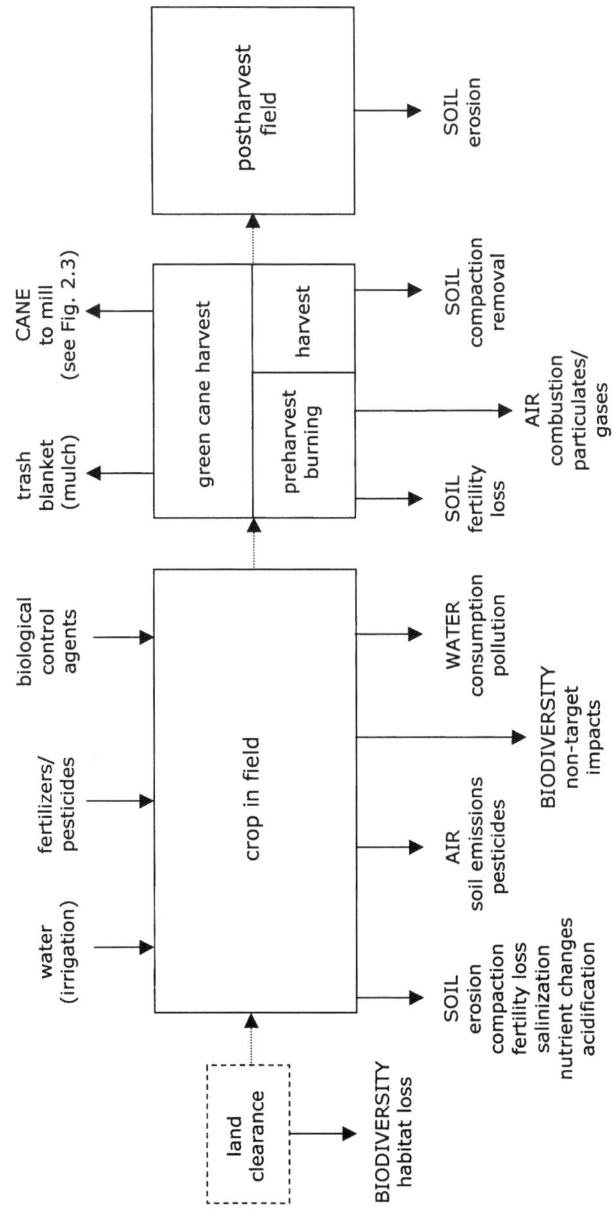

Fig. 2.1. Sources of environmental impacts relative to key processes and inputs in the cultivation of sugarcane.

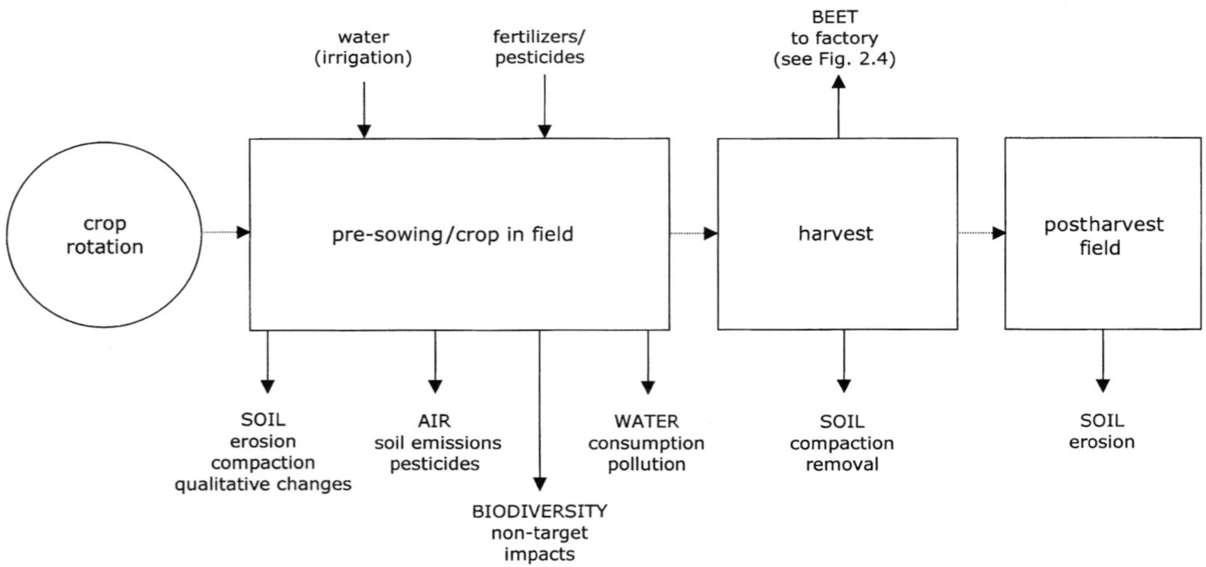

Fig. 2.2. Sources of environmental impacts relative to key processes and inputs in the cultivation of sugar beet.

Box 2.1. Environmental aspects of sugar beet cultivation in the UK.

In June 2002, following a stakeholder consultation process, a report was produced by the UK Government's Department for Environment, Food and Rural Affairs, assessing environmental aspects of sugar beet cultivation in the UK (Defra, 2002). The findings of that report are summarized below.

The main environmental impacts of beet growing were identified as those relating to biodiversity, use of agrochemicals, soil impacts and archaeological features. Environmental regulations and agri-environment schemes relevant to beet growers were reviewed, as was the state of research into beet cultivation and the environment in the UK. Sugar beet's role as a break crop in arable rotations was seen as positive, contributing to a reduction in pesticide and fertilizer inputs during other phases of the rotation (by interrupting a potential build-up in pests/diseases associated with other crops, and by contributing organic matter to the soil in the form of root fragments, leaf material and/or beet tops ploughed in following harvest). Many of the more positive aspects of beet cultivation depended heavily on individual growers adopting appropriate management practices (e.g. minimizing crop protection activities, not treating stubble from the previous crop with herbicide in the autumn, timing mechanical weeding and irrigation to reduce possible conflicts with wildlife). Encouragement of sympathetic farm management decisions was seen as the key to realizing positive environmental outcomes.

Biodiversity

Overall, the presence of sugar beet in a crop rotation was seen as a positive contribution to crop diversity (and to wider farmland biodiversity). Particular benefits were identified for farmland birds, which have experienced population declines in the UK in recent decades. These benefits arose from:

Winter stubbles from the previous crop, left in the ground prior to spring sowing of beet

Valuable as winter feeding habitat for seed-feeding birds such as finches and buntings.

The relatively open vegetation structure of the beet crop in the late spring

Valuable as habitat for ground-nesting birds such as stone-curlew, lapwing and skylark.

Late harvested crop/postharvest beet stubble

Valuable as habitat for a range of species, and as winter feeding habitat for a range of birds, including pink-footed geese, swans, skylarks, golden plover, lapwing, pied wagtail and meadow pipit. In particular, the difficulty of controlling broad-leaved weeds in this crop results in increased numbers of such plants, their seeds and associated invertebrates.

More detailed information is required on the impacts of beet cultivation on flora and fauna other than farmland birds.

Agrochemicals

Between 1982 and 1998, the total pesticide input to UK beet cultivation fell from around 11 kg/ha to just over 5 kg/ha (a reduction of well over 60%). The reduction in insecticide inputs, for example, was partly due to a shift away from spraying towards seed treatment, such that 70% of UK beet crops received no insecticide in spray form. Further details of these trends in pesticide inputs are provided elsewhere in this review.

Use of nitrogen fertilizer in UK beet cultivation was found to have declined substantially, from an average of 150–160 kg/ha in the 1970s to 100–105 kg/ha by 2000 (a fall of around 33%).

continued

Box 2.1. *Continued.*

These reduced agrochemical inputs were partly a consequence of increased adoption of a more integrated approach to crop management. Despite the considerable reduction in inputs, concern remained over a lack of detailed knowledge on pesticide impacts on watercourses and aquatic species, eutrophication of ground and surface waters (due to N and P inputs), potential hazards of increased seed treatments, and the effects of herbicide drift and runoff into adjacent habitats. It was noted that even more could be done to reduce agrochemical usage in UK beet cultivation. The role and impact of genetically modified beet (see Royal Society, 2003) were still under investigation.

Soil impacts

Erosion (by wind, water and removal of soil at harvest) was of some concern. However, it was noted that wind erosion was increasingly controlled by the planting of cover crops. Water erosion, leading to silting up of waterways, was seen as having potentially serious consequences, although quantities of soil lost were much less than those attributable to removal at harvest. This factor accounted for 350,000 t of soil removed from UK beet fields each year. Although substantial, this figure (equivalent to 6.5% soil tare) represented the lowest rate of soil lost with beet at harvest in Europe, and all soil thus removed was returned to agricultural land or used in other applications.

Although measures had been taken to reduce soil compaction risk in UK beet cultivation, this was considered to remain a potential problem area requiring further improvements.

Archaeological features

As in other cropping systems involving deep ploughing, beet cultivation was seen as posing some risk of damage to buried archaeological features.

Irrigation

Although identified as a measure generally reserved for periods of severe drought (with less than 5% of the UK beet crop normally receiving irrigation), this was seen as an area requiring future monitoring in the light of future water availability and climate change.

Environmental regulations/agri-environment schemes

A range of legislation was identified as relevant to the potential environmental impacts of UK beet growing, including regulations on the discharge of dangerous substances into groundwaters, nitrate pollution of ground and surface water and various aspects of the use of pesticides. Existing agri-environment schemes were seen as providing relatively limited opportunities for encouraging environmentally focused management practices amongst most beet growers.

State of research

Although relatively little research was devoted primarily to environmental objectives (excepting the farm-scale evaluations of genetically modified beet – see Royal Society, 2003), many broader research programmes were seen as contributing to the development of more 'environment-friendly' beet cultivation in the UK. It was also recognized that there was a good communications network within the industry, accelerating the uptake of research outputs.

including changes to hydrology and increased soil erosion. Particular concern has been expressed over impacts on wetland habitats, and the resultant effects on associated ecosystems such as rivers and coastal zones. It is likely that only relatively very small areas have been cleared specifically for sugar beet, which was adopted as a widely grown crop relatively recently and would often have been grown on areas that were already under some form of cultivation.

Impacts of ongoing cultivation

Some environmental impacts of sugar crop cultivation have the potential to affect biodiversity well away from the farmer's field. Such impacts include the sedimentation of waterways as a consequence of soil erosion, or eutrophication arising from the leaching and runoff of nutrients. In relation to the farmer's field, areas under cultivation generally support fewer indigenous species than adjacent natural habitats, although the diversity of invertebrates amongst the crop and in the soil can be considerable, and microorganismal biodiversity in these situation is often overlooked. The use of pesticides can have a direct impact on biodiversity by killing non-target organisms, but can also have indirect effects (e.g. by removing species that provide food or shelter for other organisms). The inappropriate use of biological control in cane cultivation has had negative biodiversity impacts, notably through introductions of the mongoose in the Caribbean and elsewhere, and the introduction of the cane toad to Australia. Such cases can distract from the potential environmental benefits of responsible biological control programmes. There has also been much concern expressed recently over the potential biodiversity impacts of the cultivation of transgenic crops, including sugar beet.

Impacts on water

Excessive water consumption

Although entirely rain-fed in some localities, cane cultivation relies on irrigation in many other areas, and the quantities of water used are a cause for increasing concern. This is particularly apparent in the literature from India, where cane cultivation is seen as placing a strain on available groundwater resources in a number of areas (e.g. Dhillon and Panshi, 1987; Singh and Sankhayan, 1991). There are also concerns that water extraction for cane irrigation has resulted in over-commitment and degradation of river systems, for example in Australia (Arthington *et al.*, 1997; Meyer, 1997). In South Africa, there is even debate over whether non-irrigated cane consumes so much rainwater that it should be classified as a stream flow reduction activity (e.g. Schulze *et al.*, 2000).

Sugarcane is amongst a group of crops noted for their heavy water consumption (along with rice and cotton, for example), and it has been estimated that a cane crop of 100 t/ha would be expected to consume in total approximately 7.5 Ml/ha water. In areas where this demand cannot be met by rainfall, the water requirement translates into substantial applications of irrigation water. For example, estimates of irrigation requirements for cane growing in parts of north-west Australia have ranged from 15 to 54 Ml/ha/year (Wood *et al.*, 1998). Despite their importance to the industry, irrigation systems have often been found to be inefficient, leading to wastage of water.

Sugar beet is relatively insensitive to soil moisture, but around one-fifth of the world's beet cultivation is irrigated (Dunham, 1993). In drier localities, this may be essential for successful cultivation, but in other areas the benefits of irrigation appear to be marginal. In addition to concerns over excessive water consumption for irrigation, it is also worth noting that irrigation may exacerbate other cultivation impacts (particularly salinization of soils).

Water pollution

Watercourses and aquatic habitats can be polluted by agrochemicals and sediments arising from the cultivation of sugar crops. Groundwaters can be affected by leaching of nutrients from fertilizers applied to the crops. These impacts may extend to downstream ecosystems, such as coastal zones. Examples of impacts of cane cultivation on water quality come from many areas where the crop is grown, including Australia, South America and the USA, although there may be difficulties in separating the effects of cane growing from other land uses. Similar concerns have been raised in beet growing areas, but unambiguous assignment of water pollution to the cultivation of this crop is even more difficult, as it is generally just one component of a wider crop rotation.

Impacts on soils

Erosion

In many areas, cane is cultivated on slopes, and beet is often cultivated in such a way that fields are left bare over winter; both activities exacerbate erosion risks. The extent of erosion problems is heavily dependent on local conditions. Estimates of erosion soil losses under sugarcane range from around 15 to > 500 t/ha/year (e.g. Lugo-Lopez *et al.*, 1981; Prove *et al.*, 1995). Because they are often left bare over winter, beet fields can be vulnerable to wind erosion as well as water erosion. Estimates of soil losses to wind erosion under sugar beet range from 13 to 49 tons/acre/year in the USA (Fornstrom and Boehnke, 1976), and estimates of soil losses to water erosion range from 0.3 to 100 t/ha/year in beet growing areas in Europe (De Ploey, 1986; Morgan, 1986).

Soil lost at harvest

In addition to erosion, soil is also removed from the field with the harvested crop. In cane cultivation, extraneous material (including soil) probably makes up about 1–15% of the material delivered to the mill. However, the nature of beet harvesting results in large quantities of soil being removed with the crop. Estimates suggest a soil 'tare' of 10–30% for harvested beet (Elliott and Weston, 1993), with studies suggesting figures such as 9 t/ha per harvest soil lost (Poesen *et al.*, 2001). Over large areas, these losses become substantial, with published estimates including 3 Mt/year for the EU (Elliott and Weston, 1993) and 1.2 Mt/year for Turkey alone (Oztas *et al.*, 2002).

Reduction in soil quality

Compaction of soils can occur under cane and beet, increasing bulk density and reducing porosity, as well as producing negative effects on the soil fauna. Reduced porosity (and consequently water infiltration rate) results in increased runoff, which is likely to exacerbate erosion problems. Compaction risks associated with cane or beet cultivation differ to some extent, according to the differing cultivation systems which are generally used (monoculture and rotation, respectively). In beet cultivation, the number of field operations (and therefore vehicle passes) used in field preparation and the fact that soils are often wet during harvesting, increase compaction risk and are particular sources of concern.

Other soil quality impacts commonly associated with sugar crop cultivation include loss of soil organic matter, changes in nutrient levels, salinization and acidification. Loss of organic matter and changes in nutrient levels are demonstrable under both cane and beet (as well as other crops). Salinization (associated with poor water management, particularly drainage) and acidification (mostly as a consequence of the application of inorganic fertilizers) appear to be more prevalent in certain cane growing areas than under beet cultivation.

Combined impacts on soil quality can lead to a loss of fertility, which is a particular risk under cane, which is generally grown as a continuous monoculture. Loss of soil fertility appears to be having a negative impact on cane yields in a number of countries.

Impacts on air quality

Air pollution

The practice of burning cane prior to harvest creates air pollution (and contributes to soil impacts). The use of fertilizers can exacerbate nitrogenous emissions from fields.

A Summary of the Environmental Impacts of Sugar Processing

The environmental impacts of the processing of sugar crops are summarized below, and their sources relative to key processes and inputs are illustrated in Figs 2.3 (for cane) and 2.4 (for beet).

Impacts on biodiversity

Most impacts of sugar processing on biodiversity are secondary effects from environmental pollution, such as the discharge of effluent into waterways.

Impacts on water

Excessive water consumption

Processing of sugar crops is a relatively water intensive activity, involving a number of stages that consume water. The water consumption of beet processing is exacerbated by the need to wash off the considerable quantity of soil removed with the roots at harvest.

Water pollution

Sugar processing produces effluents that can cause pollution when discharged; polluting effects being exacerbated by the high oxygen demand of the effluents and the use of agents such as lime in processing operations. Published accounts from many cane growing areas, including parts of Africa, South America and the Caribbean, report environmental impacts of sugar-processing effluents. There is a particular body of literature from India, demonstrating pollution of groundwaters and surface waters (including rivers). Reports of pollution from beet sugar-processing effluents in Europe include impacts on coastal ecosystems.

Impacts on soils

Reduction in soil quality

Soils can be negatively affected by poorly managed application of wastes (by-products) from sugar processing, or from poorly managed irrigation with processing waste waters. However, there is also much evidence that processing wastes can be used as beneficial soil amendments, if appropriately applied.

Impacts on air quality

Air pollution

The practice of burning bagasse to fuel cane processing operations can result in undesirable emissions. However, this represents the utilization of a by-product (see below) and may be less polluting than alternative arrangements. The wastes generated by the processing of cane and beet can result in significant odour problems, from the release of noxious gases.

Positive Environmental Aspects of Sugar Production

Crop characteristics

Sugarcane is a highly productive plant in terms of yield per unit area and yield per unit water consumed. The relatively large quantities of atmospheric carbon fixed by cane have led to interest in its cultivation primarily as a renewable fuel source, either as biomass or as a source of alcohol. The use of waste fibre (bagasse) to fuel cane processing in many parts of the world already contributes to cane sugar production being a relatively carbon neutral activity (e.g. see Beharry, 2001). There has also been interest in sugar beet as a source of biofuel, although the plant does not benefit from the high level of productivity associated with cane. However, its deep, spreading root system does make it an efficient scavenger of

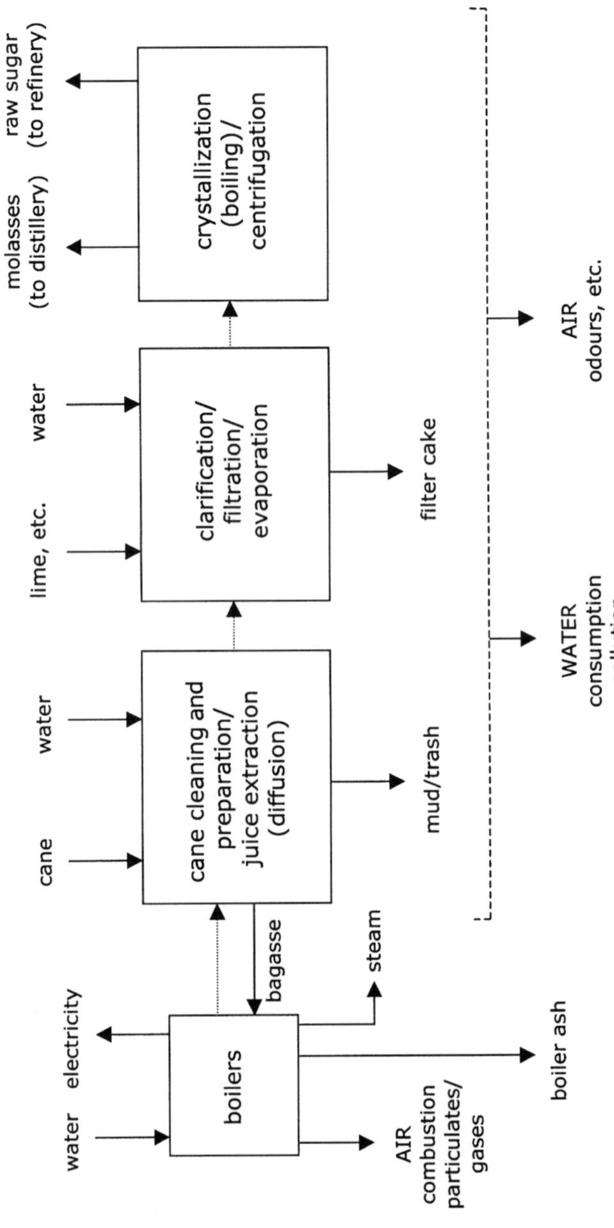

Fig. 2.3. Sources of environmental impacts relative to key processes and inputs in the processing of sugarcane.

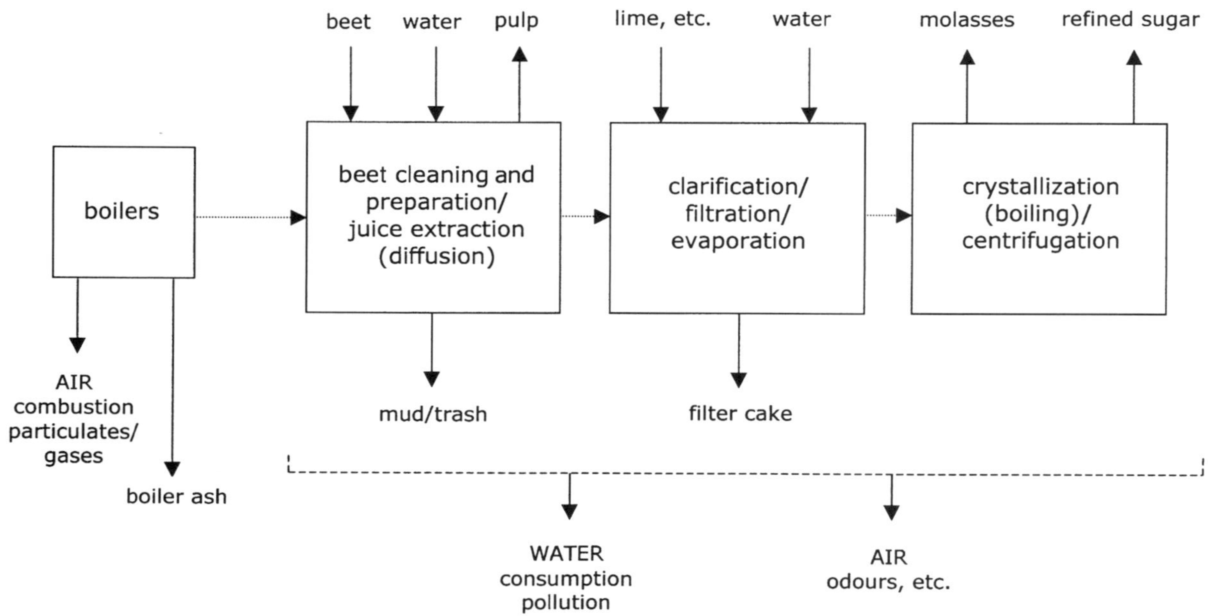

Fig. 2.4. Sources of environmental impacts relative to key processes and inputs in the processing of sugar beet.

soil water and nutrients, relative to many other crops.

By-products of sugar production

The waste materials arising from sugar cultivation and processing (particularly from cane) are, in many cases, utilized as by-products. This reduces the absolute output of waste materials, and has the potential to ameliorate negative impacts of other sugar production activities (e.g. through the use of waste materials as mulches or beneficial soil amendments). By-product utilization can also prevent, by substituting for, more environmentally damaging activities. This is most clear where bagasse (a renewable fuel) is burnt to provide power, where otherwise fossil fuels might have been consumed. However, it should be noted that the utilization and further processing of by-products can, in itself, result in negative environmental impacts. In some cases, this results in a very complex overall cost–benefit analysis. Such issues extend to wastes and by-products of secondary processing activities. For example, the consumption of molasses as a feedstock for alcohol production generates a further waste material (vinasse), which also has the potential to be either a pollutant or a useful by-product, depending on its handling.

Environmental Health Considerations

There appear to be no particular environmental health issues associated with sugar beet, although agrochemical and factory safety considerations, for example, cannot be disregarded. UNEP (1982) suggested that there were relatively few major environmental health problems associated with sugarcane. As with beet, the extent of health problems arising from, for example, the mishandling of agrochemicals and factory accidents, is difficult to quantify. Other hazards may include diseases associated with cane cultivation systems, and bagassosis from handling dried bagasse (see also Whitaker, 1978). Bilharzia and hookworm often occur naturally in areas where sugarcane is grown, and (although they are not a direct result of the sugar industry) may become established in cane fields and associated water bodies. Depending on the location of the plantation/mill, other insect- and snail-borne infections of humans and domesticated animals may also become a problem. These include fascioliasis and paramphistomiasis. Sugarcane irrigation systems that result in pooling of water on and around the fields (e.g. flood and furrow) may also serve to harbour schistosomiasis and malaria vectors. Schistosomiasis in particular is considered further under environmental impacts associated with water consumption.

SASA (2002) recognizes that the organization of settlements where cane farm workers are accommodated (i.e. farm village sites) has implications for human health and the wider environment. It makes recommendations in this respect, summarized as follows:

- Settlements should be appropriately sited for accessibility and convenience.
- Settlements should be appropriately designed, taking account of the health, education, cultural, aesthetic and recreational requirements of the community.
- Responsibility for settlement management should be delegated to designated responsible individuals.
- Water supplies must be adequate for domestic use, and should be monitored for quality in relation to health considerations.
- Adequate facilities should be provided for disposal of refuse and human waste.
- Health awareness programmes should be instigated.
- Environmental education programmes should be instigated.
- Appropriate facilities should be provided for storage and maintenance of farm vehicles, equipment, chemicals, etc.

Energy Efficiency

One means by which the environmental dynamics of an industry can be assessed and enhanced is in relation to its energy efficiency. Although this is not a primary focus of this review, the following paragraphs outline some aspects of the energy efficiency of sugar production.

Paturau (1989) illustrates how energy consumption in the processing of sugar crops showed a marked downward trend in the latter part of the 20th century. Example data from Mauritian cane factories show a fall in energy consumption from around 1.9 to 1.4 GJ/t between 1960 and 1985, and those from French beet factories show a fall from around 1.65 to 1.0 GJ/t over the same period. This is in sharp contrast to data for cane sugar production in Puerto Rico over a similar period (1950–1980) provided by Alexander (1985), which show an increase in energy consumption from around 0.55 to 1.05 kW/ton/day. Increased energy consumption here was ascribed to increased mechanization in cultivation and the energy demands of new mill technologies including scrubbers, cleaning plants and other environmental protection facilities. In this case, increased energy consumption was accompanied by a decline in energy produced (by bagasse-fired boilers), resulting in a shift from an overall energy surplus to an energy deficit.

The beet sugar industry is generally seen as being well ahead of the cane industry in terms of energy efficiency, with energy use per unit product around half that for cane, even though the final product is white rather than raw sugar (e.g. Fry, 1997). The drive for energy efficiency has been less in the cane industry, which tends to have access to a free fuel source (bagasse), whereas the beet industry has been very sensitive to external energy prices. A range of methods have been suggested for making cane processing more energy efficient, many of them relating to improved methods for generation of energy from cane-processing by-products. Bagasse is routinely used as fuel in cane mills, and co-generation (the combined production of electricity and thermal energy, as steam) has been a long-standing feature of cane sugar factories, which have been well placed to exploit the technique. This is principally a consequence of the ready supply of bagasse, and the high level of usage of low pressure steam in cane sugar processing. The latter largely accounts for the higher levels of thermal efficiency seen in many cane sugar factories, relative to utility powerplants. Co-generation can allow cane factories to sell excess electricity, providing an additional economic incentive for energy efficiency. The production of ethanol and biogas from cane and its by-products has also been explored. Otorowski (1990) suggests that a beet sugar factory, like a cane mill, can be energetically self-sufficient, by exploiting beet pulp as an energy source. Even where fossil fuels remain the main source of power for beet factories, there are economic as well as environmental reasons for improving energy efficiency (e.g. Urbaniec, 1996). Klemes *et al.* (1999) argue that adoption of new methods in beet sugar processing, such as cooling crystallization of concentrated raw juice, as opposed to the traditional method of evaporating crystallization, has the potential to improve energy efficiency and to reduce atmospheric emissions, water consumption and the polluting potential of effluents.

Reducing the Environmental Impacts of Sugar Production

Measures to reduce environmental impacts may be of greatest value if they are part of a broad, holistic and pragmatic system of sustainable management. Measures that can be taken to reduce the multiple environmental impacts of various activities are discussed in subsequent sections. In the context of the principles of ecologically sustainable development adopted by the Australian authorities, Johnson *et al.* (1997) outline what could be a 'manifesto' for increasing overall sustainability in the sugar industry:

- Natural resource management in the sugar industry is undertaken as a fundamental part of, and necessary precondition for, ecologically sustainable development and the implementation of the precautionary principle.
- Responsibility for natural resource management is shared between government,

community and sugar industry in a transparent manner.
- Appropriate incentives are put in place to encourage the protection of natural resources and to encourage their use only in ways which are ecologically sustainable.
- Appropriate mixes of incentives are developed and appropriate weighting given to motivational, voluntary, property-right, price-based and regulatory instruments in ways which vary according to the issue and local, regional and social characteristics.
- The community as a whole – as well as users and beneficiaries of natural resources – contribute towards the provision of incentives to [sugar producers], whose primary responsibility is for the protection of the environment.

Of course, there are a number of requirements that need to be met before such measures can be effectively implemented. Those noted by Johnson *et al.* (1997) include:

- Government at all levels must show unambiguous commitment to ecologically sustainable development, and develop the necessary supporting institutional capacity.
- Local communities and industry must be informed, empowered and enabled to manage natural resources.
- Sugar producers and other stakeholders must be involved in the establishment and operation of relevant decision-making and advisory bodies.
- Sugar producers must ensure that their own internal processes allow for involvement of other stakeholders.
- Visible monitoring and accountability mechanisms must be developed.
- Research must be undertaken, and knowledge systems developed to support informed decision making and monitoring.

Johnson *et al.* (1997) note that the cost of controlling and preventing environmentally damaging processes arising from sugar production should be borne by the industry itself, but that others (including government) should accept responsibility for protecting the natural environment when the costs cannot be recovered using market mechanisms.

Many environmental impacts of cane and beet cultivation could be reduced by the adoption of what would widely and generally be regarded as good agricultural practice. Such measures often have associated economic benefits to the grower (through a reduced need for costly inputs, for example). Similarly, in relation to sugar processing, general aspects of good practice encapsulate key measures that can be taken to reduce environmental impacts, a number of which may also generate cost savings, for example, through reduced water or energy consumption, or more effective exploitation of wastes as by-products.

Overall, the following appear to be particularly important considerations in the reduction of environmental impacts of sugar production:

- Reducing excessive water consumption (particularly in the cultivation of cane, and also in the processing of cane and beet).
    - Adoption of more appropriate irrigation practices in relation to sugar crops:
        - improved scheduling of irrigation to enhance water use efficiency;
        - adoption of water-saving irrigation methods (drip irrigation systems are attractive in this respect, although there may be technical and economic constraints on their installation in particular situations);
        - better assessment of whether irrigation is strictly necessary (particularly for beet);
        - adoption of general measures to conserve soil moisture (such as mulching).
    - Improvement of water efficiency in sugar-processing operations:
        - technologies are available, and there is an increasing adoption of systems which recycle processing water.

- Reducing soil impacts (declining soil quality and erosion under cane, and the loss of soil at harvest and erosion under beet).
    - Adoption of generic soil conservation methods in cultivation, such as terracing on slopes, reduced tillage and not farming marginal land.
    - Improved management for soil fertility, including monitoring of soil quality.
    - Reducing soil lost with beet at harvest (this may be difficult without altering the shape of the root itself).
- Reducing water pollution (from cultivation and the discharge of effluents arising from the processing of cane and beet).
    - Improved crop management (including soil conservation methods to restrict sediment release, rational use of pesticides, and better fertilizer management to reduce leaching).
    - More effective treatment of effluents prior to discharge from sugar-processing operations.
- Reducing air pollution (from preharvest burning of cane).
    - Adoption of green cane harvesting/ trash blanketing (already widely used in parts of the industry, this technique yields a range of environmental benefits in addition to providing an alternative to preharvest cane burning).

Broad Measures to Reduce Multiple Environmental Impacts

Appropriate planning can minimize environmental impacts

Land use planning is important at a range of scales, if the sustainability of sugar crop cultivation, and agricultural activities in general, is to be enhanced (e.g. King *et al.*, 1992; Johnson *et al.*, 1997; Rozeff, 1997). SASA (2002) regards the development and implementation of a farm land use plan (LUP) to be the fundamental basis for integrated conservation management, while noting that less than half of the South African sugar industry had thus far adopted this approach (but see example given by Platford, 1992). Johnson *et al.* (1997) note that the development of property management plans (PMPs) is increasingly accepted practice for new cane farms in Australia, although less so for established farms. Baseline data are critical in land use planning and for monitoring, but such information is often lacking (Johnson *et al.*, 1997). However, the increasing availability of computer based systems (such as Digital Terrain Modelling, DTM, as used in Mauritius by Seeruttun and Crossley, 1997) has made the handling (and to some extent, gathering) of data much faster and more efficient. Bakker (1999) regards topographical survey and mapping of soil types as key aspects, and SASA (2002) stresses the importance of developing strategies for soil and water conservation (exemplified by terracing and appropriate management of waterways), while listing the following as other elements to be considered:

- non-arable areas/natural habitats;
- wetlands and watercourses;
- dams;
- areas sensitive to burning (e.g. built-up areas and roads);
- field layouts (e.g. in relation to terracing and strip planting);
- road layouts and the cane extraction system;
- tillage;
- suitability of areas for trashing (and green cane harvesting);
- suitability of areas for mechanized operations.

SASA (2002) notes that the implementation of a farm LUP is likely to be a phased programme, over a period of up to 10 years, and stresses the importance of environmental audits for collecting baseline data and monitoring progress. Environmental audits have also been found to be valuable in Australia (e.g. McIlroy *et al.*, 1995; Ballantyne, 1998). The environmental impacts of agricultural systems at a range of scales can also be assessed and monitored using life cycle analysis (LCA), as applied to aspects of beet cultivation in the Netherlands by Brentrup *et al.* (2001).

A number of studies have shown that landscape or regional scale data can be collected in cane growing areas using remote sensing techniques (e.g. in South Africa – Peel and Stalmans, 1999; and Brazil – Fiorio *et al.*, 2000). Land use planning at this scale is challenging, but can be seen as a key component in the sustainability of agriculture (as in Brazil – Pinto and Crestana, 2001; and Thailand – Yamamoto and Sukchan, 2002). Remote sensing certainly provides a useful tool for monitoring changing land use at large spatial scales.

The potential benefits of landscape-scale planning in relation to systems involving sugar beet have also been explored, e.g. in relation to assessment of soil erosion risks in Bavaria, Denmark or Europe as a whole (e.g. Chisci and Morgan, 1986; Madsen *et al.*, 1986; Schwertmann, 1986), the establishment of mixed farming systems in the Netherlands (Bos and van de Ven, 2000), the optimum locations for particular crops in Andalucia, Spain (de la Rosa, 1989), for the development of sustainable soil and crop management strategies in parts of Greece (Floras and Sgouras, 2002); and for efficiency of use of arable land in Russia (Mukha and Sviridov, 1999). Landscape-scale planning using a geographical information system (GIS) has also been used to decide on the appropriate siting of beet processing factories (e.g. Polacik, 1992).

Appropriate planning, for example through the use, at the earliest possible stage, of environmental impact assessments (EIAs), can also contribute greatly to the reduction of impacts of sugar processing (UNEP, 1982).

Appropriate management can minimize environmental impacts without compromising productivity and efficiency

Reflecting on the Australian sugar industry in particular, Ballantyne (1998) notes that there is an increasing awareness that sustainability does not necessarily imply reduced productivity and efficiency. Indeed, many measures introduced to reduce environmental impacts have resulted in increased efficiency (see Box 2.2). Murty and Kumar's (2003) study of the effect of environmental regulation on the productive efficiency of water-polluting industries in India reinforces this. Their main empirical finding was that the technical efficiency of firms in the sugar industry increased with their degree of compliance with environmental regulation and water conservation efforts. Similar synergies between environmental protection and enhanced productivity in Mauritius are noted by Autrey (1999).

Regulation and the propagation of best management practices

Increasingly, a complex body of domestic legislation and administrative arrangements affect the operations of the sugar industry. Those relevant in Australia, for example, are tabulated by Johnson *et al.* (1997). SASA (2002) notes that, in addition to increasing domestic legislation on the environment, the provisions of international conventions such as the Convention on Biological Diversity, the Convention to Combat Desertification and the Convention on Wetlands of International Importance (RAMSAR) need to be considered by sugar producers where their nations are signatories. The broad range of regulations that affect the sugar industry are summarized by ISO (2001), which also provides a commentary on concerns over likely future regulations from respondents in the cultivation and processing sectors. Cane growers were concerned about further restrictions on cane burning and more stringent land use control, while beet growers were primarily concerned over further restrictions in the use of agrochemicals (including fertilizers). Concerns from the processing sector varied between countries, but further restrictions in relation to air and water pollution were common between those handling cane and beet.

While the weight of relevant legislation increases, there is a feeling within the sugar industry that sustainability can best be achieved through self-regulation and the adoption of voluntary codes of conduct (e.g. Zabaleta, 1997; Ballantyne, 1998). In relation to self-regulation, the establishment of local

Box 2.2. Reducing environmental impacts can improve efficiency of production.

The following examples of areas where measures to reduce environmental impacts in the Australian sugar industry have led to improved efficiency of production are given by Ballantyne (1998).

Use of chemicals

Over 70% of Queensland cane growers have completed a voluntary 1-day course and have been accredited in the use of farm chemicals, resulting in a marked increase in proficient use and reduced application rates and frequencies.

Green cane harvesting and trash blanketing

In 1997, 65% of Queensland cane was harvested green, compared with just 18% in 1987. This has resulted in reduced pollution from cane burning (as well as increased flexibility in harvest schedules), whilst the trash blanket left on the ground increases soil organic matter and contributes to a reduction in nitrogen inputs.

Irrigation

Soil water monitoring has allowed more precise irrigation scheduling, whilst drip irrigation technology (although not necessarily reducing total volume of water used) has increased water use efficiency. Nutrient movement in groundwater and runoff is more widely monitored than previously, and tailwater dams are commonly used to ensure that runoff is captured and reused.

Riparian management

Whilst native vegetation is still being cleared to allow an expansion of cane production, there is an increasing recognition of the importance of riparian vegetation in filtration of nutrient runoff, prevention of erosion and siltation, provision of wildlife corridors and reduction of pest problems in adjacent cane fields. Trees are increasingly retained adjacent to waterways, and have been replanted in some areas.

Acid sulphate soils

An increased understanding of the environmental problems associated with acid sulphate soils has resulted in the development of more precise land management strategies, leading to positive environmental outcomes and improved farm productivity.

In addition to these examples, Ballantyne (1998) notes that improved fertilizer management, measures to tackle soil salinization and consideration of coastal/reef water quality issues have imposed a short-term cost on the sugar industry, but have had longer-term benefits for farm efficiency and sustainability. Wood *et al.* (1997) also stress the combined economic and environmental benefits to the Australian sugar industry that would arise from an improved system of fertilizer management in cane cultivation.

stakeholder groups can assist in the propagation of a sense of democracy and local ownership in relation to environmental issues, as well as the propagation of better practice. In South Africa, for example, the structure of the sugar industry includes a network of Local Environmental Committees (LECs), which serve this type of role (SASA, 2002). Whilst recognizing the value of voluntary (and incentive based) agreements, Johnson *et al.* (1997) conclude that external regulation will remain important in ensuring appropriate management of natural resources by the sugar industry, particularly where potential environmental impacts are most severe, not least because it provides a 'safety net to protect against the recalcitrant few not persuaded by other incentives'.

Many of the impacts of the cultivation of sugar crops in any one place are significantly influenced by local conditions, such as soil type and climatic factors. Guides to best management practices (BMPs) to reduce impacts must therefore develop recommendations based on site-specific considerations, and combine these with more widely applicable, generic recommendations. At least two such guides have been developed and published by the industry itself, the South African Sugar Association's *Manual of Standards and Guidelines for Conservation and Environmental Management in the South African Sugar Industry* (SASA, 2002), and the Australian CANEGROWERS' *Code of Practice for Sustainable Cane Growing in Queensland* (CANEGROWERS, 1998; see also Christiansen, 1999). The development of such voluntary technical standards can assist in the self-regulation process. Alternatively, relevant international standards can be applied, which help to facilitate universally applicable, safe and efficient production systems, trading standards and consumer protection. For example, the International Organization for Standardization's ISO 14000 standards relate to environmental management systems that are applicable to the sugar industry (SASA, 2002). ISO standards provide one means of socio-environmental certification, which has also been explored as a method for ensuring the observation of minimum standards in the Brazilian sugar industry (Pinto and de Santis Prada, 1999).

Mechanisms for communication and information exchange (as through the *Manual* in South Africa and the *Code of Practice* in Australia) are important means of propagating better practice. The value of such compendia of information is also recognized in the beet sugar industry, as a means of developing more integrated approaches to quality management, including the reduction of environmental impacts (e.g. in Belgium – Moyart and Sneessens, 2001). Recently, electronic mechanisms have come to the fore in some areas, providing a fast and accessible mean of enhancing the flow of information and knowledge, e.g. between researchers and growers (May and Fisher, 2001). A range of electronic (mostly internet based) resources are available to sugar beet growers in Europe, for example, aiming to promote efficient, economic and 'environment-friendly' methods – see Box 2.3. Electronic information and decision support resources are also available to cane growers in some parts of the world, for example in South Africa via the website of the South African Sugar Association (Schmidt, 2000). Rapid communication and uptake of improved methods are also assisted by having a relatively well-integrated industry (as for beet sugar production in the UK – see Box 2.1).

Alternative systems of production – precision, integrated and organic agriculture

Precision agriculture recognizes that field conditions are not uniform, and encourages a more precise application of specific inputs, such as fertilizers, pesticides (e.g. Bateman, 2003) and soil amendments. The precision approach may also look to optimize the use of time, labour and energy. Correctly applied, this technique can reduce inputs, increase efficiency and reduce potential environmental impacts. Precision agriculture is recognized as having potential benefits in both sugarcane monocropping (e.g. in Australia – Wood *et al.*, 1996, 1997) and beet growing (e.g. Sperlingson, 2003). The potential for such an approach, using mechanized systems guided by satellite-based navigational technology, has been explored for cane cultivation in Mauritius (Jhoty and Autrey, 1999). It has also been explored in a theoretical case study for Brazil, where it may be economically feasible (Sparovek and Schnug, 2001b). Similarly, economic feasibility is the focus of an assessment of prospects for precision agriculture, including sugar beet cultivation, in the USA (Daberkow, 1997). The technique is also explored in relation to UK production of this crop by Holman and Miller (1995), in more general terms by Jahns and Kogl (1993), and in relation to fertilizer application specifically by Draycott and Christenson (2003).

An integrated farming system (IFS) approach was investigated for crop rotations including sugar beet by El-Titi and Landes (1990), who found that IFS resulted in significant reductions in agrochemical consumption and nitrogen fertilizer inputs (leading to lower nitrate concentrations in the soil profile) and positive effects on the abundance and

Box 2.3. Electronic information resources available to European beet growers.

A sequence of papers in the Proceedings of the 64th Congress of the International Institute for Beet Research (IIRB) provides examples of the electronic, particularly internet based, information resources available to European beet growers. These are briefly summarized below. Common features of many of these systems are their interactivity and free access granted to growers in the respective areas covered by each.

Southern Germany (Burcky and Vierling, 2001)

BISZ – an up-to-date, concentrated, interactive, internet-based consultancy and information system for beet growers.

Spain (Esteban Baselga, 2001)

RECOM – initially interactive software, now an internet-based system, built on a database of Associacion da Investigacion para la Majora del Cultivo de la Remolacha Azucarera (AIMCRA) research, providing information including general data, statistics and recommendations on crop water needs, and a range of crop management practices.

Denmark, Sweden and Germany (Sorensen, 2001)

GrowCom – an internet-based system provided by Danisco, regularly updated with news and advice to growers, using a common platform for the three countries served.

Romania (Kammerling, 2001)

LIZ – an internet-based system providing some 20,000 growers and consultants with information and assisted decision-making (decision-making software also available independent of website).

UK (Pettitt *et al.*, 2001)

The grower's guide – an internet-based information system, also available as a component of web-independent software (Decision and Encyclopaedic Support System for Arable Crops – DESSAC).

Netherlands (Maassen, 2001)

Betakwik – decision assistance software, from the Institute of Sugar Beet Research (IRS), now available with other information resources via the internet.

Hungary (Potyondi *et al.*, 2001)

An information and communication system developed by BETA-KUTATO, initially fax based, but now available as an internet-based system.

In addition to these mostly country-specific systems, more generic resources are also available electronically. For example, access to the International Beta Database (IDBB), which holds information on 11,000 accessions of *Beta* genetic resources, including 2500 of sugar beet and 4000 of its wild relatives (Germeier and Frese, 2001). There is also the International Beet Library (BIB), which contains and collates information published since 1990 on beet cropping and related agronomic and industrial issues. At the time of Legrand's (2001) account, the database contained around 20,000 records, with some 150 new records being added each month.

diversity of the soil fauna. Economic analyses showed a higher (not significantly) margin for the IFS. Integrated systems for beet cultivation have also been explored by Kovac and Zak (2000), and particularly in relation to pesticide and fertilizer strategies by Smid *et al.* (2001).

Studies suggest that the cultivation of sugar beet as part of a relatively low input (tillage, fertilizer application, irrigation, crop protection and treatment of crop residues) system has benefits in terms of reduced costs and reduced environmental impacts (e.g. Lanza, 1991). The need to consider farmer perceptions (as well as environmental and economic factors) in any proposed change to agricultural practice is noted by Wossink *et al.* (1997).

Organic production has also been investigated by the sugar industry in some parts of the world. Organic production requires adherence to strict standards on permissible chemical inputs and techniques. Consequently, a shift to organic sugar production requires modification of processing, as well as cultivation, methods (e.g. Deville, 1999; Hoi and Shum, 1999). A shift to organic cultivation generally results in a substantial decline in yield, at least initially, and there may be increased costs in some aspects of production (Gudoshnikov, 2001). However, these must be considered alongside production savings (e.g. with respect to synthetic agrochemical inputs) and the substantial premium that can be obtained on sale of the final product. There has been much discussion of the environmental implications of organic farming in general and, although results of individual studies have sometimes been contradictory, there is a widespread public perception that organic systems are 'environment-friendly'. There has been much interest in the potential of organic sugar, as a 'value added', speciality product (e.g. see Bosworth, 2000; Buzzanell, 2000a,b; ISO, 2000). However, the ultimate economic viability of organic sugar depends on future market trends, and is currently somewhat unclear (e.g. Jolly, 2002). Some authors see the potential for growth in this niche market (Gudoshnikov, 2000, 2001), while others suggest that a substantial increase in organic production is unlikely (Sperlingson, 2003).

Gudoshnikov (2001) reports that organic cane was initially cultivated in countries including Mauritius (where organic production declined after initial exploration – Deville, 1999), Madagascar, Malawi, India, the Philippines, Argentina, Colombia, Paraguay, Brazil, Bolivia, Costa Rica, the Dominican Republic and the USA. An organic approach has similarly been explored in beet sugar production, in countries including Chile (where organic beet cultivation gave similar yields to conventional methods, after a 5-year transition period – Venegas and Aguilar, 1992), Sweden (where it recently accounted for 1% of total production – Olsson and Nordstrom, 2002), Switzerland (Arnold, 2003), the UK (Jarvis *et al.*, 2001; McAughtrie, 2001), Denmark and the Netherlands (Gudoshnikov, 2001). Robson *et al.* (2002) assessed a range of break crops for use in organic cultivation of arable rotations, and found that sugar beet performed poorly in comparison with bean, lupin, soybean, hemp, potato, carrot, swede and linola, because (like oil-seed rape) it was difficult to grow organically and had a limited organic market. Draycott and Christenson (2003) consider aspects of the cultivation of organic sugar beet.

Appropriate use of fertilizers and agrochemicals

The appropriate use of fertilizers and agrochemicals in the cultivation of sugar (and other) crops can be an important factor in maintaining soil fertility and preventing devastating crop losses associated with weed infestations and outbreaks of pests and diseases. However, poorly managed applications can result in very serious environmental pollution and have consequences for human health. A range of fertilizers, pesticides and other agrochemicals are used in the cultivation of sugar crops. The specific substances used, their application rates and means of application vary considerably between localities, and their environmental fate is dependent on a range of factors, including soil type, climate and land use management practices (e.g. Knappe and Haferkorn, 2001). There is evidence that herbicide applications can influence the environmental fate of fertilizer-derived nutrients (Sotiriou and Scheunert, 1994).

In addition to fertilizers and pesticides, chemical ripeners may be applied to sugarcane crops to facilitate harvesting and increase sugar recovery. Alexander (1985) notes that, historically, these were generally broad-leaf herbicides. Studies of such agents include those by Donaldson (1990), Rufino *et al.* (2001) and Solomon *et al.* (2002). Davis (1997) notes that their introduction played a significant role in increasing sugar yields in the Guyanese sugar industry in the 1990s. There appears to be limited information on the specific environmental impacts of these practices.

Commercial sugarcane is usually grown with high levels of inorganic fertilizer and pesticide inputs, and contamination of ground and surface waters is a major concern, particularly in areas with shallow water-tables (Hartemink, 2003). SASA (2002) recognizes the potential harm to humans and the environment associated with agrochemical use in sugarcane cultivation, and makes a series of recommendations for the appropriate use of agrochemicals (including herbicides, fungicides, insecticides, nematocides, plant growth regulators and adjuvants), summarized as follows:

- Biological (rather than chemical) control of weeds, pests and diseases should be practised wherever possible.
- Where agrochemicals are used, those with the fewest side-effects should be selected.
- To minimize the risk of serious pollution due to spillage, agrochemicals should not be transported in bulk.
- Only registered agrochemicals may be used.
- Manufacturers' specifications (label instructions) and legal requirements must be observed, in relation to application, storage and disposal.
- Staff applying chemicals should be adequately trained and equipped.
- Equipment should be of an appropriate design and maintained in good working order.
- Agrochemicals should only be applied under appropriate conditions (wind speeds, temperatures, etc.).

Similar recommendations are made in relation to crop nutrition and the use of fertilizers and lime. More detailed recommendations, including aspects of pesticide formulation, toxicity, registration, packaging, labelling, storage, transportation, preparation, application and disposal, as well as record keeping and health monitoring of field workers, are made in a well-written and illustrated manual on *The Safe Use of Pesticides* produced by the cane sugar industry in Guyana (Eastwood *et al.*, 1997b).

Although designed for use in cane cultivation, these recommendations apply equally well to beet. Beet cultivation in many countries has been increasingly affected by legislation restricting the use of agrochemicals and increasing penalties for agrochemical pollution. In relation to fertilizers and manures, the main concern has been over nitrate pollution of drinking-water (leading to more emphasis on techniques like the growing of trap crops – Scott and Jaggard, 1993), while there has been a general drive to reduce the quantities of pesticides applied to crops. Scott and Jaggard (1993) suggest that a 50% reduction in total nitrate inputs to beet across the UK would result in an overall yield reduction of only some 10%, and note that the key consideration would be to manage the timing of nitrate inputs more effectively. Indeed, there is evidence of decreased pesticide and fertilizer use in UK sugar beet cultivation in recent years (Draycott *et al.*, 1997; Defra, 2002).

Specific aspects of better practice in the use of fertilizers and pesticides, along with other particular agricultural operations, are considered in the following sections.

More Specific Measures to Reduce Multiple Environmental Impacts

Across agriculture, there is an increasing emphasis on the rational application of plant nutrients such as N, by shifting focus from sheer productivity to a balance between productivity, crop quality and environmental impact. The aim is to develop soil management systems that optimize nutrient use efficiency by the crop, through methods such as cover cropping, reduced tillage and improved timing and rates of application of manures, crop residues and inorganic fertilizers (Christensen, 2004).

Tillage

The main objective of tillage is to produce suitable conditions for sowing/planting, establishment and growth of the new crop. As such, tillage may aim to loosen the soil, control weeds and incorporate into the soil plant residues, (in)organic fertilizers, pesticides and other soil amendments.

The following summary descriptions of different types of tillage operations are based on those given by Bakker (1999) for sugarcane cultivation and Draycott and Christenson (2003) for beet cultivation:

- Primary tillage – initial, major soil manipulation, generally by ploughing.
- Secondary tillage – subsequent operations (harrowing, furrowing, etc.), in preparation for sowing/planting.
- Deep tillage – disturbance (ploughing) of soil to a depth below about 25 cm.
- Conventional tillage – disturbance (ploughing) of soil to a depth of about 25 cm.
- Combination/conservation tillage – less direct soil disturbance, e.g. 30% or more of previous crop residue left on surface.
- Minimum tillage – little direct soil disturbance prior to planting new crop.
- No/zero tillage – new crop is planted without primary or secondary tillage following harvest of the previous crop.

Bakker (1999) suggests that cane crops are generally ploughed out (under conventional tillage) every 4 years, and in many cases less often, depending on ratooning potential. Old cane stools are mechanically uprooted, followed by harrowing, furrowing and planting of a new crop into the old inter-rows. Beet-growing soils experience more regular tillage, the precise operations involved being determined by the particular stage in the crop rotation. Henriksson and Hakansson (1993) and Draycott and Christenson (2003) summarize the tillage operations typically applied in preparation for the beet crop itself. Primary tillage or stubble treatment (typically mouldboard ploughing) is usually carried out in the autumn in Europe and the USA, allowing the soil to be weathered over the winter months. Primary tillage may be delayed until spring on coarser textured soils, or to incorporate green manures grown as a winter cover crop. Other operations, such as application of lime to ensure close to neutral soil pH, may involve more than one additional ploughing (Draycott and Christenson, 2003). The final stages of seedbed preparation generally involve one or two harrowings, followed by sowing. Overall, up to nine operations may be carried out, often in a random traffic pattern, leading to a total track area (for tractors and other vehicles) of three times the area of the field. Henriksson and Hakansson (1993) also note that ploughing depth for beet cultivation has gradually increased in recent decades, to around 25 cm in northern Europe and often deeper in southern Europe. This is generally deeper than the ploughing requirements for other crops in the rotation.

The effects of tillage are complex, influenced by factors such as soil type, topography and climate, and involve both positive and negative environmental impacts. For example, regular tillage enhances the breakdown of soil organic matter, by aerating the soil (stimulating microbial activity), releasing organic matter 'locked up' in soil aggregates and increasing its exposure to higher temperatures at the soil surface (e.g. Bakker, 1999; Haynes and Hamilton, 1999). Whilst this increases the availability of nutrients in the soil (potentially reducing the need for fertilizer inputs), soil structure and moisture holding capacity may be impaired if the organic matter is not replaced. Also, tillage can reduce the erosion risk (and associated runoff problems) presented by compacted soils, but can (in itself) exacerbate erosion problems in other situations. Conversely, reduced tillage may be recommended to reduce erosion risks, only to exacerbate soil compaction and weed problems. Another consideration is the retention of crop residues associated with minimum tillage, which may help to suppress pests and conserve soil moisture but inhibit soil warming and germination (Henriksson and Hakansson, 1993).

Overall, the trend in the literature is to recommend reduced (minimum, conservation or even zero) tillage in order to control a range of environmental problems associated with the cultivation of sugar crops. However, increased (deep) tillage has also been recommended in a number of cases, to address particular issues. For example, where soils are compacted, deep tillage may be practised in order to loosen the subsoil. However, this has the potential to do more harm than good (according to soil type and local conditions), impairing soil quality in the surface layers and possibly even resulting in greater compaction problems in the longer term (Henriksson and Hakansson, 1993). In some systems, deep tillage has been recommended to enhance water conservation and reduce soil erosion, as in cane cultivation in Sulawesi (Subagio and Mumwandono, 1992).

Reports of the environmental benefits of reduced tillage include the following:

- Improved water conservation in cane cultivation (Holden *et al.*, 1998).
- Improved water conservation in beet cultivation (Stout *et al.*, 1956; Papesch and Steinert, 1997).
- Reduced soil erosion in cane cultivation (Hadlow and Millard, 1978, 1981; Prove *et al.*, 1995; Glanville *et al.*, 1997; Ferrer and Nieuwoudt, 1998).
- Reduced soil erosion in beet cultivation (Simmons and Dotzenko, 1975; Graf *et al.*, 1983; Merkes, 1983; De Ploey, 1986; Sidiras *et al.*, 1988; Sommer, 1989; Opanasenko, 1998; Hao *et al.*, 2001).
- Reduced impact of cane cultivation on soil quality (Armas *et al.*, 1991; van Antwerpen and Meyer, 1996a,b; Blair, 2000; Dominy *et al.*, 2001; Hammad and Dawelbeit, 2001; Grange *et al.*, 2002).
- Reduced impact of beet cultivation on soil quality (Sommer, 1989; Hao *et al.*, 2001).
- Reduced fuel consumption (Dey *et al.*, 1997; Antonelli *et al.*, 2001; Draycott and Christenson, 2003).

Reduced tillage can promote weed populations (as demonstrated in beet fields in the Ukraine – Korytnik and Malienko, 1994), although it has also been suggested that minimal cultivation systems discourage weed emergence (Hopkins, 1992). Whilst weeds play an important role in the biodiversity of sugar cropping systems (as in beet fields in the UK - Defra, 2002), measures that actively promote their spread and growth may not be desirable from an agronomic perspective. Some systems of reduced tillage may, therefore, involve increased applications of agrochemicals, such as herbicides (e.g. Hadlow and Millard, 1978, 1981; Opanasenko, 1998), which may have to be offset against environmental benefits such as soil conservation. It is in this area (reduced soil erosion) that reduced tillage appears to be particularly valuable. Graf *et al.* (1983) estimated that adoption of no tillage cultivation of sugar beet on 0.5% slopes in Wyoming (USA) resulted in an 85% reduction in soil erosion. In the Ukraine, Opanasenko (1998) reported at least a halving of soil erosion (and increased beet germination) in fields with slopes of 3° or less. In cane cultivation systems in Australia, Prove *et al.* (1995) found that zero tillage reduced soil loss rates on conventionally cultivated slopes of 5–18% from an average of 148 t/ha/year to < 15 t/ha/year, and proved more effective than trash mulching.

So it is that reduced tillage can be particularly valuable on the sloping lands where sugarcane is grown in countries like South Africa. SASA (2002) concludes that mimimum tillage must be practised on slopes greater than 11% on erodible soils, 13% on moderately erodible soils and 16% on erosion-resistant soils. Bakker (1999) goes further, concluding that minimum tillage should be a general feature of cane cultivation, even on flat lands (except on heavy clay soils).

Mulching and the cultivation of catch/cover crops

Like reduced tillage, mulching has been identified as having a range of environmental benefits in the cultivation of sugar crops. There has been particular interest in the use of cane trash as a form of mulch, often in combination with a switch from preharvest burning to green cane harvesting (see Box 2.4). Probably because of fundamental differences

Box 2.4. Green cane harvesting and trash blanketing.

One particular agricultural practice in sugarcane cultivation that appears to have a range of environmental benefits is the shift from preharvest cane burning to green cane harvesting and trash blanketing.

The adoption of green cane harvesting, in itself, eliminates a range of negative environmental impacts arising from preharvest burning, including release of combustion pollutants (Scandaliaris *et al.*, 1998; Cock *et al.*, 1999) and negative impacts on the soil (see Chapter 6). In combination with the retension of cane trash in the field as a form of mulch (a trash blanket of 3–10 t/ha of crop residues – Ng Kee Kwong *et al.*, 1987; Thind, 1996; Swamy *et al.*, 1998; Thanki *et al.*, 1999), these benefits are increased. For example, instead of CO_2 being released in a single burning event, it is retained in the trash and released slowly, partly through emissions to the atmosphere (which are greater from trash blanketed fields than from bare fields), and partly through assimilation into soil organic matter (Weier, 1998). This increase in organic matter enhances soil quality, and is seen as a major benefit of trash blanketing (Wood, 1986, 1991; van Antwerpen and Meyer, 1998; Graham *et al.*, 1999, 2002; Haynes and Hamilton, 1999; Hartemink, 2003; Noble *et al.*, 2003). In the long term, computer modelling by Vallis *et al.* (1996a) suggests that trash blanketing can raise soil organic matter content by around 40% after 60–70 years. In addition to increased soil organic matter, Yadav *et al.* (1994) found that, over 3 years, trash blanketing increased available soil N by 37 kg/ha and available P by 10 kg/ha. Trash mulching also reduced optimum N fertilizer application rate from 241 kg/ha to 230 kg/ha, and increased crop yield response per kg of N (applied at optimum rate) from 263 kg to 328 kg. Wood *et al.* (1997) suggest that trash blanketing may have the potential to reduce N fertilizer input requirements by 40 kg/ha/year. Retention of trash also appears to enhance soil biodiversity, in terms of microbial and earthworm communities (Wood, 1991; Sutton *et al.*, 1996; Graham *et al.*, 1999, 2002).

In addition to improvements to soil quality, trash blanketing can reduce the risk of soil erosion (Lugo-Lopez *et al.*, 1981; Prove *et al.*, 1986; Sullivan and Sallaway, 1994). SASA (2002) recommends that trashing (mulching) should be practised on slopes greater than 15% during the wet season, to reduce the impact of raindrop action, if insufficient crop cover has developed. As well as assisting in the conservation of soil, trash blanketing contributes to the conservation of soil moisture. Some early studies questioned this (e.g. Eavis and Chase, 1973), but many subsequent experiments have indicated significant benefits (e.g. Lugo-Lopez *et al.*, 1981; Thind, 1996; Denmead *et al.*, 1997; Murombo *et al.*, 1997; Swamy *et al.*, 1998; Meier *et al.*, 2002). Yadav (1986), for example, found that trash mulching resulted in a 40% economy of irrigation water. Trash blanketing can also contribute to the conservation of natural enemies of cane pests and suppress weed development (Kuniata and Sweet, 1994), reducing the need for herbicide inputs (SASA, 2002). Other benefits ascribed to green cane harvesting and trash blanketing include reduced diurnal fluctuations in soil temperature and the facilitation of wet weather harvesting (shortening the growing season) (Garside *et al.*, 1997b).

Some disadvantages to green cane harvesting and trash blanketing have been recorded. For example, the shift away from preharvest burning in Australia contributed to the re-emergence of the sugarcane weevil borer (*Rhabdoscelus obscurus*) as a significant pest in some areas (Robertson and Webster, 1995). Hartemink (2003) suggests that it may also contribute to soil acidification (as pH-increasing ashes are no longer returned to the soil), an idea supported by the results of Noble *et al.* (2003). Where internal drainage is poor, there is some evidence that a trash blanket may produce allelopathic effects, with the potential to suppress development of cane and other plants (Wood, 1991; Garside *et al.*, 1997b). Facilitation of wet weather harvesting may increase the risk of soil compaction, and reduced cultivation for weed control reduces the alleviation of compaction associated with soil disturbance. However, experience in other systems with zero tillage and stubble retention suggest that a period of 10 years is often required before improvements in soil physical quality are recorded (Wood, 1986; Garside *et al.*, 1997b).

Other disadvantages of green cane harvesting and trash blanketing may include increased harvesting costs, complications to irrigation and fertilizer application and slowing of tiller emergence. However, these appear to be significantly outweighed by the benefits (Murombo *et al.*, 1997; Bakker, 1999), which may include increased yields. The precise effects of green (vs. burnt) harvesting on cane yield factors and the economics of cane cultivation are difficult to determine, but it is likely that any reduction in yield factors under green cane harvesting are compensated for by other cost savings (Wood, 1991; Garside *et al.*, 1997b).

Green cane harvesting and trash blanketing are now widely practised in some parts of the world, as in Australia, where uptake of the technique has increased over recent years (Garside *et al.*, 1997b). Ballantyne (1998) estimated that, in 1997, 65% of Queensland cane was harvested green, compared with just 18% in 1987. However, the shift away from preharvest burning has been more rapid in some areas than in others.

Weier (1998) estimates that, in 1994, 88% of the North Queensland crop was harvested green, whilst 96.5% of the crop in the Burdekin region was still subject to burning. Woods *et al.* (1997) discuss the complex of issues behind farmer uptake of green cane harvesting and trash blanketing in Australia, in which environmental considerations are just part of the decision-making process.

in the cultivation system, much less attention has been paid to mulching, as such, in beet growing. However, reported benefits of leaving crop residues on or near the soil surface in beet fields include improved water and soil conservation (Hagen, 1974; Simmons and Dotzenko, 1975; Sommer and Zach, 1984; Grundwurmer, 1991; Geelen *et al.*, 1995; Papesch and Steinert, 1997). A related practice more relevant to beet growing is the cultivation of catch/cover crops. Cover cropping has attracted limited interest in cane cultivation, but is seen as having potential to conserve soil moisture, reduce erosion, improve soil quality and suppress weed growth (Scandaliaris *et al.*, 2002).

Catch/cover crops may be grown over winter in fields awaiting the sowing of sugar beet in the spring. This practice appears to have a range of environmental benefits. Where fields awaiting a beet crop are left bare over the winter, there can be an increased risk of soil erosion by wind and water and of nitrate leaching. At the simplest level, some degree of protective cover is provided simply by allowing weeds or volunteer cereals to grow over winter (Selman, 1976). Rye has attracted particular attention as a sown cover crop, specifically to reduce wind erosion (Lumkes and te Velde, 1973; Pickwell, 1974; Knottnerus, 1976a; Bastow *et al.*, 1978; Cherry, 1983; Merkes, 1983). Defra (2002) noted that wind erosion in UK beet cultivation was increasingly controlled by the planting of cover crops. A wide range of cover crops, including *Phacelia*, mustard, oil-seed radish and clover have been investigated to counter the general threat of erosion in beet fields (Marlander *et al.*, 1981; Merkes, 1983; Schmidtlein *et al.*, 1987; Sommer and Lindstrom, 1987; Sidiras *et al.*, 1988; Sommer, 1989). However, there have been suggestions that cover cropping can inhibit crop development under certain circumstances. Elliott *et al.* (1979) found that ploughing in of wheat and oats sown as winter cover crops resulted in sugar beet seedling losses of 17–30%, caused largely by phytotoxicity of decomposing cover material, but only where this was in close contact with the emerging crop plants (suggesting that residues should be kept out of beet rows). Various plants have also been found to serve as catch crops, as well as cover crops. These remove residual nitrate from the soil after harvest of the preceding crop in the rotation, reducing the risk of nitrate leaching over winter (Allison and Armstrong, 1991; Allison *et al.*, 1993, 1998a,b; Duval, 2000). Such catch crops can also be used as green manures, being ploughed into the field prior to sowing of beet. From an environmental perspective, however, it should be noted that cover crops may be killed off with an application of herbicide, prior to beet being sown.

Fertilizers

Environmental problems associated with fertilizer use in the cultivation of sugar crops include:

- Impacts on soils:
  - perturbation of soil nutrient balance;
  - soil acidification.
- Impacts on water quality (through runoff and leaching):
  - contamination of groundwater (including drinking-water);
  - contamination of surface water;
  - pollution of downstream aquatic ecosystems.
- Impacts on air quality:
  - release of gaseous emissions.

Environmental impacts typically arise because the nutrients applied with fertilizers are not entirely taken up by the crop (Neeteson and Ehlert, 1988), although deficiencies of certain nutrients may also contribute to a decline in soil quality. Nutrients that are taken up by sugar crops may later pose environmental problems, as waste products of processing. Where fertilizer applications result in an excess of soil nutrients, the threat of water pollution (particularly by leaching) varies according to the nutrients concerned, as well as their concentrations. This is because different nutrients tend to be more or less mobile in solution. For example, nitrate and sulphate tend to be relatively mobile, whilst phosphate and ammonium are relatively immobile (Draycott and Christenson, 2003). Similarly, different fertilizers pose different levels of risk in terms of atmospheric pollution.

Aspects of soil quality in cane cultivation systems, including broad coverage of issues related to fertilizer usage, are reviewed by Haynes and Hamilton (1999) and Hartemink (2003). The role of nutrients in sugar beet production is reviewed by Draycott and Christenson (2003), and general papers on the environmental aspects of nitrogenous inputs in beet cultivation systems include those by Brentrup *et al.* (2001) and Venturi and Amaducci (2002).

Patterns of nutrient usage by sugar crops

Sugarcane and sugar beet both require around 14 different chemical elements for normal growth and development, of which the most important is N. In the absence of nitrogenous fertilizers, plants incapable of symbiotic N fixation rely on net mineralization of soil organic matter and N input from atmospheric sources. Net mineralization rates may be high (around 200 kg N/ha/year) in newly cultivated soils, but tends to decline (often to < 50 kg N/ha/year) as soil organic matter content decreases. Atmospheric sources (rainfall, dry deposition of ammonia, non-symbiotic N fixation) generally contribute around 20–40 kg N/ha/ year (Keating *et al.*, 1997). In the absence of fertilizer, few arable soils can provide more than 100 kg N/ha during the growing season, and beet generally requires twice this amount for maximum production (Draycott and Christenson, 2003). Similarly, cane requirements exceed the N available naturally in most soils.

Because of the importance of N in crop nutrition (and fertilizer inputs), and because N is relatively mobile in soil solution, creating a particular threat of nitrate impacts on water quality, the level of N recovery by sugar crops is an important consideration. Estimates of N recovery by sugarcane are generally of the order of 20–50%, although these may be underestimates, failing to account for gaseous N losses from the plant (Hartemink, 2003). Sugar beet has a reputation as a very effective scavenger of soil N. Estimates of soil or fertilizer N recovery by beet vary, with reported values ranging at least from 8 to 80%, although 50–60% seems to be representative, with much of any remaining fertilizer N incorporated into soil organic matter (Draycott, 1993; Draycott and Christenson, 2003; Sotiriou and Scheunert, 1994). The nitrogen dynamics of a typical sugar beet field (after Draycott and Christenson, 2003) are summarized in Fig. 2.5. In a sugar beet crop, a greater proportion of N is found in the leaves than in the roots. Consequently, beet tops can be left in the field after harvest as a form of organic fertilizer, but this can increase the risk of leaching.

In relation to other major nutrients, it appears that levels of P tend to increase in both cane and beet growing soils following regular inputs from fertilizers (Draycott and Christenson, 2003; Hartemink, 2003), partly because it is relatively immobile when compared to other nutrients (and is therefore less likely to be leached than, for example, N). Sugarcane is a relatively heavy consumer of K, and cane growing soils can become depleted of this nutrient (Hartemink, 2003). Beet also consumes much soil K, but is a relatively heavy consumer of Na (which it can partially substitute for K); this is related to its halophytic origins (Draycott and Christenson, 2003).

Inorganic fertilizers

Inorganic fertilizers typically supply N, P and/or K in mineral form, and some also supply S. The specific form in which they are applied can influence their environmental fate. In many areas of the world, nitrogenous fertilizers are routinely applied in cane cultivation at rates of around 50–200 kg/ha/year (Ruschel *et al.*, 1982; Haynes and Hamilton,

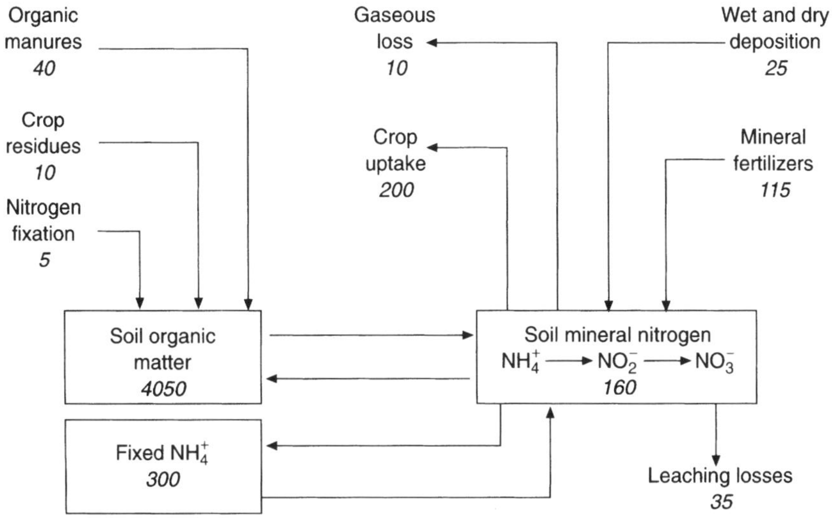

Fig. 2.5. The nitrogen dynamics of a typical sugar beet field (after Draycott and Christenson, 2003).

1999). Smith (1997) suggests that annual application rates of 120–240 kg N/ha, 30–50 kg P/ha, 40–100 kg K/ha and 15–25 kg S/ha are common. Whilst these figures appear to be representative of typical application rates, greater ranges of rates are undoubtedly used worldwide. In many cane growing areas, detailed accounts of the pattern of fertilizer use and its environmental impacts are not readily available. However, such issues have been well studied in Australia, where fertilizer application rates have shown an upward trend in the last 50 years (see Box 2.5). Draycott and Christenson (2003) report wide variation in the application rates of inorganic fertilizers in beet cultivation worldwide, with ranges of 76–240 kg/ha for N, 37–315 kg/ha for P_2O_5 and 0–290 kg/ha for K_2O. They also report substantial decreases in the use of NPK fertilizers in sugar beet cultivation in western Europe since the 1960s (see Fig. 2.6).

Fry (1997) suggests that, globally, environmental restrictions and quality payment systems have encouraged a reduction in the use of nitrogenous fertilizers in sugar production. However, as with the use of other technologies, patterns of fertilizer usage are very variable from country to country. While the Australian sugar industry has embraced mechanization and accelerated its use of fertilizers (Box 1.1 and Box 2.5), the industry in India, for example, is much less mechanized and uses only limited amounts of inorganic fertilizer (Fry, 1997).

Organic fertilizers

Although their use has potential benefits, organic fertilizers (such as manures and sugar-processing wastes) can also generate environmental problems. Use of such organic fertilizers is complicated by the variable quantities of individual nutrients that they contain, and by unpredictable nutrient releases from the breakdown of their organic content (Deville, 1999; Draycott and Christenson, 2003).

Draycott and Christenson (2003) report on the use of a range of organic fertilizers (including poultry manures and animal slurries) in beet cultivation. Although reliable figures are lacking, they conclude that inputs of such materials have probably declined overall in the last 50 years. Biosolids (sludge from sewage treatment) are widely used on beet fields in the EU and USA prior to sowing, but the use of this material is heavily regulated, because of the risk of pathogens and heavy metals entering the food chain. Historically, and in recent years, it appears that growers regarded organic inputs as having principal value as soil conditioners. However, lack of attention to the nutrient content of such

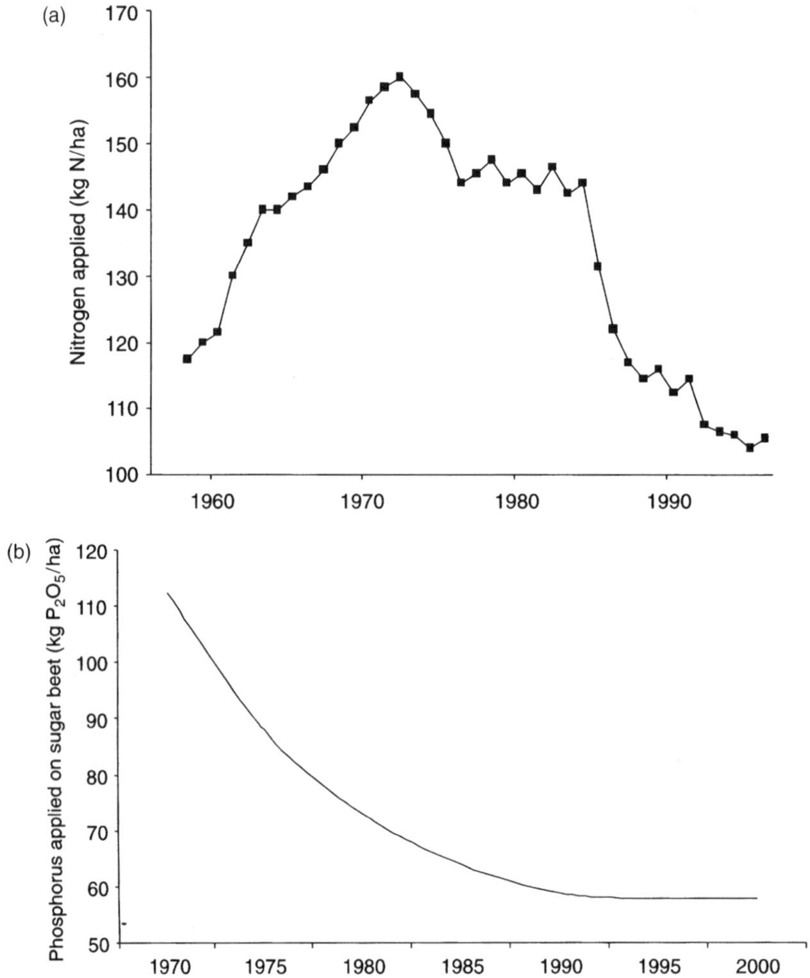

Fig. 2.6. Patterns of inorganic fertilizer use in UK sugar beet cultivation: (a) N inputs and (b) P inputs (after Draycott & Christenson, 2003).

Box 2.5. Inorganic fertilizer use in cane cultivation in Australia.

Whilst enabling increased production, increased inorganic fertilizer use in Australia has contributed to degradation of cane growing soils, contamination of ground and surface water and enhanced greenhouse gas emissions; application rates are high by world standards, increased substantially in the postwar years and often exceed industry recommendations and levels required to maximize yields (Garside *et al.*, 1997b; Keating *et al.*, 1997; Verburg *et al.*, 1998). Mitchell *et al.* (2001) estimate that, in the Tully River catchment, fertilizer N use increased by 130% between 1987 and 1999. Thorburn *et al.* (2003b) concluded that, for more than a decade, N fertilizer use had been almost double that required to produce the actual cane (and sugar) yields obtained by the industry. Garside *et al.* (1997b) quote figures which suggest typical N application rates of around 210 kg/ha in the early to mid-1980s, followed by a temporary reduction before applications increased again, to as much as 300 kg/ha in some areas. Heavy inputs of other elements, notably P, have further contributed to an imbalance in soil nutrients under cane in Australia (Garside *et al.*, 1997b). Trends in inputs of N and P fertilizers, and cane yields, in the Herbert River district during the 20th century are illustrated in Fig. 2.7.

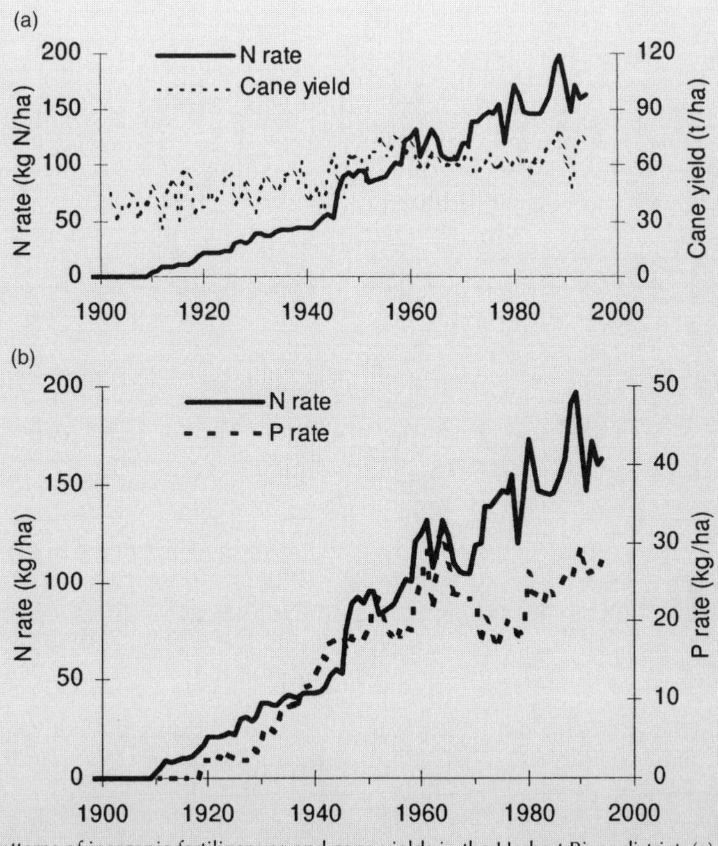

Fig. 2.7. Patterns of inorganic fertilizer use and cane yields in the Herbert River district: (a) N inputs and cane yields (after Keating *et al.*, 1997) and (b) N and P inputs (after Wood *et al.*, 1997).

Wood *et al.* (1997) note that, in Australia, around 20–25% of N is commonly applied as NPK fertilizer at planting or soon after ratooning, with the remainder applied around the time of canopy closure, usually as urea. Urea is by far the cheapest and most popular form of N fertilizer, but can result in substantial (around 40%) losses of N through volatilization when applied to the surface of a trash blanket (see Chapter 7).

Thorburn *et al.* (2003b) explored why Australian cane growers tended to over-apply N fertilizers, and concluded that uncertainty about the size of the forthcoming crop and possible long-term effects of reduced applications were the main issues. This was despite considerable evidence that greatly reducing N inputs for a single crop did not significantly reduce production, that only sustained under-application of N fertilizer was likely to reduce profitability and later evidence that over-application of N could reduce the sugar content of cane (Muchow *et al.*, 1996). Even in the late 1990s, Wood *et al.* (1997) suggested that the misconception that cane yields could be increased simply by applying more fertilizer may have been widespread in Australia. However, these authors also noted a shift in the philosophy behind fertilizer use, away from maximizing production (by focusing on supply to the crop rather than managing the soil resource), towards maximizing efficiency of use (and minimizing soil nutrient imbalance), within the context of maintaining soil health in the longer term.

materials increases the risk of over-fertilization, and of environmental impacts from runoff and leaching. Increasingly, recommendations and regulations are being developed to counter such risks in beet growing countries. In the context of cane cultivation, SASA (2002) considers that cane processing wastes (filter cake, effluent) and other organic fertilizers (chicken manure, kraal manure, abattoir wastes) should be used as soil amendments only with caution and after obtaining professional advice and guidance. None the less, there is evidence of the benefits of application of such materials in cane cultivation from many parts of the world.

Draycott and Christenson (2003) also report on the use of green manures in beet cultivation, where leguminous or non-leguminous crops are grown and ploughed in prior to sowing. Such operations improve soil structure, provide cover to the soil over winter and decrease the risk of nitrate leaching.

Rationalizing fertilizer use

Maximizing crop nutrient use efficiency requires a complex calculation of crop requirements, based on an understanding of nutrient (particularly N) cycles and particular characteristics of cultivars, soils and climate and the effects of management practices. Assessment of appropriate fertilizer input levels can be based on yield response functions, soil analysis and plant tissue analysis.

A number of strategies are available for rationalizing fertilizer inputs in cane cultivation systems:

- Improved crop and soil monitoring to assist decision-making (Keating *et al.*, 1997).
- Improved advisory services for growers (Wood *et al.*, 1997).
- More site-specific assessment of fertilizer requirements (Wood *et al.*, 1997; Thorburn *et al.*, 2003b).
- More frequent application of fertilizers, but in smaller quantities (Weier, 1998).
- Greater emphasis on organic matter and 'minor' soil nutrients (Wood *et al.*, 1997).
- 'Replacement' strategy (grower aims to replace the N lost from the previous crop, rather than aiming to fertilize the coming crop) (Wood *et al.*, 1997; Weier, 1998; Thorburn *et al.*, 2003b).
- Slow release fertilizers and nitrogen stabilization packages, including nitrification and urease inhibitors (Keating *et al.*, 1997; Wood *et al.*, 1997; Weier, 1998; Hallmark *et al.*, 1999).
- Cultivation of cover crops during fallow periods to remove residual nitrate, possibly followed by ploughing in as a green manure (Garside *et al.*, 1996; Weier, 1998; Bakker, 1999).
- Green cane harvesting/trash blanketing (see Box 2.4).
- Rotations of cane with other crops (Wood *et al.*, 1997).
- Biofertilizers (Bangar *et al.*, 1993; Dobereiner *et al.*, 1995; Hunsigi and Shankariah, 2001; Shankaraiah *et al.*, 2001; Kannaiyan, 2002).
- Drip fertigation (Keating *et al.*, 1997; Ng Kee Kwong *et al.*, 1999b; Thorburn *et al.*, 2000).

In some sugarcane cultivation systems, there is a direct economic incentive for reducing fertilizer inputs, as fertilizer represents a significant cost to the farmer (e.g. Dobereiner *et al.*, 1995). Where fertilizers do not represent a significant cost, economic incentives for reduced N inputs depend on negative effects of excessive applications on the crop. Excessive N does not appear to reduce biomass production (e.g. cane response curves tend to remain flat once the optimum N application rate is reached – Keating *et al.*, 1997). However, there is evidence that excessive N reduces the quality of both cane and beet in terms of sugar yields (Muchow *et al.*, 1996; Draycott and Christenson, 2003). None the less, it seems likely that reduced N inputs to sugar crops are more likely to be driven by environmental imperatives than by economics (e.g. Keating *et al.*, 1997).

Drip fertigation is perhaps of particular interest from an environmental perspective, in that it combines the increased water use efficiency of a drip irrigation system with the potential to manage more effectively (and thereby reduce) fertilizer applications. Keating *et al.* (1997) see this as a particularly promising technology in its potential to control N leaching in sensitive areas. The results of Ng Kee Kwong *et al.* (1999b) and Thorburn *et al.* (2000) suggest that drip fertigation allows N fertilizer inputs to be reduced by 25–50% without impairing cane productivity. Studies of the biofertilizer approach are also of interest, suggesting that combinations of nitrogen-fixing microorganisms such as *Azotobacter chroococcum*, *Azospirillum brasilense* and *Acetobacter diazotrophicus* and organic amendments could reduce fertilizer requirements by 20–25% (Bangar *et al.*, 1993; Hunsigi and Shankariah, 2001) and reduce the risk of nitrate leaching (Shankaraiah *et al.*, 2001).

A range of measures similar to those above have been investigated for rationalizing the use of fertilizers in sugar beet cultivation systems. These include:

- Choice of appropriate fertilizer application equipment (Tugnoli and Maini, 1992).
- Choice of appropriate crop rotations (Ceotto, 2001).
- Choice of appropriate form of fertilizer (Zak *et al.*, 2002; Draycott and Christenson, 2003).
- Modified timing, rate and method of application (Hills *et al.*, 1978; Becker and Bruss, 1996; Henriksen *et al.*, 1998; Draycott and Christenson, 2003).
- Improved crop and soil monitoring to assist decision making (Draycott and Christenson, 2003).
- Improved information and advisory services for growers (Bartocci *et al.*, 2001; Ver Elst *et al.*, 2001; Draycott and Christenson, 2003).
- More site-specific assessment of fertilizer requirements (Draycott and Christenson, 2003).
- 'Replacement' strategy (grower aims to replace the P lost from the previous crop, rather than aiming to fertilize the coming crop) (Draycott and Christenson, 2003).

Christensen (2004) suggests that cultivation of crops such as sugar beet and potatoes in cereal-dominated crop rotations presents particular challenges in 'tightening the nitrogen cycle'. Problems with beet include late harvesting (not allowing for the sowing of autumn cereals or cover crops), heavy soil disturbance at harvest and the potential of beet tops (left in the field after harvest) to add to nitrate leaching losses. Draycott and Christenson (2003) note that, despite 100 years of research, it is still difficult to make specific recommendations for fertilizer application rates to beet crops at the field scale.

Scott and Jaggard (1993) consider that restrictions on inputs of organic N sources (manure and slurry) in many sugar beet growing areas will lead to improved yields. Many farmers fail to consider the nitrates from these organic sources when designing their fertilizer regime, resulting in excess nitrates that only serve to reduce beet quality. For example, 20% of beet fields in the UK are estimated to receive soil dressings of manure or slurry prior to sowing, and half of these receive inorganic fertilizers (sometimes at rates in excess of 100 kg/ha) as well. Scott and Jaggard (1993) suggest that a 50% reduction in nitrate inputs to beet across the UK would result an overall yield reduction of only some 10%, and note that the key consideration would be to manage the timing of nitrate inputs more effectively. Indeed, there is evidence of decreased fertilizer use in UK sugar beet cultivation in recent years.

Pesticides

Agricultural food production in general strongly depends on the use of pesticides, with herbicides representing about 50% of pesticides used in many countries (Lanchote *et al.*, 2000). Quite apart from impacts on the wider environment, the negative impacts on human health are considerable. Specific figures for sugar crops are not readily available, but, in relation to pesticide use in agriculture in general, the World Health Organization (WHO) estimates that there are 25 million cases of acute chemical poisoning in developing countries each year (Dent *et al.*, 2003). Despite widespread concern over pesticide misuse, the total value of world sales has increased 2.5 times in the last 20 years, to US$30 billion (Bateman, 2003). Agrochemical companies, which often provide farmers with most of their information on synthetic chemical inputs, are unlikely to develop or promote techniques that reduce pesticide use. However, it is in their interest to promote practices that maintain the longer-term viability of their business (Dent *et al.*, 2003), rather than becoming associated with negative impacts on environment and health. A range of methods is available for optimizing pesticide use in agriculture and minimizing undesirable impacts (e.g. see Wilson, 2003).

Environmental problems associated with pesticide use in the cultivation of sugar crops include:

- Impacts on soils:
  - accumulation of pesticides in soils.
- Impacts on water quality (through leaching and runoff, involving soluble residues and those bound to sediments):
  - contamination of goundwater (including drinking-water);
  - contamination of surface water;
  - pollution of downstream aquatic ecosystems.
- Impacts on biodiversity:
  - non-target effects (in the field and in adjacent areas owing to spray drift).
- Impacts on human health.

In addition, as with poorly managed fertilizer application, the environmental impact of pesticides can manifest itself in effects on sugar crop yields. In long-term agrochemical, microbiological and ecological experiments on the use of pesticides on sugar beet, Mineyev *et al.* (1993) demonstrated accumulation of toxic substances in roots and aerial parts of the crop plants, and found that maximum doses of pesticides resulted in retardation of growth and decrease in sugar content.

A wide variety of pesticides (herbicides, insecticides, fungicides, nematocides) is used in the cultivation of sugar crops. In cane cultivation, chemical agents may also be used against vertebrate pests, notably rats. In environmental terms, the most persistent compounds are the most problematic, although a range of factors influence the rates at which particular pesticides break down (e.g. see Pussemier, 1991; McMartin *et al.*, 2003). The quantities and specific agents used in any given case will depend on a range of factors, including the identities of the weeds and pests involved, availability and cost of pesticides and equipment to the farmer, systems for registration of pesticides in individual countries and enforcement of relevant regulations. Although there has been a shift towards more rigorous regulation of pesticide use and adoption of less persistent agents, degrees of enforcement of regulations and preferred practices of growers inevitably vary considerably between different localities. In addition to pesticides themselves, Wevers (1997) notes that concern is growing about the environmental effects of other ingredients in agrochemicals, like solvents and emulsifiers.

Along with inorganic fertilizers, pesticide (particularly herbicide) use increased in Australia in the post-war years, as the economics of cane production justified their application (Garside *et al.*, 1997b). Hamilton and Haydon (1996) reported that pesticide usage in the Queensland sugar industry was dominated by the use of herbicides (notably atrazine), which exceeded insecticide use by a factor of ten. None the less, control of pests relies heavily on insecticide use, although there is an increasing move towards alternative and integrated methods. These already play a significant role in disease control (Allsopp and Manners, 1997); fungicides represent only a small proportion of the pesticides used in Australian cane cultivation (Hamilton and Haydon, 1996). The shift towards more integrated methods of pest control has been driven by a recognition that old pesticide-based strategies are unlikely to remain viable in the long term, as environmental pressures have increased, and agents like thallium sulphate and organochlorines have been withdrawn (Allsopp and Manners, 1997). Although fewer persistent agents are now used in Australia, there is evidence that compounds such as DDT and dieldrin may continue to pose a threat through bioaccumulation in some cane growing areas (Johnson *et al.*, 1997).

In some other parts of the cane growing world, there is a greater emphasis on non-chemical control methods, particularly for insect pests. In Guyana, for example, chemical pest control has largely been abandoned (SAC, 2000), owing to a number of factors, including the success of biological control against some key pests. It is likely that the threat of long-term pesticide impacts has reduced in recent years in many cane growing areas, as less persistent agents have replaced older formulations. Reduced impacts associated with less persistent agents depend, of course, on new formulations being adopted by growers.

Recent studies have shown a pattern of reduced pesticide use in beet cultivation in the UK (see Box 2.6). Elsewhere in Europe, there is also evidence of a reduction in quantities of pesticides used in beet cultivation, for example, in Belgium (Eeckhaut, 2001). In some beet growing areas, conventional practice involves only limited use of chemical pesticides. For example, herbicide use against weeds in Moroccan beet is still very limited, manual and mechanical control being the dominant methods (El Antri, 2001).

Rationalizing pesticide use

> Indiscriminate use of pesticides creates a number of problems, such as development of resistance in pests, upsurge of secondary pests because of elimination of natural enemies, pollutes the environment making it hazardous for human beings and animals. Moreover, they are expensive and increase the cost of crop production. (Dr Zafar Altaf, Chairman of the Pakistan Agricultural Research Council, in the foreword to Mohyuddin *et al.*, 1994)

Reduced pesticide inputs have been found to have a range of environmental benefits (increasing biodiversity in a range of taxa in sugar beet cultivation systems, for example – Esbjerg, 1998). Various strategies are available for rationalizing and reducing pesticide use in sugar crop cultivation systems. These include:

- Development of more selective pesticides.
- Use of appropriate application equipment (Hopkins, 1992; Tugnoli and Maini, 1992).
- Use of reduced pesticide concentrations (Muchembled, 1992; Eeckhaut, 2001; Hermann *et al.*, 2001).
- More precise application (Hopkins, 1992; Rudolph and Klee, 1993; Scott and Jaggard, 1993; Wijnands and van Asperen, 1999).
- Use of trap crops (Hafez and Sundararaj, 1999; Held *et al.*, 1999).

Box 2.6. Pesticide use in beet cultivation in the UK.

In examining environmental aspects of beet cultivation in the UK, Defra (2002) identified a number of trends and considerations with respect to pesticide use. Overall, between 1982 and 1998, the total pesticide input to UK beet cultivation fell from around 11 kg/ha to just over 5 kg/ha (a reduction of well over 60%). The reduction in insecticide inputs, for example, was partly due to a shift away from spraying towards seed treatment, such that 70% of UK beet crops received no insecticide in spray form. Use of nematocides (seen as, generally, the most toxic group of agrochemicals – see Held *et al.*, 1999) had fallen by around 50% between 1994 and 2000, in terms of both quantities applied and area treated. In 2000, around 7.25 t nematocides were used to treat approximately 10,500 ha (7% of the total beet cultivation area in the country). Levels of molluscicide use, a relatively small component of pesticide usage overall, varied considerably according to prevailing conditions each year, but typically involved treatment of 3000–10,000 ha. Fungicide use had not shown the recent decline found in application rates of other pesticides, but remained low in beet relative to other crops (a single application was often made, in comparison with three in cereals and up to seven in potatoes). Herbicide use tended to be greatest in the earlier stages of beet crop development, with four to five applications typically made in spring/early summer. However, development of more efficient and low-dose sprays had led to a reduction in herbicide inputs of more than 60% over 20 years. The reduced agrochemical inputs to beet in the UK identified by Defra (2002) were partly a consequence of increased adoption of a more integrated approach to crop management. Despite the considerable reduction in inputs, concern remained over a lack of detailed knowledge on pesticide impacts on watercourses and aquatic species, the potential hazards of increased seed treatments and the effects of herbicide drift and runoff into adjacent habitats. It is noted that even more can be done to reduce pesticide use in UK beet cultivation, possibly through an increased focus on integrated pest management.

- Use of non-chemical methods (Becker *et al.*, 1989; Hopkins, 1992; Scott and Jaggard, 1993; Ceccatelli and Peruzzi, 1995; Peruzzi *et al.*, 1995; Barberi, 1997).
- Increased use of seed dressings for beet (Eeckhaut, 2001; Pigeon *et al.*, 2001; Defra, 2002).
- Expert systems and computer software to assist decision-making (Wevers, 1997, 2001; Simpson *et al.*, 2003).

Perhaps the greatest potential benefits arise from the adoption of integrated management strategies for pests and weeds (Allsopp and Manners, 1997; Fernandez-Quintanilla *et al.*, 1999; Defra, 2002), drawing on rational pesticide use in combination with other (e.g. cultural and biological) control methods.

Integrated pest management and biological control

Integrated pest management (IPM) aims to combine a variety of appropriate control methods (including rational pesticide use) towards a more holistic and sustainable approach to pest control. Hence, IPM strategies may draw on, for example:

- enhanced knowledge of pest ecology;
- improved decision support systems;
- increased crop monitoring;
- resistant cultivars;
- cultural control methods;
- pheromones and other chemical deterrents/baits;
- a variety of trapping methods;
- biological control.

Cooperative approaches in preventing the spread of pests can also be an important component of IPM (e.g. see Mauremootoo, 2001). Maredia *et al.* (2003) review the application of IPM in a global context, including reference to IPM successes in sugarcane cultivation in a number of countries. Koul *et al.* (2003) review some of the constraints and practicalities associated with IPM programmes in general.

Biological control is an important component of many IPM programmes and has significant potential environmental benefits, in reducing the need for applications of chemical pesticides. However, poorly executed biological control programmes also carry environmental risks, based on potential impacts of biological control agents on (particularly indigenous) non-target species. It is unfortunate that two of the most widely cited examples of 'biological control gone wrong' are associated with early, very poorly judged attempts to control pests in sugarcane. These cases tend to overshadow the substantial contribution that biological control has made to the reduced use of pesticides in sugar crops, particularly cane.

Integrated pest management and biological control in sugarcane

Pesticide inputs can be reduced through the adoption of IPM strategies in sugarcane cultivation. Even where formal integrated strategies are not employed, practices intended to reduce environmental impacts in one area (such as soil erosion) may have beneficial effects in relation to pesticide application and impacts. Peng and Twu (1980) demonstrated, for example, that application of the soil conservation agent Curasol AH reduced problems associated with redistribution of herbicides. Trashing (mulching) can also reduce the amount of herbicide needed for weed control (e.g. SASA, 2002).

There are numerous examples of successful IPM programmes in sugarcane throughout the literature. One specific example is given here (Box 2.7), from Papua New Guinea, where IPM is seen as a high priority in the sugar industry (Hartemink and Kuniata, 1996). IPM programmes in the Asia–Pacific region (including those used in cane cultivation in countries such as Papua New Guinea, Fiji, Taiwan, India and Pakistan) are discussed in Ooi *et al.* (1992). Maredia *et al.* (2003) review the application of IPM in a global context, including reference to IPM successes in sugarcane cultivation in Brazil (Hoffmann-Campo *et al.*, 2003), India (Singh *et al.*, 2003; see also Madan, 2001; Pandey, 2002), Peru (Palacios Lazo *et al.*, 2003) and South Africa (Charleston *et al.*, 2003). Although the Australian sugar industry relies heavily on chemical agents for the control of pests, there is an increasing

Box 2.7. IPM against the sugarcane stem borer *Sesamia grisescens* in Papua New Guinea.

Papua New Guinea is considered to be the centre of origin of sugarcane. Consequently, the relatively recently established sugar industry here (see Box 1.2) is particularly afflicted by pests, diseases and weeds, most of which are native and may have co-evolved with the ancestors of the crop plant (Hartemink and Kuniata, 1996; Kuniata *et al.*, 2001; Magarey *et al.*, 2002).

The most important insect pest is the stem-boring larva of the noctuid moth *Sesamia grisescens*, control of which requires an integrated approach (Kuniata and Sweet, 1994). Such a programme has been developed, involving identification of resistant cultivars and optimum planting times, combined with rational pesticide use and biological control, based on close monitoring of the situation in the crop (Kuniata, 2000; Lloyd and Kuniata, 2000). The programme has been built on a sound knowledge of the biology of the pest species (Young and Kuniata, 1992), and has also investigated the use of pheromones for trapping or mating disruption (Whittle *et al.*, 1993, 1995). Because the weevil borer *Rhabdoscelus obscurus* (another significant indigenous pest) tends to preferentially attack cane already afflicted by *S. grisescens*, measures to control the moth larva also result in reduced damage from the weevil (Kuniata, 2000; Lloyd and Kuniata, 2000). The biological control element of this IPM programme has developed from studies of the natural enemy complex associated with the pest (Kuniata and Sweet, 1994). The most important natural enemies are hymenopterous parasitoids: the indigenous *Cotesia flavipes* (widely used in biological control programmes around the world, e.g. see Polaszek and Walker, 1991) and *Enicospilus terebrus*, and the exotic *Pediobus furvus*. Appropriately timed releases of *C. flavipes* and *P. furvus* have been incorporated into the *Sesamia* control programme (Kuniata, 2000; Lloyd and Kuniata, 2000). Kuniata and Sweet (1994) demonstrated that inappropriate chemical control aimed at *S. grisescens* in 1987 and 1988 resulted in a rapid build-up of cicada pests, which were then responsible for considerable cane damage from 1988 to 1991.

movement towards greater use of IPM systems. The implications of this and options for major pests of Australian cane cultivation are reviewed by Allsopp and Manners (1997). Integrated methods for the control of cane pests in Pakistan are reviewed by Mohyuddin (1992) and Mohyuddin *et al.* (1994). Biological control as a component of IPM programmes against cane pests (particularly stem borers) in Africa and neighbouring islands, including Mauritius, are briefly reviewed by Greathead (2003) and Overholt *et al.* (2003). Other relevant studies include those by Liu (1985) for Guangdong, China.

It is unfortunate that two of the most widely cited examples of 'biological control gone wrong' come from sugarcane cultivation systems. The small Indian mongoose *Herpestes javanicus* was introduced to Mauritius, Fiji, Hawaii and islands in the Caribbean (Jamaica, initially) in the late 19th century (Cock, 1985; Lowe *et al.*, 2001). Despite having relatively little impact on the target pests (rats in cane fields), the mongoose has become a significant predator of small, indigenous vertebrates, and has been implicated in the extirpation of native biodiversity. The cane toad *Bufo marinus*, released into Australia in the 1930s as a biological control agent against insect pests of sugarcane, spread rapidly and has become a major environmental problem species; it has subsequently been considered as a potential target for biological control programmes (e.g. Speare, 1990; Wittenberg and Cock, 2001; Hazell *et al.*, 2003).

Both these examples represent relatively early attempts at biological control and involve the release of generalist vertebrate predators. Such releases would never be made as part of officially sanctioned biological control programmes today, as environmental safety standards are now very rigorous (Wittenberg and Cock, 2001). In the wake of such high-profile 'horror stories', it is easy to overlook the fact that biological control has contributed a very great deal to reducing pesticide inputs, whilst protecting crops, in sugarcane in many parts of the world, often as part of wider IPM strategies.

Forms of biological control in sugarcane are most commonly practised against exotic or indigenous insect pests, either through the importation and release of exotic natural enemies, or through measures to encourage the action of indigenous predators and parasitoids (see Box 2.8). In India, for example,

Box 2.8. Biological control in sugarcane in Guyana.

The Guyanese sugar industry has all but abandoned chemical control of insect pests, partly due to biological control successes and a relatively low pest diversity, and there has been a shift towards low-volume application technologies in herbicide use (Eastwood *et al.*, 1997a; SAC, 2000). The system of cane cultivation in Guyana also allows for controlled flooding of fields, which can assist in non-chemical pest control strategies. Specific examples of the use of biological control against insect pests in Guyana are outlined below.

Historically, two of the most important pests in Guyana have been cane-boring larvae of the pyralid moths *Diatraea saccharalis* and *Diatraea centrella*. In the early 20th century, their population levels were approximately equal, although *D. saccharalis* was considerably more damaging to the crop. The introduction of the Amazon fly *Metagonistylum minense* from Brazil in the 1930s and a subsequent programme of mass rearing and release into cane fields proved successful in reducing numbers of *D. saccharalis* (Cock, 1985; Dasrat *et al.*, 1997; SAC, 2000). Although *D. centrella* (now the dominant species, numerically) was only responsible for economically significant damage locally, biological control attempts were also targeted at this pest. Particular efforts have concentrated on the hymenopterous parasitoids *Allorhogas pyralophagus* and *Cotesia flavipes* (Cock, 1985; Quashie-Williams, 1991; Dasrat *et al.*, 1997), although largely without sustained success in the field.

Perhaps a particularly valuable lesson regarding the use of insecticides can be learned from experience with the froghopper *Aeneolamia flavilatera* in cane cultivation in Guyana. Populations of this pest declined to low levels after attempts at chemical control were discontinued, even without the release of specific biological control agents. This reduction in pest numbers was apparently due to the recovery of existing natural enemy populations following the withdrawal of insecticide treatment (Dey *et al.*, 1997; Eastwood *et al.*, 1997a; SAC, 2000). Natural enemies such as predatory syrphids (hoverflies) and cocinnelid beetles also appear to provide adequate control of the aphids *Sipha flava* and *Longiunguis sacchari* (SAC, 2000).

After its early successes, work to enhance biological control of pests in the Guyanese sugar industry has continued, with interest not only in insect natural enemies, but also other potential agents, including *Metarhizium anisopliae* (SAC, 2000). This fungus has been used in biological control programmes (as a 'biopesticide') against cane pests in other areas, including Trinidad and Brazil (Cock, 1985; Dent *et al.*, 2003; Hoffmann-Campo *et al.*, 2003). The potential use of entomopathogenic nematodes against pests such as *Castniomera licus*, against which chemical control attempts have yielded indifferent results, has also attracted attention in Guyana (SAC, 2000; Dasrat, 2001).

a wide range of such natural enemies have been investigated for their biological control potential in sugarcane (Ashok Varma, 2002). Biological control has also been used successfully against weeds in cane growing areas, for example, in Papua New Guinea, against the giant sensitive plant *Mimosa invisa*, which is a problem in cane fields and adjacent areas (Kuniata, 1994; Kuniata and Korowi, 2001). Having been successful as a biological control agent against *M. invisa* in Australia, the psyllid *Heteropsylla spinulosa* was released at the Ramu Sugar Estate in Papua New Guinea in 1993. The insect provided good control of the weed, allowing for reduced herbicide applications, although N fertilizer is applied to *M. invisa* plants, as this increases the numbers of the biological control agent (Kuniata, 1994; Kuniata and Korowi, 2001). Prospects for biological control of weeds in the neighbouring South-east Asia region are discussed by Waterhouse (1994). In some cases, forms of biological control may even be attempted against vertebrate pests. For example, Ballantyne (1998) notes that some Australian cane farmers have erected nest boxes and artificial roosts on their property to encourage the control of vermin by owls.

There are numerous further reports of biological control in sugarcane throughout the literature. In a number of cases, publications are available which catalogue releases of biological control agents in cane cultivation and other systems (e.g. Cock, 1985, for the wider Caribbean region). Further examples here are confined to those given in Box 2.8, which describe activities in Guyana, where chemical control of insect pests has been all but abandoned, partly due to biological control successes.

Integrated pest management and biological control in sugar beet

The adoption of an increasing focus on an IPM approach is one of the factors underlying the reduction in pesticide usage seen in UK beet cultivation in recent years (Defra, 2002 – see above and Box 2.1). Row crops like sugar beet provide good opportunities for more precise agrochemical applications, such as band spraying of herbicides. For example, Rudolph and Klee (1993) demonstrated a 60% reduction in herbicide application, an average reduction in spray drift of 50%, and increased beet yields in trials of this method. Non-chemical methods, such as the use of flame weeders in place of herbicides, have been investigated for their potential in beet cultivation in Europe and the USA (Ceccatelli and Peruzzi, 1995; Peruzzi *et al.*, 1995; Barberi, 1997). Overall, however, in comparison with cane, it appears beet cultivation systems tend to place a greater emphasis on combinations of resistant varieties and rational pesticide use, rather than on biological control and cultural methods. None the less, there are examples of biological control in beet growing systems, generally as part of broader IPM programmes (e.g. see Samersov and Skur'yat, 1985). For example, Zarrabi and Ganbalini (2001) outline a programme of IPM against sugar beet weevil *Bothynoderes obliquifasciatus* in Iran, which aims to bring together biological control (with entomopathogenic fungi), rational pesticide use and cultural methods.

Araji and Hafez (2001) examined the economic and environmental benefits of nematode biological control methods and IPM in potatoes, sugar beet and lucerne in the Pacific North-West (USA). Given that nematode pests in these crops are at present managed with expensive and toxic soil fumigants, and alternative methods show significant efficacy, these authors estimated that investment in the development of nematode biological control methods could eliminate 6.17 million kg of active toxic material from the environment in Idaho, and that farmers could see a gross annual benefit of $43 million ($29 million net).

Other areas of interest in relation to biological control in sugar beet include the control of plant diseases with antagonist microorganisms, often in the form of seed dressings (e.g. Kiewnick and Jacobsen, 1998; Moenne-Loccoz *et al.*, 1998; Weiergang *et al.*, 2001).

Appropriate harvest operations

Harvest operations have a number of potential environmental impacts. In both cane and beet cultivation systems, harvest results in soil removal from the field, and soil compaction is a risk. These problems are more acute for beet than for cane, as the root crop has a relatively high soil tare and is often harvested in wet conditions. Erosion risk may also be increased by the soil disturbance accompanying harvest operations. Preharvest burning is a particular feature of cane cultivation in many areas, which has implications for air and soil quality, although a shift towards green cane harvesting and trash blanketing is occurring in many areas, with consequent environmental benefits (see Box 2.4).

SASA (2002) makes general recommendations for good practice in cane harvesting, summarized as follows:

- Planning should take account of topography, soil characteristics, weather, extraction routes, waterway crossings and loading zone sites.
- Relatively wet and poorly drained areas should be harvested during dry periods.
- Staff should be appropriately trained (and, where appropriate, accredited).
- Vehicles should be appropriately set up and deployed (e.g. with respect to loads on axles, tyre pressures and patterns of in-field traffic).
- Trashing rather than burning should be practised wherever possible, particularly on slopes and relatively erodible soils.

Again, while designed for use in cane cultivation, many of these recommendations apply equally well to beet harvesting. For example, the importance of selecting appropriate agricultural machinery is noted by Spiess and Diserens (2001), controlling patterns of in-field traffic in beet fields (use of tramlines) is recommended by Brunotte and Sommer (1993) and Draycott and Christenson (2003), limiting axle loads (to less than 6 t) is recommended by Henricksson and Hakansson (1993), and the benefits of increased tyre size are noted by Spiess and Diserens (2001) and Defra (2002).

3

Water Consumption

Human activities that have an impact on water availability are a growing concern. For example, as Rao (2000) notes, the World Watch Institute of Washington and the Birla Foundation forecast that, by 2000, 60% of groundwater would be depleted and that, by 2025, 40% of the world's people will face chronic water shortage. The general trend of global warming, coupled with low rainfall and higher evaporation levels have already created drought conditions in the USA, eastern Canada, the Pacific islands and portions of Australia. In India, low groundwater levels have led to drought in Haryana, Gujarat, Madhya Pradesh, Rajasthan and Andhra Pradesh. Dilip Yewalekar (1998) reported an estimated decline in total water availability per person per year in India from 1353 to 910 m^3 between 1981 and 2001. It has been estimated that water demand will exceed supply in many parts of South Africa by 2020 (Schmidt, 2000). Human activities that lead to over-consumption and contamination of water resources have severe impacts on the wider environment. For example, the biodiversity of freshwater ecosystems is considered to be more threatened globally than the biodiversity of terrestrial systems (e.g. Darwall, 2003; McNeely, 2003). Such threats have further knock-on effects for human well-being and livelihoods, for example amongst communities that rely on fisheries (e.g. Dugan, 2003).

Concern over water availability and quality is reflected in recent discussions on these subjects at major international meetings, and in global awareness-raising initiatives. For example, these issues received much attention at the World Summit on Sustainable Development (WSSD) in Johannesburg in 2002 and at the 3rd World Water Forum in Kyoto in 2003, and 2003 was declared the International Year of Fresh Water (Steiner, 2003). Those concerned with the conservation and sustainable management of water resources have urged an ecosystem approach, consistent with the Convention on Biological Diversity (e.g. Steiner, 2003). Such an approach is ambitious and complex, in aiming to integrate the needs of human communities with those of the wider environment across, for example, entire river basins. For such an approach to work, it must be tailored to local conditions (economic, social, political, cultural and environmental), and must provide the necessary powers and resources to local communities, preferably within a robust legal framework (e.g. Masundire, 2003; Scanlon, 2003). There have been recent moves towards whole catchment management approaches in some major sugar producing countries, e.g. Australia (Johnson *et al.*, 1997).

Agriculture, irrigation and water consumption

In most areas, there are three main natural sources of water: rainfall, surface water and groundwater. Where irrigated agriculture is practised (drawing on surface and groundwater reserves), a major objective should be to maximize crop yield per unit of water consumed (e.g. WMO, 1998), although different frameworks for maximizing water productivity may operate at larger than farm scales (Kijne *et al.*, 2003). The enhancement of water use efficiency and water productivity in agriculture is particularly important, as in many parts of the world this activity is a dominant consumer of water. For example, irrigated agriculture is the largest user of water in South Africa (Schmidt, 2000; Ascough and Kiker, 2002), Australia (Lisson *et al.*, 2003) and Pakistan (Azhar Javed and Tariq, 2003). There is also increasing demand for water from the agricultural sector in many countries. In India, Dilip Yewalekar (1998) suggested that expansion of the area under irrigation and improved irrigation systems were essential for increased agricultural production. McNeely (2003) estimates that more than 250 million ha of agricultural land are currently being irrigated globally, around 75% being in developing countries (particularly India, China and Pakistan). This equates to around 70% of all freshwater consumption being directed to irrigation, and over 85% in low-income countries where the extent of irrigation is typically increasing (Herschy, 1998a; McNeely, 2003). Consequences include decreased river flows and falling groundwater levels, and wider environmental impacts on hydrological processes and biodiversity (McNeely, 2003). In addition, irrigation has been a key driver in the salinization of agricultural soils, in both ancient and modern times (Ghassemi *et al.*, 1995).

With these general considerations in mind, irrigation in the cultivation of sugar crops is an issue of particular interest. In some areas of the world, either cane or beet is cultivated as a purely rain-fed crop, in other areas they are irrigated. The need for irrigation varies according to local conditions, notably climatic factors such as rainfall. Cane in particular is noted as a relatively heavy consumer of water, and (being grown largely in tropical conditions, where available sunlight is not limiting) water availability can be expected to be one of the main factors limiting growth. None the less, there is evidence that much could be done to reduce water consumption and improve the efficiency of water use in many cane irrigation systems, and this has been recognized in relation to best management practices recommended, for example, for South African cane growers (SASA, 2002). The main concerns that have been expressed in relation to irrigation in the cultivation of sugarcane relate to the over-exploitation of water resources, and environmental impacts (notably salinization) related to soil waterlogging. Major infrastructural projects related to irrigation may also be a source of environmental concerns. There is also some evidence that cane irrigation systems have contributed to human health problems, notably schistosomiasis, in local communities, although cases where a connection can be confidently determined are fewer than anecdotal evidence would suggest.

There are also some areas where irrigation is essential for the effective cultivation of beet. However, with this crop, there may be questions over whether irrigation is actually necessary in some of the other areas where it is practised. Such questions are based on at least three considerations: (i) beet is not particularly sensitive to water availability (e.g. Dunham, 1993); (ii) the crop is principally grown in temperate areas, where sunlight rather than water availability would be expected to limit plant growth (e.g. Scott and Jaggard, 1993); and (iii) the economics of beet production in Europe (based on quotas) mean that maximum financial returns are not necessarily based on maximizing potential yields from a given cultivated area (e.g. see Box 3.3). In contrast to cane cultivation, the main concerns that have been expressed in relation to water use in the cultivation of sugar beet relate to impacts on soil and water quality, notably through enhanced leaching as a consequence of irrigation.

As for other crops, the development of more efficient and effective irrigation systems for sugarcane and beet requires consideration of technical aspects of the operation. A key consideration is the accurate estimation of crop water requirements, based on soil water balance (see Box 3.1). This often involves calculations using various forms of the Penman combination equation or the Blaney–Criddle method (e.g. see WMO, 1998), and can assist in assessing the amounts of irrigation water

Box 3.1. Parameters for estimating crop water use and irrigation requirements.

The following summarized account is based on Bakker (1999) for sugarcane, but the principles are the same for beet (e.g. see Dunham, 1993).

Summary of terms used:

Et = evapotranspiration (or moisture consumptive use)
Ept (or Et_pot) = potential evapotransiration
Ea (or Et_act) = actual evapotranspiration
Eo = evaporation from an open Class A pan
TAM = total available moisture content (of the soil)
FAM = freely available moisture content (of the soil)

Potential evapotranspiration (Ept) is the amount of water used by the plant at its maximum potential rate of growth. Actual evapotranspiration (Ea) is the amount of water actually used by the plant, which may be limited by a lack of available soil moisture. For example, rainfall may be sufficient to meet all of the plant's water requirements, such that Ea = Ept. Otherwise (and in many cases), irrigation may be required to top up the soil moisture reservoir at a sufficient rate to allow the plant to transpire (and grow) at its maximum potential rate.

Effective rainfall is that proportion of total rainfall (say, 70%) that infiltrates the soil and becomes available to the root system of the cane. Ultimately, the availability of water to the root system depends on the overall soil moisture balance, a 'profit and loss' account of soil moisture derived from the amount of water entering the soil (effective rainfall, irrigation), leaving the soil (evapotranspiration, drainage) and stored in the soil (determined by the soil's water holding capacity).

The key to effective irrigation management is to know when to irrigate and how much water to apply. In other words, how to ensure that Ea = Ept. Insufficient water will not allow the plant to grow at its maximum potential rate, and excess water may also impede the plant's development (as well as wasting resources and leading to possible environmental problems such as soil waterlogging).

Estimating Et (particularly Ept) is an important part of determining when to irrigate and how much water to apply. Penman developed an equation for estimating Et, based on environmental factors (such as sunshine hours and temperature). It has also been shown that Et can be estimated by measuring the evaporation of water from an open water surface, such as that in a US Weather Bureau Class A Pan (Eo). When the crop has developed a full canopy, Et = Eo (in other words, the Et/Eo ratio = 1), and it is reasonable to assume that Ept = Et = Eo. At earlier stages of crop development, Et depends on a combination of evaporation from the bare soil surface and transpiration from the sparse canopy. In this situation, Et is less than Eo (Et/Eo ratio < 1), although it increases in a more or less linear fashion as the canopy becomes fully developed.

Total available moisture content (TAM) and freely available moisture content (FAM) are standard measures of the water available to the root system of the plant, based on soil water content and rooting depth, and are expressed in mm units. Regular assessment of such measures (e.g. based on laboratory analysis of soil samples using suction devices), along with estimates of Et factors, allows the irrigation requirements of the crop to be estimated using a meteorological approach, based on estimates of the soil moisture balance. Alternatively, methods such as tissue analysis can be used to assess the irrigation requirements of the crop.

A record of TAM/FAM measures and assessments of Et factors through the growing season allow estimates of the total water consumed by the crop to be maintained (e.g. in a moisture balance book – MBB). This allows a farm manager to assess the amount of water consumed over the lifetime of the crop, compare crop performance between fields and (on the basis of predictive equations) estimate likely yields. In assessing the overall efficiency of water use (and water management), it is appropriate to calculate a form of water utilization factor, for example the amount of water used to produce 1 t of crop or the yield of crop per 100 mm of effective water used.

needed and the appropriate timing (scheduling) of its application. Such calculations are most easily undertaken using computer models, which can also incorporate many other relevant parameters.

Other important considerations in the development of efficient and effective irrigation systems include the individual characteristics of available systems, sources of irrigation water and non-technical aspects (economic and social factors). While cane may remain in the field for a number of years, influencing the suitability of particular irrigation methods, beet irrigation must be seen in the context of the annual rotation within which this crop is typically grown. The principal irrigation systems available are variations on three basic types: surface (e.g. flood, inundation, furrow) methods, overhead sprinklers (e.g. dragline, centre pivot) and surface/subsurface drip (trickle) systems. In general (e.g. WMO, 1998), surface irrigation tends to result in oversupply (with substantial drainage losses) of water, sprinkler irrigation results in losses of water intercepted by above-ground parts of the crop and drip/trickle methods have greatest potential for maximizing irrigation efficiency but are relatively expensive to install. Particular irrigation systems may be more or less suitable in given situations, and there has been much interest in the use of waste water (including that from sugar processing) for irrigation.

Consumption of water in sugar processing

Although a large proportion of the raw crop material itself is water, the processing of sugarcane and beet involves relatively high levels of water consumption. Sucrose is typically extracted using diffusion into water, and water-cooled condensers are used in later stages of processing. None the less, methods are available for significantly reducing water consumption in sugar processing, particularly through recycling of water between stages in processing operations.

SUGARCANE CULTIVATION

Water availability is likely to represent a limiting factor to plant growth in the tropics, where there is generally a reliable supply of sunlight (and of carbon dioxide – together the three fundamental resources for photosynthesis). The likelihood of water resource limitation only increases for *Saccharum* and its near relatives, with their exceptional potential for rapid and vigorous growth (e.g. Alexander, 1985). Whilst specific responses vary between varieties, Rao (2000) notes that drought conditions (water stress) can affect aspects of cane development such as germination, root growth, cane elongation and tiller formation, and a wide range of physiological processes (including enzyme activity, translocation of photsynthate and sucrose accumulation). In addition, water stress has been found to result in increased damage from some diseases, such as red leaf sheath spot *Mycovellosiella vaginae* (Nass *et al.*, 2001). However, the water requirements of a sugarcane crop are not simple. Particular growth stages are especially vulnerable to water stress, and (to some extent) water inputs can be used to regulate the development of the crop. Gaudin (1999) considers water supply in sugarcane cultivation. Watering can theoretically be used to control the agricultural calendar (adjusting the growth cycle, facilitating harvest after 12 months), while satisfying plant water requirements during the so-called 'main growth' phase between 4 and 9 months. Water shortage at this stage results in reduced yields, and the only way to make up the shortfall is to extend the growth cycle (provided the weather is warm enough). This biological flexibility is due to the way the plant grows, and its C4 type photosynthesis. Particular care has to be taken during ripening, when requirements (water, nitrogen and heat) are reversed in relation to the main growth stage. Muchow *et al.* (2001) note that irrigation of sugarcane (where practised) is usually suspended prior to harvest to dry the field for harvesting operations and to increase the relative sucrose content of the cane.

The water requirements of crops such as rice, cotton and sugarcane are relatively high when compared to other crops (e.g. Dhillon and Panshi, 1987; Singh and Sankhayan, 1991; Munir Ahmad, 2001). However, as in terms of productivity per unit area, sugarcane has been noted as being a very productive crop per unit of water consumed (see Table 3.1). In order to feed its substantial water requirements, sugarcane is able to extract soil water to depths well below 1 m (Inman-Bamber *et al.*, 1998). In the Ord River Irrigation Area in Western Australia, Plunkett and Muchow (2003) found that sugarcane extracted water down to

Table 3.1. Productivity per unit water depleted for four crops in the Indus Basin 1993/94 (after Bastiaanssen *et al.*, 2003).

Crop	Evapotranspiration (mm)	Crop yield (kg/ha)	Productivity per unit water consumed (kg/m^3)	Gross value of production per unit water consumed (US\$/m^3)
Cotton	579	1,293	0.22	0.43
Rice	414	1,756	0.42	0.13
Wheat	357	2,276	0.64	0.10
Sugarcane	965	47,929	4.97	–

1.8 m, 1.6 m or 1.0 m, according to soil type. Moisture absorption by sugarcane doubles in sunshine relative to the rate under a cloudy sky, and wind increases rates of transpiration by driving moisture away from the above-ground parts of the crop – the latter effect can be sufficient to cause water stress under some conditions, undesirable except in the ripening phase of growth (Bakker, 1999). A sugarcane plant comprises approximately 75% water (and 25% organic matter), but consumes around 250 parts of water for every one part of dry matter produced, with the excess water being lost through transpiration. On this basis, a cane crop of 100 t/ha would be expected to consume approximately 7.5 Ml (750 mm) water/ha (Bakker, 1999). This may be a conservative estimate, or may fail to take into account water lost by evaporation from the soil as well as by transpiration, as Bakker (1999) later suggests that a crop produced under more or less ideal conditions might be expected to yield 10 t of cane per 100 mm of moisture consumed and that < 7 t of cane per 100 mm is more realistic.

In some areas, sugarcane cultivation operates primarily as a rain-fed system, as in the uplands of the north-eastern part (the Isaarn Region) of Thailand (Anecksamphant and Wichaidit, 2002). However, in many places supplementary irrigation is practised, and in some areas cane cultivation is fully irrigated (as in Swaziland – Magwenzi, 2000).

Irrigation in the Cultivation of Sugarcane

The need for irrigation in the cultivation of sugarcane varies between localities, according to factors such as topography and soil type, but particularly climatic factors such as rainfall. As well as influencing the need for irrigation, local climatic conditions affect the precise pattern of cane development and the most appropriate timing for particular operations (planting, harvest, etc.). However, the seasonality of rainfall may not match the seasonality of crop water requirement, resulting in a need for water storage and irrigation, even in areas of relatively high rainfall. Bakker (1999) gives a number of detailed examples of the influence of climatic patterns on the nature and timing of cane cultivation operations, citing parts of Kenya and Papua New Guinea where cane can be grown without irrigation, parts of Sri Lanka where irrigation may be necessary and parts of Zambia where irrigation is required. As well as local conditions, the need for irrigation varies between sugarcane varieties. A large number of published research papers deal with responses of sugarcane varieties to management variables including irrigation. Some recent examples are provided in Appendix 1. In addition to its direct effects on cane development, irrigation can influence the relationship between the crop, its pests and diseases and their control. For example, Parsana *et al.* (1994) and Mrig *et al.* (1995) studied the effect of irrigation on the incidence of insect pests (*Chilo infuscatellus*, *Emmalocera depressella* (*Polyocha depressella*), *Melanapis glomerata* and *Saccharicoccus sacchari*) on sugarcane in India. Mrig and Chaudhary (1993) studied the efficacy of different insecticides, combined with different irrigation levels, for control of *C. infuscatellus* and *E. depressella*.

Irrigation is important in many areas where sugarcane is grown, in localities as far apart as Jamaica (Strohl, 1985) and Côte d'Ivoire (Pene, 1999). In Mauritius, 22% of the total cane cultivation area is irrigated (Jhoty *et al.*, 2001). Availability of water is one of the principal constraints on expansion of the sugar industry in Australia, where 60% of the crop is fully or partially irrigated, and where irrigation is essential for consistently profitable cane cultivation and accounts for around a third of production costs (Chapman and Milford, 1997; Ballantyne, 1998). Garside *et al.* (1997b) cite figures suggesting that irrigation of cane in Australia increased by 6% between 1970 and 1990. Examples of the scale of irrigation of sugarcane, even from specific localities, illustrate the huge quantities of water involved. Reportedly, 3.8 Ml/ha/year is consumed in the Bundaberg region (Kingston, 1994), whilst consumption of 10 Ml/ha/year is reported from the Burdekin region (Raine, 1995). In the Ord Irrigation Area in north-west Western Australia, where commercial sugarcane production only started in 1995 (Muchow and Keating, 1998), estimates of irrigation requirements have ranged from 15.3 to 53.8 Ml/ha/year (Wood *et al.*, 1998). Muchow *et al.* (2001) subsequently recommended less frequent irrigation of Ord cane, particularly during late growth, and Plunkett and Muchow (2003) noted the need for irrigation management strategies to support profitable production with minimal environmental consequences such as rising water-tables.

South Africa provides many examples of the importance and scale of irrigation in sugarcane cultivation. Here, Schmidt (1998, 2000) reported that approximately 21% (87,000 ha) of the 412,000 ha under production was irrigated in 1996/97. In northern areas (comprising some 47,000 ha, and typically producing around 16% of South Africa's sugarcane) irrigation is a prerequisite for cultivation. In 1996/97 production in the northern irrigated areas was worth R0.6 billion in miller and grower revenues. The remaining 40,000 ha of irrigated areas are in the coastal and midland regions of KwaZulu-Natal, where irrigation is generally supplementary to rainfall. Efficient management of irrigation is becoming increasingly important in catchments with limited water, where there is competition amongst users for resources, and this is reflected in recent water legislation in South Africa (Schmidt, 1997, 2000, 2001). One aspect of this legislation is the concept of levying land uses that represent stream flow reduction activities (SFRAs). Measurement of stream flows is an attractive means of estimating the prevailing conditions in a catchment, as stream flows represent the combined results of all climatological and hydrological factors, and can be relatively easily measured while water is confined in well-defined channels such as rivers and streams (e.g. see Herschy, 1998b). There has been considerable debate in South Africa over whether rain-fed sugarcane should be classified as an SFRA (Schulze *et al.*, 2000).

Particularly where water is scarce, the importance of effective and efficient irrigation systems is increasingly recognized. None the less, Schmidt (2000) notes that cane irrigators in South Africa are generally perceived to be inefficient users of water, and poorly managed irrigation systems elsewhere in the world have been shown to use water inefficiently and ineffectively (Robertson *et al.*, 1997). For example, Agbossou *et al.* (1995) studied cane irrigation in the Save area of central Benin, and found that established scheduling caused severe water stress during critical periods, and Proag (1998) found that inappropriate irrigation methods were amongst the factors confounding accurate assessments of sugarcane yields in the north of Mauritius. In a study in the North-West Frontier Province (NWFP) (Pakistan), Tahir Sarwar *et al.* (2001) concluded that, when water for irrigation is abundant and farmers have full control over its supply, they tend to over-irrigate their fields. Studies in Australia suggest that cane growers are often unaware of the quantities of water that they are applying, leading to wastage (Shannon *et al.*, 1996). Hence, training of farmers in better water management may also be important (even in areas where water supply is not limiting) if problems arising from poor practice are to be avoided. Pressure for the adoption of better practices may come from a range of sources. For example, in Australia, Attard *et al.* (2003) note that increasing community awareness of environmental issues has raised expectations for irrigators to improve practices, minimize off-site impacts and maximize productivity from scarce water resources. In Swaziland, McGlinchey (1998) reports that irrigation was highlighted as one of the most important and costly agricultural inputs in the sugar industry and that the need for industry expansion with limited water resources, combined with the likely introduction of new water laws, prompted renewed efforts to fine-tune water management.

There is evidence that effective best management practice (BMP) recommendations can be developed for sugarcane irrigation. Klok *et al.* (2003) report on field trials in the Burdekin Delta region in Queensland, Australia, comparing conventional and BMP irrigation systems. Trials were conducted on different soil types, and the BMP applied was site specific (determined by consultation with extension officers and farmers). BMP irrigation reduced the amount of water applied by an average of 15% and increased yields on three of the six sites by approximately 6%. Irrigation is also amongst the issues covered in an impressive set of generic BMP recommendations for conservation and environmental management in the South African sugar industry (SASA, 2002).

Problems Arising from Sugarcane Irrigation

Over-exploitation of water resources

India provides a number of examples of the way in which sugarcane cultivation can place a strain on available water resources. Fry (1997) suggested that problems in the Indian sugar industry have been exacerbated by heavily subsidized electricity, which has encouraged excessive use of groundwater from great depths. Problems are particularly severe where areas under cane cultivation have increased (or continue to increase), as in parts of the Indian Punjab during the post-Green Revolution period (Singh and Sankhayan, 1991). Dhillon and Panshi (1987) reported plans to increase the area under sugarcane in Indian Punjab from 80,000 to 136,000 ha, although Joshi and Tyagi (1991) noted that the rate of change in production and yield here and in Haryana had slowed. In some areas, the demand for water for cane cultivation has outstripped groundwater supply, causing significant ecological problems. For example, increasing irrigated cane cultivation in the Indian Punjab contributed to a situation in which water demand in many areas exceeded levels of annual groundwater recharge, such that the water-table started to fall in almost the entire water zone of the state (Dhillon and Panshi, 1987; Singh and Sankhayan, 1991). Selvarajan and Subramaniam (1988) estimated that water application to cane cultivation in the Amaravathy River basin, Tamil Nadu was 28% higher than the recommended levels (excess applications to rice were even greater). Problems have also been reported in Maharashtra, where the strain that cultivation of sugarcane (the most important commercial crop in the state) placed on water supply was noted by Shiva and Bandyopadhyay (1986), and where Inamdar *et al.* (1995) reported that productivity per unit of area and of irrigation water had shown a continuous decline due to limitations on water availability and farm-level inefficiencies in irrigation management. Limited access to water is a major consideration for cane growers in some areas. Chawla *et al.* (1989) concluded that only 43% of small and marginal farmers in western Uttar Pradesh (whose main crops were sugarcane and wheat) had access to groundwater.

Concerns over water supply and the potential for over-exploitation of available resources for cane cultivation are not confined to India. Availability of water has been identified as a major limiting factor to cane cultivation in areas as far apart as Peru (Iguiniz, 1987), Malawi (Panje *et al.*, 1987) and southern China (Lu, 1991). Sugarcane cultivation is also recognized as a major consumer of water elsewhere, as for example in Mauritius (Ramjeawon and Baguant, 1995). In many areas, cane cultivation is dependent on irrigation, which has to compete with other demands on scarce available water resources, as in central Sudan (Hussein, 1999) and parts of South Africa (Schmidt, 1998). There are concerns in Australia that irrigation of sugarcane has contributed to over-commitment and degradation of river systems and over-exploitation of groundwaters in the Burdekin Delta and Bundaberg regions (Arthington *et al.*, 1997; Meyer, 1997).

Waterlogging of soils

Cane cultivation on flat or gently sloping land is generally preferable to cultivation on slopes, and the rich soils of alluvial valley bottoms and coastal plains may be particularly suited to agriculture. However, cultivation in such areas may result in problems associated with waterlogging (e.g. Bakker, 1999), not all of which arise from direct effects on crop development. In Guyana, for example, poor drainage can contribute to populations of the froghopper *Aeneolamia flavilatera* reaching serious pest levels (SAC, 2000). Waterlogging has been blamed on under-utilization of groundwater (as in parts of India – Joshi and Tyagi, 1991), but is essentially a consequence of poor drainage (e.g. Bakker, 1999). Some soils are naturally poorly draining, and the need for improved drainage is often exacerbated by local rainfall patterns (Garside *et al.*, 1997b) and particularly by irrigation. The urgency of this need is only increased by the threat of soil salinization which accompanies rising water-tables (Ghassemi *et al.*, 1995).

SASA (2002) notes a range of indicators of poorly drained areas, including the presence of particular plant communities (such as those including sedges, reeds, bulrushes, algal mats) and particular soil forms and surface characteristics (ponding, capping, salt encrustation). They suggest that waterlogging is often characteristic of the following situations:

- In areas where water accumulates naturally (such as footslopes, valley bottoms).
- Adjacent to large areas of open water (such as estuaries, lakes, dams).
- Where surface water management is poor.
- Where there is leakage of water storage or transport systems (such as dams, canals, pipes, irrigation infrastructure).
- Around human-made obstructions (such as embankments, bridges, drifts).
- Where over-irrigation has taken place.
- On soils with impervious horizons close to the surface.

Water storage and major infrastructural projects

In reviewing the early history of agriculture in Australia, Meyer (1997) concludes that major irrigation projects were seen as being positive aspects of development, but that realistic economic analyses were rarely carried out. Consequently, major infrastructural projects were undertaken, on the basis of promises of increased productivity, but early returns were invariably insufficient to service the capital debt or cover running and maintenance costs. Nor were environmental consequences taken into account, such as over-commitment of supplies and problems related to drainage. Johnson *et al.* (1997) note that recent government reforms in Australia have included increased recognition that investment in publicly owned resources such as dams and irrigation systems should generate satisfactory rates of return, as well as taking ecologically sustainable development principles into account. Even at smaller scales, infrastructure for water storage needs to be appropriately managed, as the breaching of even a farm-scale dam can have a major impact on human safety and the downstream environment (SASA, 2002). Consequently, in South Africa, storage of more than 10,000 m^3 of water per farm requires registration of the dam or storage facility with the appropriate authority.

Very large infrastructural projects have been undertaken in some areas to enhance water supply for sugarcane growing areas. Examples include government approval of plans for construction of the Driekoppies Dam on the Lomati (Mluwati) River, to guarantee a water supply for increased cane cultivation in eastern Transvaal (Bekker, 1992). In some cases, projects perceived to be primarily for the benefit of the sugar industry have been controversial. Damandeep Singh (1994) reports that the Narmada Dam (which had been proposed as a means to provide irrigation, drinking-water and power for local communities) was strongly opposed by those who argued that the water was not meant for drought-prone villagers, but for the burgeoning industrial sector and the sugar industry of already rich central Gujarat. A range of environmental and social impacts are associated with dam construction (including the flooding of villages and displacement of communities – Damandeep Singh, 1994). The Chittaurgarh irrigation project, involving the construction of a dam in the outer Indian Himalayas, was expected to increase agricultural productivity of crops, including sugarcane (Ahmad and Singh, 1991). However, anticipated environmental effects included serious impacts on flora and fauna, a considerable loss of forest land, rising water-tables, waterlogging, soil salinization, fodder and fuel wood crises following deforestation, and increased levels of crop pests and human diseases. After one decade, siltation was anticipated to increase 20 times due to the poor vegetation in the catchment area.

Irrigation and human health

UNEP (1982) noted that bilharzia and hookworm often occur naturally in areas where sugarcane is grown, and (although they are not a direct result of the sugar industry) may become established in cane fields and associated water bodies. Depending on the location of the plantation/mill, other insect- and snail-borne infections of humans and domesticated animals may also become a problem. These include fascioliasis and paramphistomiasis. Sugarcane irrigation systems may also serve to harbour schistosomiasis and malaria.

Grosse (1993) provided a comprehensive review of international literature on schistosomiasis in relation to water resource management and development projects, including a brief section on irrigated sugarcane. This concludes that hard evidence for the promotion of schistosomiasis by the introduction of sugar irrigation systems (as with water management projects in general) is harder to find than anecdotal reports would suggest. However, there are clear examples, as in Puerto Rico. Here, studies revealed a substantial increase in the prevalence of schistosomiasis following the establishment of irrigation systems to allow sugarcane cultivation, particularly the South Coast Irrigation System. This was constructed in 1914, and local prevalence of *Schistosoma mansoni* infection rose from zero before 1910 to around 25% by 1930 (Jobin, 1980). There appears to be no evidence of similar problems elsewhere in the Caribbean. In Africa, a small number of studies appear to clearly link sugarcane irrigation with increased rates of schistosomiasis. For example, on the Wonji estate in the upper Awash valley of Ethiopia, rates of infection rose from 7.5% in 1968 to 20% in the 1980s (Kloos *et al.*, 1988), and both the disease and its *Biomphalaria* host snails were unknown in the area prior to irrigation. In 1988, the prevalence of *S. mansoni* infection in children at one Wonji labour camp reached 82% (Simonsen *et al.*, 1990), partly due to poor maintenance of water and sanitary facilities. Another example comes from the Richard-Toll sugar complex in the lower Senegal valley, where an outbreak of *S. mansoni* infection in the 1980s was linked to *Biomphalaria pfeifferi* host snails found only in irrigation canals, and not in natural water bodies (Talla *et al.*, 1990; Diaw *et al.*, 1991). The ecology of this disease system has since been studied further by Sturrock *et al.* (2001). However, Grosse (1993) concluded that a number of other reported cases of sugar irrigation schemes promoting schistosomiasis in Africa were inconclusive.

Ijumba *et al.* (2002) examined malaria transmission risk variations in different agroecosystems in the Lower Moshi irrigation area, northern Tanzania. Malaria vectors and human populations associated with an irrigated sugarcane plantation, a smallholder rice irrigation scheme and savannah with subsistence crops were compared. Irrigated sugarcane cultivation resulted in water pooling, but this did not produce more vectors. The study suggested that malaria transmission risk was lower for villagers near the irrigated rice scheme (despite greater vector potential) than for the other communities, largely as a consequence of socio-economic factors leading to greater use of antimalarials and bednets.

Technical aspects of improved irrigation systems

Robertson *et al.* (1997) suggest that the development of strategies to improve the efficiency of cane irrigation can concentrate on: (i) increasing the application efficiency (the proportion of water applied that is stored in the root zone); or (ii) varying the timings and quantities of water applied so as to maximize the crop response. In addition to the influence of local climatic and other environmental factors, a range of technical aspects need to be considered when developing irrigation management strategies for sugarcane. A lack of technical information (and its application) will only inhibit moves towards more efficient water use. Relevant technical aspects of sugarcane irrigation systems are discussed by a range of authors, including Combres *et al.* (1996), Torres *et al.* (1996), Robertson *et al.* (1997), Muchow and Keating (1998), Singels *et al.* (1998, 1999), Schmidt (2000), Attard *et al.* (2003) and Plunkett and Muchow (2003). Important considerations include:

- estimation of crop water requirement;
- monitoring of water use;
- irrigation scheduling;
- irrigation system characteristics;
- source of irrigation water;
- maintenance of irrigation infrastructure;
- availability of knowledge and training.

Assessment of the effectiveness of cane irrigation strategies can be based on an analysis of water use efficiency (WUE), the ratio of cane yield to water consumed by the crop. WUE is often used (at least in Australia and South Africa) for estimating crop response to irrigation and the amount of irrigation required to attain a given yield, and a 'rule of thumb' has been developed, based around a benchmark WUE figure of 8–10 t/Ml (Thompson, 1976; Kingston, 1994). Although this appears to be reasonably reliable under fully irrigated cultivation, variability in WUE between sites can range more widely (around 6–16 t/Ml), and it is generally a less useful measure under rain-fed or partially irrigated conditions (Robertson and Muchow, 1994; Robertson *et al.*, 1997).

In addition to WUE, potential environmental impacts should be considered (e.g. Robertson *et al.*, 1997) as well as linkages to other aspects of farm management, such as drainage (Meyer, 1997) and soil conservation measures (SASA, 2002).

Tools to assist in estimation of crop water requirement and irrigation scheduling

As Hussein (1999) and Chinnusamy and Jayanthi (2000) note, a key aspect in the accurate prediction of the water requirements of a sugarcane crop is the reliable estimation of evapotranspiration rates (see Box 3.1). As Attard *et al.* (2003) note, irrigation water is used most efficiently when crop requirement is defined accurately and water is applied to meet this demand both fully and at precisely the right time. The most effective irrigation sceduling is ideally based on day-to-day monitoring of field conditions (Agbossou *et al.*, 1995). However, at a regional scale, optimum scheduling is substantially influenced by climatic conditions across a range of sites (as demonstrated for India by Rajendra Gupta and Tripathi, 1998a,b). A particular question over irrigation scheduling of relevance to sugarcane growers in parts of South Asia is the extent to which warabandi-type irrigation schemes (in which irrigation water is supplied to a group of farmers in rotation, rather than on demand) can effectively serve the needs of the crop (e.g. Qureshi, S.A. *et al.*, 2002a,b).

Appropriate scheduling can produce significant savings of irrigation water. Ellis *et al.* (1985) reported that severe water restrictions in Zimbabwe led to a review of irrigation strategy and a revised system of scheduling that provided a potential saving in water use of 32% when compared with conventional practices. Comparisons of estate sugar production in successive years showed that adoption of this strategy did not reduce estimated recoverable sugar yield and that it resulted in a 20% saving in total water applied.

With so many parameters to be considered in developing optimum irrigation strategies for sugarcane, farmers can easily be discouraged by the time and paperwork required to carry out the calculations (Torres, 1998). It is not surprising, therefore, that a number of computer models and other tools have been developed to assist in decision making. Robertson *et al.* (1997) and Schmidt (2000, 2001) note the particular value of a modelling approach in its ability to deal with a complex set of variables, drawing on baseline information gathered in long-term data sets. However, it is also noted that models invariably fail to take account of fine-scale, site-specific considerations, and may overemphasize biophysical parameters when socio-economic factors are also a major feature of decision making. Examples of computer models developed to assist in the refinement of sugarcane irrigation strategies are listed in Table 3.2. New technology can also be harnessed in other ways to assist in the management (including irrigation) of sugarcane crops. For example, Baran *et al.* (1999) describe how a geographical information system (GIS) is thus used in northern Côte d'Ivoire.

Where farmer access to computer models (or the requisite baseline soil and climatic data) is constrained, the development of simpler decision support tools may be possible. For example, Torres (1998) suggests a simple device which serves as a visual aid for sugarcane irrigation scheduling. This consists of a plastic bucket, which acts simultaneously as a pluviometer and as an evaporimeter. Once calibrated, there is no need for human intervention beyond checking the position of the water level in relation to the irrigation control marks. The importance of the availability of baseline data and simple tools for irrigators is also noted by Holden *et al.* (1998) and Culverwell *et al.* (1999). In this context, Olivier and Singels (2001) provide a database of crop water use coefficients for irrigation scheduling of sugarcane in South Africa and note the value of automated weather data logging techniques.

Simulation modelling, in combination with field trials, may be able to contribute to the formulation of generic recommendations for irrigation best management practices. Care

Table 3.2. Computer models for the refinement of sugarcane irrigation strategies.

Model	Source – locality
CANEGRO	Inman-Bamber *et al.* (1993) – northern Natal
	McGlinchey *et al.* (1995) – South Africa
	McGlinchey (1998) – Swaziland
Thompson	McGlinchey (1998) – Swaziland
PAWCER	Zund and McDougall (1997) – Bundaberg, Australia
SEPI	Lidon *et al.* (1999) – Kenana sugar growing area, Sudan
IRRICANE	Combres *et al.* (1996)
	Singels *et al.* (1998, 1999) – South Africa
DAM EA$Y	Lisson *et al.* (2003) – Bundaberg, Australia
APSIM-SUGARCANE	Robertson *et al.* (1997) – Australia/South Africa
	Muchow and Keating (1998) – Ord, Australia
	Cheeroo-Nayamuth *et al.* (2000) – Mauritius
CROPWAT	Rajendra Gupta and Tripathi (1998b) – India
SWAP93	Qureshi, S.A. *et al.* (2002a,b) – Sindh, Pakistan
Other models	
CANESIM	Schmidt (2000), SASA (2002) – South Africa
A soil moisture deficit (SMD) model	Inosako *et al.* (1995) – Miyako island
An atmospheric evaporative demand (AED) model	Attard *et al.* (2003) – Ord, Australia
A model to predict changes in total evaporation under deep contour tillage	Gardiner and Cazalet (1991) – steep slopes on the Natal coast
A model for assessing cost–benefit of on-farm water storage	Lisson *et al.* (2003) – Australia
A model for optimal irrigation scheduling	Chaudhry and Leme (1996) – São Paulo, Brazil
A model to simulate effects of different irrigation regimes on yields	Ah-Koon *et al.* (2000a,b) – Mauritius

must be taken in interpretation, as so many factors in irrigation effectiveness are site specific. However, it is interesting to note results like those obtained by Ah-Koon *et al.* (2000a,b), who combined field trials and simulation modelling in Mauritius and found that optimum cane yield would be achieved by irrigating a larger area at 0.50 ETc, whilst lowest yield would be obtained if all the water were to be concentrated on a small area at 1.0 ETc, with the remainder of the perimeter rain-fed. Chaudhry and Leme (1996) used an optimal scheduling simulation model to assess supplemental irrigation of sugarcane in São Paulo, Brazil, and found that the elimination or partial reduction of irrigation was possible without affecting economic returns.

The accumulation of baseline data sets is important if irrigation strategies are to be 'benchmarked' and generic recommendations for best management practices developed. Wood *et al.* (1998) describe the benchmarking of irrigation practices in the Ord River area, Australia, and indicate how their survey results represent one of the most comprehensive data sets on irrigation practices gathered anywhere in the Australian sugar industry. The data set has substantial potential, in combination with field experimentation and crop simulation modelling, to underpin a research programme to develop the most profitable and sustainable irrigation strategies for the Ord area. Similarly, Jhoty *et al.* (2001) describe a survey of irrigation practices for sugarcane in Mauritius, collecting baseline data for a GIS land management database.

Irrigation system characteristics

Application efficiency (the proportion of water applied that is stored in the root zone) varies considerably between (and within) the different irrigation systems used in cane cultivation (Robertson *et al.*, 1997). Factors to be considered include the uniformity with which the system applies water (e.g. Lidon *et al.*, 1999; Ascough and Kiker, 2002).

In any given area of sugarcane cultivation, a variety of irrigation systems tend to be in place, and these change over time. A useful overview of different irrigation methods and their use in cane cultivation is provided by Bakker (1999). In many areas, traditional irrigation methods are based on the flooding of small cultivated plots (as described for Nigeria by Phillips-Howard, 1996), and the same inundation method has been widely used in larger cane fields. Modern techniques, such as the use of sprinkler or drip systems, are replacing traditional inundation methods in many areas (e.g. in India – Dilip Yewalekar, 1998; Jamaica – Strohl, 1985). In some areas, the adoption of such techniques has been identified as a key factor in the potential expansion of the area under sugarcane cultivation, for example in the Costa del Sol region of Spain (Fernandez Lavandera and Pizarro Checa, 1983) and Egypt (Fawzy and Amer, 2002). In a survey of irrigation practices in Mauritius, Jhoty *et al.* (2001) found that three main irrigation systems were being used (overhead, drip and surface) and that low-pressure overhead irrigation systems, such as the centre pivot and dragline, were gradually replacing high-pressure overhead systems.

A number of studies examine the responses of sugarcane varieties to different irrigation methods and other management variables. Some recent examples are provided in Appendix 3. In addition to these, a number of more generic comparisons of different sugarcane irrigation systems have been made. For example, in the northern areas of KwaZulu-Natal and Mpumalanga, Schmidt (2000) considered the small proportion of sugarcane under pivot (6%) and drip systems (4%, but increasing) to be more efficient than the 90% under overhead sprinkler systems. In Swaziland, Magwenzi (2000) estimated that water application efficiencies were 72–89% under drip and centre pivot systems, 49–88% under dragline and 48–75% under furrow irrigation. Using agronomic and economic data, Qureshi, M.E. *et al.* (2002) compared profitability (based on net present value, NPV) of growing sugarcane under different irrigation systems in the Burdekin Delta, Australia. Furrow irrigation on well-draining soil had the highest NPV, followed by use of centre pivot systems on moderately draining soil, then furrow irrigation on poorly draining soil and drip irrigation on well-draining soil. However, when volumetric water charges were used instead of area-based charges, the ranking changed. Use of the centre pivot on moderately draining soil had the highest NPV, followed by furrow irrigation on well-draining soil, furrow irrigation on poorly draining soil and drip irrigation on well-draining soil. Under the volumetric water charging option, the overall NPVs for each irrigation system were lower than the NPVs for area-based water charges. Ascough and Kiker (2002) compared cane irrigation systems in South Africa, and estimated average application efficiencies of 83.6% for centre pivot, 73.5% for dragline, 76.7% for floppy sprinklers and 78.9% for semi-permanent sprinklers. These authors also examined uniformity of water application, which increased through semi-permanent sprinklers, dragline, floppy sprinklers, drip and centre pivot systems, and concluded that good maintenance and correct operation were key considerations. Overall, Robertson *et al.* (1997) concluded that application efficiency increases from furrow to sprinkler to surface drip to subsurface drip irrigation systems.

SURFACE IRRIGATION. In surface irrigation, the cane field is inundated, and the soil serves as a medium for infiltration and conveying water from the upstream to the downstream end of a field. Considerable improvements in irrigation efficiency can be achieved by flooding furrows between the crop rows rather than the whole field. It can be difficult to achieve consistently high levels of water use efficiency with furrow irrigation. For example, Raine and Bakker (1996) found that application efficiency of furrow irrigation varied between 14 and 90% for individual water applications and between 31 and 62% over the course of a season. However, this method has the considerable advantages of being inexpensive and easy to operate. The main technical considerations are the water holding capacity and infilration rate of the top 60–100 cm of the soil, the gradient, shape and length of the furrow, and the volume of water to be discharged into it (e.g. Bakker, 1999; Lidon *et al.*, 1999). A key objective is to obtain even application of irrigation water along the furrow. Common practice (as in Mauritius – Ng Cheong *et al.*, 1996) is to stop the inflow of water as soon as it has reached the end of the furrow (or preferably before – Bakker, 1999) and to bund the latter to allow *in situ* water percolation.

As the studies in Appendix 3 indicate, there are further potential efficiencies to be achieved by flooding alternate (rather than all) furrows. Yadav (1986) achieved water savings of 36% using this method, and Ved Singh (2001) achieved a water saving of 30% using alternate furrow irrigation with rotation (odd/even) of irrigated furrows. Alternate furrow irrigation can also be combined with trash mulching, yielding water savings of 36% (Thanki *et al.*, 1999). Pandian *et al.* (1992) achieved increases in water use efficiency of 43–66% by using alternate furrow irrigation, with the greatest increases attained in combination with mulching.

Raine and Shannon (1996) studied further possible modifications to alternate furrow irrigation in the Burdekin, Australia. They found that modifications to the furrow shape to produce a narrow 'v' shape with surface compaction could reduce water use throughout the season by 45%. This represented a saving of Aus$218/ha/year to the grower and a potential saving of Aus$1.74 million annually to the Burdekin sugar industry. Where furrow lengths of 300 m were used instead of 600 m, the volume of irrigation water applied was reduced by 42%. These shorter furrows were found to produce a net return of either Aus$132 or Aus$210/ha/year after the capital and production costs were assessed, depending on the nature of the water delivery system installed. Such modifications of furrow shape were also found to be beneficial in studies by Holden *et al.* (1998).

Bakker (1999) suggests that land preparation is a key element for effective furrow irrigation, which relies on uniform slopes (generally of 1 : 200 or less). Although contour furrows may make surface irrigation practicable on more steeply sloping areas, Bakker (1999) recommends that, where land levelling cannot be undertaken, overhead irrigation is more appropriate. SASA (2002) suggests that furrow irrigation is not suitable where soils are less than 0.4 m deep, underlain by an impervious layer, or on slopes of greater than 6%. They also suggest specific standards for irrigation furrow gradients on particular soil types.

SPRINKLER IRRIGATION. Sprinkler irrigation systems can provide for application efficiencies of up to 85% (Shannon *et al.*, 1996), although this is rarely attained. Sprinkler systems come in a number of forms. For example, Zadrazil (1990) reports on dragline irrigation, an overhead sprinkler system where sprinklers are connected by portable hoses and (permanent or semi-permanent) pipes to a pressurized water supply. The dragline system, which has been in operation for many years in Swaziland and South Africa, was derived from the conventional sprinkler system of portable pipes, in order to reduce the labour requirement for its operation. This was achieved with a modest increase of capital costs, well below the comparative costs of other, more automated irrigation systems. Dragline systems can be adapted to the irrigation of large estates, or to smallholder plots of 1 ha or less. They can also be successfully integrated with contour planting, as used for erosion control on slopes. The dragline sprinkler system used for the first 15 years of cane cultivation on the Simunye sugar estate in Swaziland provided very good service, being low in capital cost, well suited to the topography, simple to operate, and highly visible (allowing faults to be detected easily). However, the potential for increased productivity and water use efficiency prompted a shift to a subsurface drip system in the late 1990s (Merry, 2003; see Box 3.2).

Sprinkler irrigation can produce substantial water savings in comparison with furrow irrigation (e.g. 37.5% – Shrivastava *et al.*, 1993). However, the necessary infrastructure is considerably more expensive and may be significantly more labour intensive to operate.

Bakker (1999) notes that the key elements of an overhead sprinkler system are the layout of the infrastructure, the spacing between portable sprinkler lines (laterals), the maintenance of appropriate water pressure in the system and the design of the nozzles. SASA (2002) suggests that soils of a minimum depth of 0.45 m and a total available moisture (TAM) of 50 mm are suitable for irrigation with overhead systems, noting that where TAM is low and a very short (3–5 days) irrigation cycle is required, centre pivot or drip systems may be preferred.

DRIP IRRIGATION. Bakker (1999) notes that drip (trickle) irrigation is designed to deliver water directly to the root system of the crop, through frequent applications of relatively small amounts of water. Typically, this is achieved through a network of porous plastic tubing, running along rows (or some multiple of alternative rows), either on the surface or buried at a depth of around 5–10 cm. Bakker (1999) considers this system to be much less intensive in use of labour, energy and water than overhead or surface irrigation, but other authors have noted problems with the cost and maintenance of the infrastructure. Hodnett *et al.* (1990) conclude that drip irrigation methods have the greatest potential for maximizing water use efficiency in cane irrigation, but that subsurface drip systems may not be cost effective in many circumstances.

Schmidt (2000) summarizes the perceived advantages and disadvantages of drip irrigation in South Africa (relative to the widely used overhead systems), as shown below:

Advantages

- Higher water application efficiency (increasing yield per unit water and/or reducing water applied per unit area).
- Lower operating pressure, reducing pumping costs.
- Lower operating labour costs.
- Potential for full automation.
- Better control over daily applications (useful on 'problem' soils with low moisture holding capacity or low infiltration rates).
- Lower wetting area, enhancing rainfall efficiency and restricting weed growth.

Disadvantages

- Higher capital cost.
- Greater management and maintenance requirement.
- Better repair and maintenance backup required for more automated systems.
- Potential for damage to dripper lines (e.g. by rats or at harvest).

Schmidt (2000) also notes some of the perceived advantages and disadvantages of subsurface versus surface drip systems:

Advantages

- Cheaper, thin-walled dripper tape can be used.
- Lower operating labour costs (dripper line does not need to be retrieved at harvest or extracted from lodged cane).
- Higher water and fertilizer (fertigation) application efficiency.

Disadvantages

- System must be planned and operated with particular care.
- Greater difficulty in monitoring performance.
- Greater difficulty in effecting repairs.
- Root intrusion.
- Poor germination on rapidly draining soils.
- Obstructs pest and disease control by deep tillage.

Studies clearly indicate that even surface drip irrigation systems can provide substantial water savings over other irrigation systems, particularly inundation methods. Water savings of around 50% are commonly reported under drip irrigation of various forms (e.g. Cho and Kuroda, 1987; Hapase *et al.*, 1990, 1992; Parikh *et al.*, 1992; Shinde and Jadhav, 1998, 2000; Raskar and Bhoi, 2001). Water use efficiency can also be enhanced substantially, from around two to three times the level under furrow irrigation (Hapase *et al.*, 1990, 1992; Shinde and Jadhav, 1998, 2000). Yields and quality factors may also be enhanced, e.g. by figures of around 20% (Hapase *et al.*, 1990, 1992; Shinde and Jadhav, 2000; Raskar and Bhoi, 2001). As with furrow irrigation, combining mulching with drip irrigation can improve results further (e.g. an additional water saving of 16% – Shinde and Jadhav, 1998). Thorburn *et al.* (1998) review the agronomic and environmental benefits of drip irrigation for sugarcane. Early research showed few benefits, but studies published since the mid-1980s have shown yield increases of 5–20%, although these increases do not necessarily persist through ratooning. A number of studies found that irrigation water use efficiency had increased by 50–80% under trickle irrigation. One of the studies reviewed suggested that nitrogen applications could be reduced under drip irrigation, but studies on other crops, such as citrus, showed that nitrogen leaching is substantially enhanced where too much water and/or nitrogen is applied. Camp *et al.* (2000) review the use of subsurface drip irrigation (SDI) in agriculture in the USA, and reach similar conclusions. Early problems were followed by improvement of systems such that crop yields were equal to or better than those under other irrigation methods (including surface drip systems). SDI water requirements were equal to or lower than surface drip, and fertilizer requirements were sometimes lower than for other irrigation methods. Camp *et al.* (2000) conclude that SDI is a very precise irrigation method, both in the delivery of water and nutrients to desired locations and in the timing and frequency of applications for optimal plant growth.

As noted below, however, non-technical (including economic) issues must also be considered when assessing the benefits of an improved irrigation system. This is particularly relevant to drip irrigation, which is perceived to have relatively high associated costs. Whilst some studies (e.g. Cho and Kuroda, 1987; Hapase *et al.*, 1992) conclude that a drip irrigation system represents a good investment, others stress the constraints to its adoption. Analyses of the economic and social aspects of adopting drip irrigation for sugarcane are particularly readily available from India. Gurav *et al.* (2003) describe constraints experienced by Maharashtra sugarcane farmers in the adoption of drip irrigation, and their suggestions for overcoming these. Based on interviews with 102 farmers who had adopted drip irrigation, reported problems included high initial cost, requirement for regular maintenance and lack of technical guidance and locally available spare parts. Farmers suggested that adoption of drip irrigation would be enhanced by timely provision and availability of subsidies and loans, reduction in total initial cost of the drip unit, provision of regular after-sales service and technical training. Inamdar *et al.* (1995, 1996a,b) examined the economic efficiency of biwall drip irrigation in sugarcane production, using a case study in Ankalkhop village in Sangli District of Maharashtra. The benefits of the new irrigation system were a saving of farm labour days, decreased manure and fertilizer requirements, lower water usage, reduced soil erosion and weed growth and increased sugarcane yield and quality. The cost of installation of the biwall drip irrigation unit was Rs36,423 per farm, with a government subsidy (also noted by Chauhan, 1995) providing a major incentive for farmers to adopt the new technology. Per hectare, the output and return from sugarcane were 24.32 t and Rs13,989 higher under biwall drip than surface irrigation, and the biwall drip system was estimated to have a benefit/cost ratio of around 1.5. It was concluded that the biwall drip irrigation system is an important technique for increasing crop production and on-farm economic efficiency. However, a later study by Kumar *et al.* (2000) suggests that microtube systems are more efficient than biwall. Srivastava and Upadhayaya (1998) also studied the factors influencing the economics of drip irrigation in India, and highlighted the importance of yield gain ratio, electricity charges, irrigation requirement and depth of groundwater.

Berthelot and Robertson (1990) conducted a comprehensive comparison of the financial and economic viability of drip and overhead irrigation of sugarcane in Mauritius, and concluded that drip irrigation provided higher net returns. The drip experience in Swaziland is outlined in Box 3.2. Other studies examining drip irrigation in sugarcane include Robertson *et al.* (1997), Ullman (1999) and Sanchez-Roman (2000), and further studies examining its application in India include Dua (1995), Magar (1995), Dilip Yewalekar (1998), Kareem (1999) and Soman (2002).

Source of water for irrigation

The source of water for irrigation is an important consideration. Surface and groundwater sources may be in limited supply, and the location of cane fields may affect the ease with which water can be imported from a particular source. Another important factor is the quality of water available from particular sources. SASA (2002) suggests that quality of irrigation water should be kept within certain limits, in order to prevent soil degradation effects (such as salinization) and impacts on cane yields. A possible classification of water quality and its suitability for cane irrigation is given (see Table 3.3; cf. Table 3.4), and it is suggested that irrigation water quality should be monitored by an accredited laboratory. Despite the need for caution in using water of questionable quality, there is considerable interest in using waste water (from cane processing or other sources) and saline water for cane irrigation.

IRRIGATION WITH CANE PROCESSING (AND OTHER) WASTE WATERS. A number of studies suggest that waste water from sugarcane mills is suitable for irrigation, e.g. in Andhra Pradesh (Srimannarayana and Sudheer, 2000), Taiwan (Lu and Chen, 1991; Wang *et al.*, 1999; Tzeng *et al.*, 2001), Hawaii (Yang *et al.*, 1991), Cuba (Inklan, 1991; Arzola Pina and Yera Martin, 1995c) and Brazil (Scaloppi *et al.*, 1989). In most cases, some treatment of the effluent is recommended before its use in irrigation. However, it is also suggested that (non-water) waste materials remaining in the effluent can have agriculturally beneficial effects, substituting for mineral fertilizers (Arzola Pina and Yera Martin, 1995c). In some cases, sugarcane yield increases could effectively cover the costs of disposing of waste water via irrigation systems (Scaloppi *et al.*, 1989). However, the effects on crops of irrigation with cane processing waste water are not universally positive. Tewari and Archana Tripathi (2001), for example, found that sugar factory effluent suppressed germination of peas (*Pisum sativum*) in a study conducted in Balrampur (India), and Kumar Arindam (2001) found a range of negative effects of irrigation with sugar mill effluent on barley. Scaloppi *et al.* (1989) note that a major benefit of using cane mill waste water for irrigation is the reduced environmental impact to surface waters, into which the effluent would otherwise have been discharged. However, there is concern that untreated (and in some cases, treated) effluent used in irrigation has the potential to pollute soils and groundwater (Inklan, 1991).

Deliberate irrigation of sugarcane with waste water from other sources has also been explored. Again, this can be an attractive proposition, potentially reducing impacts on aquatic environments where effluent was previously discharged (Braddock and Downs, 2001). Lau (1979) showed that sewage effluent could be used to supplement water for furrow irrigation of sugarcane without detriment to groundwater quality and sugar yields, suggesting that the technique could be

Box 3.2. Conversion from sprinkler to drip irrigation – a case study from Swaziland.

Except where otherwise indicated, this summary is based on Merry (2003).

Irrigation is vital to sugarcane production at the Simunye sugar estate in Swaziland, which grows around 11,000 ha of cane. Effective rainfall is about 440 mm/year, and around 750 mm/year (net) of additional water is required for optimum cane growth. Irrigation water is obtained primarily from the Mnjoli Dam and elsewhere on the Mbuluzi River. When full commercial production began here in 1982, around 75% of land was irrigated using a dragline sprinkler system, the remainder by furrow (flood) irrigation. Subsequent expansion of the estate led to experiments with surface drip systems (see Pollok and Bosua, 1986). By the mid-1990s, further expansion, drought and the deterioriation of existing sprinker infrastructure prompted complete redevelopment of the irrigation system. A cost analysis of seven different options led to conversion from dragline sprinkler to subsurface drip. Details of the design used are provided by Merry (2003), but the construction of 'cluster houses', to group together irrigation control valves, is a notable feature. This system design also incorporates water filtration, and mechanisms to allow fertigation and chemigation.

Performance of drip vs. sprinkler systems (see Ndlovu, 2000; Ndlovu *et al.*, 2001)

Assessments indicate overall water savings of around 25% under drip, and an increase in water use efficiency of 29% under plant cane and 18% under the first ratoon. Cane yields were greater under drip, by around 17% for plant cane and 14% for first ratoon. Percentage sugar content was also greater under drip, resulting in a final sugar yield increase of nearly 25%.

Post-investment audit

The cost of conversion to a subsurface drip system was US$2542/ha, versus US$868/ha to retain the sprinkler system and replace worn-out parts. Benefits from the new system were principally labour cost savings (US$219/ha/year), an increase in sucrose yields (worth US$91/ha/year) and water savings. In this case, water savings did not produce substantial, direct cost savings, because there is no bulk water charge, and the main supply is gravity fed. However, the opportunity value of water saved, in terms of potential returns had it been used to irrigate cane, was estimated at US$162/ha/year. Together with a small saving in the cost of power, these figures indicated an incremental rate of return, on the additional investment required for conversion to drip, of around 30% per year.

Part of the success of this project was in the early recognition of the challenges involved, both in the installation of thousands of kilometres of pipelines and drip laterals, and in changing operating practices. Detailed forward planning was critical, as was the careful recruitment of reliable contractors to handle design, supply and installation. It was also possible to reduce costs by incorporating some of the infrastructure (pumping stations and mainlines) of the previous sprinkler network into the new system.

Table 3.3. A classification of water quality and its suitability for sugarcane irrigation (after SASA, 2002).

Class	EEC (mS/m)[a]	ASAR	Suitability for irrigation
A = good	< 50	< 5	On all soils, except those with extremely low permeability or excessive salinity/sodicity
B = moderate to poor	< 120	< 10	On well-drained soils only
C = poor	< 150	< 10	Only where no other water is available, and then in combination with appropriate drainage
D = very poor	> 150	> 10	Not under normal circumstances

[a]Total concentration of soluble salts, adjusted for local rainfall leaching.
ASAR, Adjusted sodium adsorption ratio (concentration of sodium, calcium, magnesium and bicarbonate).

used to augment natural water resources, provide supplemental or alternative fertilizer, alleviate ocean water pollution and reduce costs of engineering systems in Hawaii. Wu (1996) reported that the Taiwan Sugar Corporation (TSC) owned 30 pig farms with 0.49 million pigs, accounting for 5% of the pigs in Taiwan. Treated effluents from these farms are recycled for irrigation in sugarcane fields. Concerns over accumulation of heavy metals (particularly Cu and Zn) in soils under long-term irrigation with effluents appear to have been overcome by modified application models (Liu *et al.*, 1996a,b). As the effluents are also rich in N, P and K (Liu *et al.*, 1996a,b), their use in irrigation not only contributes to solving waste disposal problems in the pig industry, but also reduces the use of chemical fertilizers (and enhances soil organic content) in sugarcane cultivation in Taiwan (Wu, 1996).

Palaniswami and Ramulu (1994) studied the effects of continuous sugarcane irrigation with paper factory effluent. They found that organic carbon, pH, electrical conductivity (EC), CEC, exchangeable Na, Ca, Mg and K, available P, K, Fe, Mn, Zn and Cu, and activity of urease and acid and alkaline phosphatase enzymes all increased in the 0–15 cm soil layer after 15 years, and that soil nitrogen levels fell. However, soil physicochemical changes were negligible in subsurface samples, and no visible toxicity symptoms were observed in the crop. Working in Cuba, Arzola Pina and Yera Martin (1995a) found significant increases in soil sodium, bicarbonate and soluble salt following irrigation of sugarcane with paper mill effluents. Also in Cuba, Arzola Pina and Yera Martin (1995b) studied the effects on soil and sugarcane yield of irrigation with effluent from torula yeast production, diluted with water, or with sugar liquid from cane mills. They concluded that this effluent represented a valuable soil amendment, especially useful for potassium deficient soils. Linedale (1998) reports on irrigation of sugarcane in Bundaberg, Queensland, with urban waste water from the Thabeban treatment works. Studies showed virtually no beneficial accumulation of nutrients (except S) at irrigated sites, and there was some accumulation of Na. Leachate analyses found only minor levels of heavy metals, nitrates and phosphates in deep drainage waters, suggesting a low likelihood of significant groundwater pollution.

Despite reports of successes arising from irrigation with waste waters, there are environmental risks where the activity is undertaken without due care. SASA (2002) notes that, in South Africa, waste water may not be used for irrigation within an area liable to flooding, or within 10 m of a watercourse or borehole. Irrigation with industrial waste water (or from waterworks) is subject to controls from a regulating authority, the Department of Water Affairs and Forestry, with whom any such irrigation involving the application of more than 10 m^3/day must be registered. However, a licence to irrigate with waste water is not required where certain quantitative and qualitative conditions are met (shown here in Table 3.4). None the less, SASA (2002) urges that, where waste waters are used for irrigation, regular water and soil samples should be taken to monitor and control any likely detrimental environmental effects from accumulation of salts, nutrients or trace elements. Such monitoring should be undertaken by an accredited laboratory, and general good practice should be implemented to prevent waterlogging, nuisance (from flies, mosquitoes, odours or secondary pollution), pollution of watercourses and unreasonable degradation of the soil.

IRRIGATION WITH SALINE WATER. Rising water-tables have led to problems of soil salinity and sodicity in a number of areas under sugarcane cultivation (see Chapter 6). Saline soils will tend to produce saline drainage waters, which may (directly or indirectly) be chosen for irrigation purposes. Nelson *et al.* (2002) note that sodicity and related properties of soils and irrigation water restrict sugarcane yields and cause environmental problems such as turbid runoff. Meyer and van Antwerpen (1995) reported preliminary results of a survey of water quality for 12 selected rivers in the South African sugar industry. Considerable spatial variability was observed in salinity and sodicity levels within and between rivers. The Mkuze River was

Table 3.4. Quantitative and qualitative standards for unlicensed irrigation with waste water in South Africa (after SASA, 2002).

	< 500 m³ domestic or biodegradable industrial waste water on any day	< 50 m³ biodegradable industrial waste water on any day
Electrical conductivity (EC)	< 200 mS/m	< 200 mS/m
pH	6–9	6–9
COD	< 400 mg/l	< 5,000 mg/l
Faecal coliforms	< 100,000 per 100 ml	< 100,000 per 100 ml
Sodium adsorption ratio (SAR)	< 5	< 5

COD, chemical oxygen demand.

found to be moderately saline with a fairly high sodicity hazard, and salinity levels in the lower Crocodile River had more than doubled since the previous (1976) assessment. Of the other rivers, the Pongola, Umfolozi and Mhlatuze showed a moderate sodicity hazard, suggesting that use for irrigation in the long term could lead to soil degradation and eventual yield decline on sensitive soils.

Except where over-irrigation and drainage are used specifically as a remedial measure in saline soil areas, or where soils are naturally saline, saline drainage waters are symptomatic of poor irrigation management (Meyer, 1997). Consequently, they should be seen less as a potential resource for use in irrigation, and more as an indicator of the need to alter existing irrigation strategies. None the less, lack of available fresh water has led to studies of the effects of sugarcane irrigation with saline waters (e.g. Thomas *et al.*, 1981; Rozeff, 1998a) and the investigation of more salt-tolerant cane varieties (e.g. Sundara and Reddy, 1994; Lingle *et al.*, 2000).

Non-technical aspects of improved irrigation systems

Successful design and implementation of improved irrigation systems must consider non-technical aspects, including economic and political considerations and farmer involvement. A number of authors stress the importance of economic evaluations in the design and implementation of improved irrigation systems, including Carruthers (1987), Schmidt (1996, 2000), Robertson *et al.* (1997), Frizzone *et al.* (2001) and Magwenzi (2002). It may be appropriate to evaluate the economics of investment in research (e.g. Wegener *et al.*, 2000), as well as particular aspects of irrigation systems themselves (e.g. Brzesowsky and van Vilsteren, 1988; Chawla *et al.*, 1989). The recognition that the aim of improved irrigation must be to maximize the increase in crop production per unit of water, not per unit of irrigated land, should be central to economic evaluations (e.g. Mitra, 1989; Magwenzi, 2002). While the potential for increased efficiency of water use from changing irrigation systems may be very clear, there may be little incentive to change methods when economic factors (e.g. capital, operating, maintenance, labour and water costs) are considered. Whilst it may be economic to improve the application efficiency of a system that is already in place, it may not be viable to switch to an entirely different system. An example of one case where an economic analysis revealed the viability of switching systems is outlined in Box 3.2.

A significant economic factor is likely to be the price of water itself; when this is low, the benefits to the grower of upgrading the irrigation system are unlikely to justify the investment required. Meyer (1997) notes that area-based water allocations (as practised in Australia from the early days of irrigation) are undesirable when compared to systems that charge for the amount of water actually used, but, once such allocations are enshrined as a 'right' of the farmer, there is likely to be resistance to change. Schmidt (2000) notes that inappropriate water pricing in South Africa has led to greater misuse of irrigation water, but that moves towards charging for full cost recovery will encourage more efficient use. Selvarajan and Subramaniam (1988) noted that excessive use of irrigation water in Tamil Nadu had no effect on water cost, as water was charged on a per area (rather than per volume) basis in the canal irrigation systems. Carruthers (1987) notes that improved irrigation systems will often need to be accompanied by a shift in willingness of farmers to pay user fees for services, and Meyer (1997) stresses that water prices must cover long-term maintenance costs, or delivery systems will inevitably degrade. In Australia, Johnson *et al.* (1997) note a recent move towards water (and other natural resources) being priced according to the full social cost of their use. Meyer (1997) stresses that pricing structures must separate fixed and variable costs and be in a form that irrigators can understand. Where this is not the case and where there is insufficient political will to maintain improved systems, high overhead costs may result in poor management (e.g. Ahmed, 1991). Ghassemi *et al.* (1995) also consider water pricing as a means to encourage the development of more efficient irrigation systems, in the context of avoiding soil salinization problems.

As with technical aspects, the development of decision support tools may be useful for assessing the complex economics of irrigation in sugarcane (e.g. Magwenzi, 2002). One such system is the IRRIECON model developed in South Africa, which allows the return from potential yield increases to be compared with the costs of purchasing and operating different irrigation systems (Singels *et al.*, 1999; Schmidt, 2000).

A number of authors stress the importance of farmer involvement in the development and implementation of improved irrigation strategies (e.g. Anon., 1984; Meyer, 1997). Problems have arisen in projects where farmer involvement has been limited, for example, in Nigeria's Kano River Project (covering about 16,000 ha under surface irrigation), where there was no involvement of local farmers at the initial design and construction stage, and subsequent system operation and management was handled entirely by the government (Ahmed, 1991). Other projects have ascribed their success largely to the active involvement of farmers, for example, the Water Check irrigation extension project in Queensland, which engaged grower groups in on-farm demonstrations and trials, employed participatory, action-learning methods and achieved an increase in water use efficiency of 0.9 t/Ml across Bundaberg district (Holden *et al.*, 1998). One way in which to engage farmers, and ensure their ongoing participation in the development and management of irrigation projects, is through the establishment of water user associations, autonomous groups of farmers able to manage and operate a scheme for redeveloping their own region (e.g. White, 1996).

Increasingly, forms of regulation in the supply and use of water for irrigation are being introduced which affect cane farmers. In this context, Meyer (1997) concludes that water authorities need to avoid the conflict of interest of being supplier and regulator. SASA (2002) notes that, in South Africa, cane farmers who use greater than specified quantities of water (or who use waste water) in irrigation must register with the Department of Water Affairs and Forestry (DWAF), and that licensing restrictions apply. It is then important that water consumption is metered, or otherwise recorded, to ensure that licensed allocations are respected. SASA (2002) also notes that a permit is required from the Department of Agricultural and Environmental Affairs before any virgin/new areas are converted from dryland to irrigated land.

Reducing quantity of water used in sugarcane cultivation

In addition to (or in combination with) appropriate management of irrigation systems, a range of measures has been explored for conserving water and soil moisture in sugarcane cultivation. Aspects of mulching (including trash retention) and modified tillage are discussed in Chapter 2. Other suggested methods include:

- Improvement of on-farm water storage systems (Lisson *et al.*, 2003).
- Cultivation of drought-tolerant varieties (Hellmann, 1977; Rao, 2000; Olaoye, 2001).

- Management of fertilizer application (Yadav, 1986).
- Soil amendments (including potash, coir waste, farmyard manure, cane processing wastes, hydrophilic polymers) (Bishop and Kruger, 1979; Durai *et al.*, 1996; Rao, 2000).
- Foliar applications (potash, urea, $CaCO_3$) (Thind, 1996; Rao, 2000).
- Weed control (Thind, 1996).
- Other aspects of crop management (planting time, row spacing) (Yadav, 1986).

Measures to reduce runoff (e.g. in relation to soil conservation) are considered in Chapters 2 and 6.

Measures to reduce waterlogging – drainage

Where irrigation is practised, it must be accompanied by effective and appropriately managed drainage. Otherwise, there is a substantial risk that environmental degradation will occur, and irrigated cultivation will become uneconomic (Meyer, 1997). Simple surface drainage systems, such as a network of shallow ditches intercepting drainage flow in the cane furrows (row drainage), can assist in removing excess water (e.g. Smedema, 1983). However, the subsurface drainage characteristics of soils are critical. Soils with poor drainage characteristics are less suitable for crops with the greatest irrigation requirements (such as sugarcane), as they are particularly prone to waterlogging and salinization (Gajja *et al.*, 2000; see also Chapter 6). Shallow water-tables can provide substantial contributions towards meeting a cane crop's water requirement. However, irrigation scheduling in areas affected by shallow water-tables should be modified to avoid excess irrigation and to promote increased water use efficiency. Shallow water-tables are generally associated with areas of low elevation in the landscape, and there is evidence that they are common throughout the sugar industry (Sweeney *et al.*, 2001a,b).

SASA (2002) notes that there are limitations on the establishment of drainage systems in South Africa, including legislative requirements to protect natural wetlands and regulations that require approval for the excavation of new drains. The economic benefits of draining poor (wet) agricultural land may not exceed the costs, and it may only be in situations where irrigation is practised (and methods are required to regulate waterlogging and leaching of salts) that new subsurface drains are not just appropriate but essential. A range of issues should be considered in the establishment of a new drainage system, including:

- Prior to establishment:
  - the mapping of soil forms and types in the target area;
  - the establishment of an effective surface water management system, into which subsurface drainage can be integrated;
  - the positioning of a main drain at the lowest point on any gradient;
  - the development of a drainage plan (showing the position of all main drains, subsurface drains, outlets, etc.);
  - the establishment of cut-off drains to prevent seepage into wet areas;
  - the stabilization of waterways;
  - land-planing to prevent ponding and promote free movement of water across soil surfaces;
  - the selection of an appropriate type of drainage system (many are available) to match the soil type in the target area.
- Following establishment:
  - protection of drain outlets to prevent erosion and collapse;
  - alignment of cane rows to facilitate free draining;
  - the establishment of an annual maintenance programme;
  - the installation of inspection boxes in a system of subsurface drainage pipes, to enable regular cleaning of silt, roots or blockages;
  - the possible capture and re-use of drainage water for irrigation (depending on water quality);
  - allowing areas in which soil conditions remain permanently wet to revert to natural wetlands.

SUGARCANE PROCESSING

Relatively large volumes of water can be consumed in the cane sugar factory, in cleaning (washing) of cane, extraction of juice and subsequent processing and in the production of steam for processing and generation of electricity (co-generation – see Box 8.2). UNEP (1982) reports that harvested cane may require 3–10 m^3 of washing water per tonne. Payne (1991) estimates that 4000 l of washing water may be consumed per ton of clean cane per hour. This estimate is based on operations in Hawaii, where cultivation and harvesting methods result in unusually large amounts of extraneous material having to be cleaned from the cane delivered to the factory. However, the estimate also accounts for the reduced consumption that results from recycling of water in successive stages of the cleaning process. With reference to water consumed in the extraction and subsequent processing of juice, Albert-Thenet in Payne (1991) observes that:

> Water is the material most extensively used in sugar manufacture. It is normally freely available to the operator and for this reason is often used indiscriminately. But limiting water use is of critical importance to the conservation of energy, for all water added to the process will have to be evaporated before crystalline sugar can be produced.

Measures for Reducing Water Consumption in Cane Processing

Treatment and recycling of water in cane processing reduces the total volume consumed and can reduce the volume and pollution potential of effluent ultimately generated for disposal (e.g. Srimannarayana and Sudheer, 2000). A key feature in recycling of cane factory water is minimizing contamination with organic materials, as clean water can be used for a wider range of operations than contaminated water. The cleanest recycled water should be 'primary' condensate, steam which has condensed without coming into contact with organic material – turbine exhaust steam, for example. This should be suitable for use in high pressure boilers in the power plant, which need to be fed with very clean water. 'Secondary' condensates (from later stages in juice processing) are likely to be contaminated with organic volatiles and are unsuitable as feedwater, except in lower pressure boilers after appropriate treatment (Payne, 1991). Condenser cooling water is generally produced in large volumes with only low levels of contamination (provided that operations are managed carefully). It can be recycled for use in cooling towers or spray ponds, reducing total factory water usage. As Snoad (1995) notes, such measures not only contribute to reduced environmental impacts, but can facilitate more effective mill operations, possibly even delaying or removing the need to expand the capacity of processing equipment (with associated capital expenditure). Modifications of equipment and processes to allow better recycling of waste water need not be expensive, although some measures may involve more significant costs (Aso *et al.*, 2001). In some cases, financial assistance may be available to support the introduction of new or improved systems. For example, Mangal Singh (1996) reports that cheap loans available from the central government and financial institutions have promoted modernization projects at Indian sugar factories.

Specific measures that can be taken to analyse and reduce water usage in cane processing are examined by Ragen (1992), Wright (1992), Snoad (1995), Mangal Singh (1996) and Aso *et al.* (2001).

SUGAR BEET CULTIVATION

Dunham (1993) suggests that water supply is not a major consideration in the cultivation of sugar beet in many areas, as the plant is relatively insensitive to soil moisture conditions. Sugar beet is moderately resistant to drought, and appears to be a relatively efficient user of water, even when compared to a C4 species like maize. The beet plant develops over a long growing season, and develops a deep, dense network of root fibres, penetrating up to 2 m below the surface (Dunham, 1993;

Defra, 2002). This root system makes the plant a relatively efficient scavenger of soil water and nutrients (Bailey, 1990). Although sensitive to severe waterlogging of soils, beet can withstand a water-table at about 1 m. In addition, its relatively high value has meant that the crop tends not be cultivated in marginal areas. Consequently, resistance to drought or wet soils has not been a primary focus of selective breeding programmes for this crop. In many areas, it is available sunlight (not water availability) that ultimately limits growth of the beet crop (Scott and Jaggard, 1993). None the less, beet growth and quality are influenced by soil water availability, either directly or through effects on uptake of soil nutrients, which in turn can affect the impacts of fertilizer application (e.g. Rover and Buttner, 1999; Ruzsanyi, 2000; Kenter and Hoffmann, 2002). The water consumption of beet varies considerably according to where it is grown. In Finland (with a short, cool growing season) water use by sugar beet during its development is about 400 mm, whereas in Morocco or southern California (with longer, hotter growing seasons) water use might be as great as 1500 mm (Dunham, 1993). Despite its relative insensitivity to soil moisture, dry areas are generally less suitable for sugar beet (e.g. Zimmermann, 1974; Papesch and Steinert, 1997). Even where conditions are not consistently arid, water stress can be a constraint on beet and/or sugar yields, at least in some years (e.g. in France – Richard-Molard and Cariolle, 2001; in Poland – Choluj *et al.*, 2001). Local conditions in some beet growing areas (for example, on soils with particularly poor water-retention capacity) mean that irrigation is essential (e.g. Zimmermann, 1974). Also, the lack of a serious need for additional water has not prevented irrigation from being applied elsewhere.

Irrigation of Sugar Beet

Detailed accounts of irrigation in the cultivation of sugar beet are given by Cavazza *et al.* (1976), Dunham (1993) and (for the UK in particular) Bailey (1990). A number of published research papers deal with responses of sugar beet varieties to management variables including irrigation. Some recent examples are provided in Appendix 2.

While irrigation of sugar beet is primarily undertaken to increase rates of plant growth during the main phase of plant development, irrigation water is sometimes applied to a beet field as part of seedbed preparation (Dunham, 1993; Henriksson and Hakansson, 1993) or at harvest time, to assist lifting of the crop (Bailey, 1990; Dunham, 1993). Irrigation may also be applied to assist seedling emergence through a dried soil cap (Bailey, 1990) or, in extreme cases, to dilute soil salts at germination time to overcome the saline sensitivity of beet at this one stage in its development (Dunham, 1993). The irrigation requirements of beet grown for seed also represent a minor consideration in the wider cultivation of the plant (Dunham, 1993).

Overall, but with particular reference to the UK, Bailey (1990) concludes that irrigation of sugar beet has relatively limited benefits, because the plant is only moderately sensitive to drought and because irrigation generally results in no marked increase in beet quality (sugar yield). For a more detailed consideration of the UK situation, see Box 3.3. None the less, lack of water availability can compromise yields in parts of the former Soviet Union (e.g. Turkistan, Kazakhstan, Kyrgyzstan, Republic of Georgia, Armenia and southern Ukraine), the Mediterranean (Spain, Italy, Greece, Turkey) and parts of the former Yugoslavia, where irrigation may be a prerequisite for satisfactory cultivation of beet (Winner, 1993; Kolomiets *et al.*, 1998; Tognetti *et al.*, 2002). Irrigation is also routinely practised in other beet growing areas. Dunham (1993) estimates that about one-fifth of the world's 8 Mha of sugar beet receives irrigation. As with quantities of irrigation water, the proportion of irrigated land varies between growing regions: in dry areas like parts of the USA and Mediterranean, the Middle East (notably Iran) and Chile, 80–100% of beet growing areas may be irrigated; in the western Mediterranean, the proportion is 20–80%; in northern Europe, the former Soviet Union, Japan and China, it is less than 20%.

Box 3.3. Sugar beet irrigation in the UK.

Defra (2002) reports that beet is not normally irrigated in the UK, except in severe drought conditions (largely because the crop is relatively drought tolerant). It is suggested that less than 5% of UK beet growing areas are normally irrigated, mostly on lighter soils during the driest months. None the less, it is considered that irrigation of beet in the UK should be monitored in the light of future water availability and climate change, and it is predicted that the practice will decline as abstracting of water in eastern England becomes increasingly unviable for economic and environmental reasons.

Bailey (1990) and Scott and Jaggard (1993) suggest that UK rainfall is often inadequate to maximize beet yields. Experiments have shown that irrigation can increase sugar yield by around 3 t/ha over unirrigated fields yielding about 10 t/ha (Bailey, 1990; Scott and Jaggard, 1993; Groves and Bailey, 1994). However, Bailey (1990) anticipates smaller yield gains (< 1 t/ha sugar), except in the driest summers, and Fisher and Kerr (1998) show that responses differ between varieties.

None the less, the potential value of beet yield losses from drought stress in the UK are substantial. Jaggard *et al.* (1998) and Pidgeon and Jaggard (1998) estimate that the mean annual loss of sugar production from 1985 to 1990 was 10.5%, values for individual years ranging from 0 to 25%. This is equivalent to 141,000 t/year, worth £27.9 million. Jaggard *et al.* (1998) and Pidgeon and Jaggard (1998) conclude that drought stress is the largest single constraint on yield, but that irrigation has made little impact on drought losses. Why is this?

An important factor in the cultivation of sugar beet in the UK, and elsewhere in the European Union (EU), is the quota system (e.g. Scott and Jaggard, 1993). Farmers have an allocated quota, a predetermined quantity of beet for which they will be paid a guaranteed, fixed (and relatively high) price. Any additional beet (surplus) does not attract such a high price, so there is little incentive to maximize yield, as long as the quota is met. This substantially affects the economics of operations like irrigation. If a quota-filling crop can be grown without irrigation, there is little incentive to irrigate, even if it would increase yield still further. This is particularly true, given that irrigation itself is a relatively costly operation.

Bailey (1990) reports studies of sugar beet irrigation in the UK, suggesting that irrigation was relatively poorly matched with soil type, quantities of water applied were poorly matched with crop requirements and water use efficiency for irrigation was low (except in the driest years). He concludes that (in addition to the influence of the quota system) this is because UK farmers typically design their irrigation regimes around more drought-sensitive crops, like potatoes, and beet only tends to be irrigated when these crops are not placing heavy demands on the system. There is further evidence that UK farmers tend to irrigate sugar beet (and, indeed, other crops) at levels below the calculated plant water requirements (Bailey and Minhinick, 1993; Knox *et al.*, 1997). Whilst concluding that the economic returns from enhanced yield would justify the cost of irrigation of some crops, Morris (1994) considers that irrigation benefits for sugar beet (and some field-scale vegetables) are marginal, especially where water storage is needed.

Jaggard *et al.* (1998) and Pidgeon and Jaggard (1998) conclude that irrigation is unlikely to be used to overcome drought stress yield losses in sugar beet in the UK in the future, given likely water supply constraints. However, they note that drought stress may become a more significant problem under predicted climate change, and recommend plant breeding for drought stress tolerance.

Jafari-Darabjerdi (1993) provides an overview of irrigated agriculture in Iran, where a system covering some 2.5 Mha and delivering around 50 Mm^3 water per annum feeds rice, sugar beet, cotton and winter wheat. The infrastructure combines ancient storage dams and underground channels (qanate) with modern extraction and storage facilities. Zeng and Liu (1992) report on the optimization of irrigation of sugar beet in the Hetao irrigation area, China.

Dunham (1993) reports that the amounts of irrigation water applied to sugar beet crops vary greatly worldwide. For example, in the UK or France, where irrigation is very much supplementary to reasonable levels of rainfall, only 100–200 mm irrigation might be required under dry conditions. In hot, dry environments, such as parts of the USA and the Mediterranean, and in Pakistan, irrigation of beet is essential, and 500–1000 mm of water may be applied to the crop (Dunham, 1993).

Irrigation and sugar beet diseases

Although some studies consider the effects of irrigation on diseases of sugarcane, there is a more extensive literature on such effects in the cultivation of sugar beet. This suggests that the interactions between irrigation, plant water stress, plant pathogens, insect vectors and soil microorganisms are complex.

Dunham (1993) considers that insect pest problems are generally reduced by irrigation, and Bailey (1990) notes that irrigation can be effective in washing virus-vectoring aphids away from the beet plant.

Soil-borne plant pathogens respond differently to soil moisture levels. Fungi like *Pythium* (responsible for damping off) and *Phytophthora* possess zoospores, which are produced and move more freely in liquid water, and can therefore benefit from moist conditions. Others, such as *Fusarium*, are able to perform better in dry soils, possibly because the activity of antagonistic bacteria is reduced under these conditions (Manners, 1982). Dunham (1993) notes that, whilst root rots are generally favoured by wet conditions, roots that have previously experienced drought are more vulnerable. Hence, season-long irrigation may actually reduce root rot risk. However, some foliar diseases like *Cercospora* and *Ramularia* are greater problems in humid conditions, and may need to be controlled in irrigated fields.

A number of studies suggest that irrigation increases pathogen populations and/or disease incidence in sugar beet crops. Wang *et al.* (1995) found that soil populations of *Pythium* spp. (mainly *P. ultimum*) were increased by presowing irrigation in field experiments on sugar beet in California. In Khuzestan province, Iran, Mahmoody *et al.* (1997) studied *Urophlyctis leproides*, the causal agent of sugar beet leaf and crown wart. They compared fields under conventional irrigation scheduling with fields where irrigation was applied only when required by the crop. Not only did the additional irrigation applied under the conventional system fail to increase root yield or sugar content, but disease incidence was four times greater than in fields irrigated when required by the crop.

In studies of sugar beet root rot caused by *Fusarium oxysporum* f. sp. *betae* in Texas, USA, Harveson and Rush (1998) found that irrigation treatments had no effect on disease incidence or severity. However, Harveson and Rush (2002) later found that reduced irrigation resulted in lower disease incidence in fields supporting a complex of sugar beet root diseases (the fungal pathogens *Aphanomyces cochliodes*, *F. oxysporum* f. sp. *radicis-betae*, *Rhizoctonia solani*) and the viral pathogen beet necrotic yellow vein virus (BNYVV). They concluded that, when few alternative options are available, sugar beet growers may benefit from reducing irrigation and growing locally adapted cultivars in soils severely infested with root pathogens. Other studies also suggest that reduced irrigation can reduce the incidence of viral diseases of sugar beet. Tuitert and Hofmeester (1994) studied BNYVV in sugar beet, and found that differences in infection levels between irrigated and non-irrigated plots were apparent only at low initial inoculum levels, with irrigated plots having a greater disease incidence than non-irrigated plots. In studies involving soils infested with BNYVV and beet soil-borne mosaic virus (BSBMV), Piccinni and Rush (2000) compared plots under four irrigation regimes (every 2, 3, 4 and 5 weeks). They found that sugar beets irrigated every 4 weeks had the lowest disease severity, and yield was not significantly different from beets irrigated every 2 weeks.

Bailey (1990) and Dunham (1993) note that irrigation has probably contributed to the spread of *Rhizomania* in some countries, both because the *Polymyxa betae* host fungus prefers wet soil conditions, and because the fungus and viral disease agent (BNYVV) may be carried in runoff waters from infected soils.

Other studies show a decrease in disease incidence in sugar beet under irrigation. Vesely (1975) found that losses to black leg (*Pleospora betae*) were greatest under drought conditions, whilst Maiti *et al.* (2000) found that increased irrigation enhanced root yield and suppressed root rot caused by *Sclerotium rolfsii* (*Corticium rolfsii*).

The complexity of irrigation effects on soil microorganisms in sugar beet cultivation systems is illustrated by the work of

Piotrowski *et al.* (1996) in Poland. Over several years, these authors studied the effect of irrigation of sugar beet on the number of total soil microflora (bacteria – B, actinomycetes – P and fungi – G), the biotic relations index [(B + P)/G], bacteria from the genus *Pseudomonas*, fungi from the genera *Pythium* and *Fusarium*, and *A. cochlioides* inoculum potential. On the supplementary irrigated plots, the system of biotic relations that developed gave higher soil fertility than that found on non-irrigated plots. Actinomycete and bacterial populations, including cellulolytic and pectinolytic bacteria, and bacteria solubilizing calcium phosphate, increased as the irrigation treatment progressed. There was a parallel decrease in certain fungal populations, including those solubilizing calcium phosphate, and *Fusarium* plant pathogens, but numbers of lower fungi from the genera *Pythium* and *Aphanomyces* increased. In this study, populations of potential soil antagonists (*Pseudomonas* bacteria) decreased under the irrigation treatment.

Problems Arising from Irrigation of Sugar Beet

Dunham (1993) notes that surplus irrigation is wasteful, and can reduce yields through waterlogging, nutrient leaching, increased pest/disease problems and harvesting difficulties. None the less, in many parts of the world where irrigation is relatively inexpensive (such as the USA) there has been a tendency to use too much water in beet cultivation. In addition, irrigation systems may be ineffective and inefficient, leading to wastage of irrigation water. For example, Grigorov and Grigorov (2001) examined the use of sprinkler irrigation (the most widely used method) in the sugar beet growing region of Volgograd, Russia. They found that the effectiveness of sprinkler irrigation was compromised by inadequate planning, poor choice of irrigation priorities and regimes, low quality design, construction and maintenance of irrigation systems and inadequate crop care. As with cane cultivation, the effectiveness of irrigation in beet cultivation can be estimated using some form of water use efficiency calculation, based on the amount of dry matter produced per unit of water consumed, for example (see Box 3.1).

Technical aspects of improved irrigation systems

Tools to assist in the estimation of crop water requirement and irrigation scheduling

The calculation of crop water requirements, for example, for the accurate scheduling of irrigation, is outlined in Box 3.1. Bailey (1990) notes that the calculation of soil water balance to assess the irrigation needs of sugar beet is complicated by the fact that, relative to other crops in the UK, the root system develops so much during the growing season, substantially changing a key parameter in the calculation (rooting depth). Scheduling of irrigation is unlikely to be critical to the development of the plant, as there are no growth stages of sugar beet that are highly sensitive to water stress, unlike flowering in cereals and peas, or tuber expansion in potatoes, for example (Dunham, 1993; Groves and Bailey, 1994). None the less, scheduling is important in order to optimize the efficiency of water use. Bailey (1990) reports on studies from the UK, investigating the methods used by farmers to schedule their irrigation of sugar beet. Most used some form of water balance calculation, although use of tensiometers, degree of wilting and 'feel of soil' were also listed as practised methods. Dunham (1993) also notes that traditional 'by eye' methods are widely used in scheduling of irrigation of sugar beet, while discussing the availability of more sophisticated techniques. Examples of computer models used for simulating water dynamics and developing irrigation strategies in the cultivation of sugar beet are listed in Table 3.5. Other crop irrigation models with applicability to sugar beet are discussed by Dunham (1993). Remote sensing techniques can also aid irrigation scheduling. Infrared radiometry can be used to assess average canopy temperature in sugar beet fields, overcoming problems of small-scale variations across the

Table 3.5. Computer models used in the development of improved irrigation strategies for sugar beet.

Model	Source – locality
OPUS	Smith (1995) – Germany
DAISY	Svendsen *et al.* (1995) – Germany
MORECS	Thompson *et al.* (1981) – UK
PLANTGRO	Davidoff and Hanks (1989)
SUBGRO	Fick *et al.* (1975)
Other models	
An optimum irrigation model	Zeng and Liu (1992) – China/Mongolia
A daily water balance irrigation scheduling model	Knox *et al.* (1997) – England and Wales
A farm water balance model	Gabellini *et al.* (2001a) – Italy
A linear model for crop distribution and water resources utilization across a typically irrigated area	Chen Han *et al.* (1995) – China

site (Dunham, 1993; Roth and Rossler, 1998; Margotti, 2000).

Irrigation system characteristics

Dunham (1993) notes that practically all known methods of irrigating field crops are used for sugar beet somewhere in the world, and that methods chosen are invariably guided by local conditions and economic considerations, rather than the strict requirements of the beet crop. Thus, surface irrigation (via basins, borders or furrows) is widely used for beet in the USA, Turkey and Iran, whereas overhead sprinkler systems (using travelling rain guns, booms and centre pivots) predominate in France, Italy and northern Europe. In Spain, traditional surface irrigation methods are being replaced by sprinkler systems. Draycott and Christenson (2003) conclude that furrow and overhead sprinkler systems both tend to lead to inefficient water use, and may increase the risk of leaching. A number of published research papers deal with responses of sugar beet varieties to different irrigation methods and other management variables. Some recent examples are provided in Appendix 4.

SURFACE IRRIGATION. Narang *et al.* (1992) found that beet root yields were greater under flood irrigation than furrow (or alternate furrow) irrigation, but the relative quantities of water used are not clear. As with surface irrigation of cane, it is apparent that refinement of furrow irrigation can result in water savings. For example, Sepaskhah and Kamgar-Haghighi (1997) found that alternate furrow irrigation at 6-day intervals used 23% less water than irrigation in every furrow at 10-day intervals, maintaining yields and increasing water use efficiency by 43%.

SPRINKLER IRRIGATION. There is evidence that sprinkler irrigation does not result in greater sugar yield from beet than furrow irrigation (although it may enhance leaf growth). Indeed, some studies show lower yields under sprinklers than under furrow irrigation (e.g. Eckhoff and Bergman, 2001). However, sprinklers tend to be more efficient than furrow methods, using some 20% less water and resulting in greater uniformity of water application (Dunham, 1993).

DRIP IRRIGATION. Dunham (1993) notes that subsurface drip irrigation might be particularly appropriate for sugar beet, given its deep rooting habit. However, experiments in the UK have shown no consistent benefits for drip systems over sprinkler systems for beet cultivation. In the USA, Sharmasarkar *et al.* (2001b) found that beet root yields and sugar content were greater under drip than under flood irrigation, and that water savings and increased water and fertilizer use efficiencies were also achieved. Tognetti *et al.* (2002) recommends drip irrigation over low-pressure sprinklers for cultivation of sugar beet in semi-arid Mediterranean environments, while noting the general lack of information on the use of drip irrigation in the region (but see Tugnoli, 2001).

As drip irrigation is a relatively expensive method, costs and benefits must be weighed up carefully, not least in relation to the wider rotation of crops used in most beet growing areas, and other economic factors. Sharmasarkar *et al.* (2001a) provide an agro-economic analysis of drip irrigation for sugar beet production in Wyoming (USA). They concluded that economic returns from drip were 11% greater than with furrow irrigation, and that cultivation under drip irrigation would be most profitable for a 40 ha area, with payback periods ranging from 7 to 10 years. Despite such encouraging findings, Draycott and Christenson (2003) conclude that a greater number of studies under different conditions are required before drip irrigation can be widely recommended for sugar beet production.

Source of water for irrigation

IRRIGATION WITH BEET-PROCESSING WASTE WATER. In laboratory-based studies of seeds on paper, effluent from sugar beet factories has produced equal or improved germination for a range of crops when compared with distilled water (Klimakhin *et al.*, 1998). Wider environmental effects of irrigation with sugar beet factory waste water were studied by Izsaki *et al.* (1993).

IRRIGATION WITH SALINE WATER. As noted in relation to soil salinity (Chapter 6), sugar beet is relatively tolerant of saline conditions, except in the germination stage. Studies of the impact of irrigation with saline water on sugar beet show a range of responses, which would be expected to vary with beet variety, site-specific factors such as local soil characteristics, and whether irrigation with saline water is a short- or long-term measure. Some studies suggest that irrigation of sugar beet with saline water has little detrimental effect on the crop. For example, in the USA, Rhoades *et al.* (1988) found no significant differences in yield or crop quality in a range of crops, including sugar beet, when brackish water was used for up to 25–50% of crop irrigation requirements. Other studies suggest that, under appropriate conditions, irrigation with moderately saline water can benefit a sugar beet crop. Greenhouse pot experiments by Mekki and El-Gazzar (1999) suggest that moderate salt concentrations in irrigation water can enhance sugar beet cropping characteristics. Comparing water with 0, 2500, 5000 or 7500 ppm chloride salts, these authors found that the highest fresh root yield, root diameter and whole plant dry weight were obtained under irrigation with water with 2500 ppm chloride salts. While high salt concentration (7500 ppm) caused an increase in sucrose and total soluble sugar percentages, juice purity and sugar yield (g/plant) were reduced. In a review of water requirements and crop yields in Sweden, Johansson (1978) concluded that, on clay soils high in Ca, responses of sugar beet to irrigation with saline water were good.

Other studies show negative effects of irrigation with saline water. Kandil *et al.* (2001) studied the physiological response of sugar beet to irrigation with different levels of chloride salinization. There were differences in response between varieties, but results showed that root yield, root length and diameter, top height, dry weights of top and root, as well as the total dry weight of whole sugar beet, were significantly decreased by enhanced levels of chloride salinization in irrigation water up to 6000 ppm. However, percentage of sucrose and total soluble solids (TSS) of sugar beet roots was significantly increased by increasing the concentration of chloride salinity in irrigation water up to 6000 ppm. Other plant physiological characteristics (such as cell sap concentrations and proline content) were also affected by saline irrigation. Shehata (1999) also studied these chemical constitution characteristics, and found that increasing the salt concentration in irrigation water up to 4000 ppm significantly increased total soluble solids in sugar beet root juice, free proline concentration in leaves and sodium concentration for both leaves and roots. In this study, sucrose percentage, chlorophyll B and carotenoids were not affected by irrigation with saline water up to 4000 ppm. Kaffka *et al.* (1999) found that irrigation with saline water (ECw 6.7 dS/m) decreased beet percentage sugar, and hence sugar yield, but considered that this was due to relatively high levels of N in the saline water source.

Schleiff (1982) studied one mechanism by which irrigation with saline water influences the development of the sugar beet plant, noting that crop growth was often limited by a suboptimal water supply to the shoot. Easily soluble salts accumulated in the soil solution close to the roots, such that root water uptake was reduced more than might have been expected from the salt concentration of the average soil solution. In this study, roots of young maize plants were able to absorb water from rhizospheric soil solutions exceeding an osmotic potential of about –9 bar, while sugar beet roots did not absorb water from rhizospheric soil solutions of less than –30 bar.

Non-technical aspects of improved irrigation systems

The economics of sugar beet irrigation are complicated by the typical cultivation of this crop as part of a wider rotation. Hence, the costs and benefits of installing and maintaining an irrigation system must be considered in relation to the full range of crops produced, not just sugar beet (Dunham, 1993). The type of irrigation method used also has important economic ramifications (e.g. Avillez and Ramos Rocha, 1988).

As noted elsewhere (see Box 3.3), the quota system used for sugar beet in the European Union (EU) significantly influences the economics of crop production, as the maximum price is not necessarily secured by attaining the maximum yield. Irrigation systems are also relatively costly, so there may be no economic benefit to irrigation, even where this might enhance yields to some extent. Studies suggest that the economic benefits of irrigation of sugar beet can be marginal at best in some situations (e.g. in the UK – Morris, 1994; and parts of Italy – Valli *et al.*, 1996).

Economic instruments may be proposed as a means of reducing the volume of water used in irrigation. Berbel and Gomez-Limon (2000), for example, model the economics of irrigated farming including sugar beet in Spain, to assess the possible benefits of reducing water consumption by water pricing. They conclude that farm incomes would fall by some 40%, and there would be significant loss of employment (in agriculture and processing) before water consumption was significantly decreased, resulting in catastrophic impacts on the agricultural sector. Berbel and Gomez-Limon (2000) conclude that water pricing as a single instrument would not serve the desired purpose of reducing water consumption, but that small charges might make farmers aware of the scarcity of water resources and induce them to adopt water-saving technologies.

Reducing quantity of water used in beet cultivation

Draycott and Christenson (2003) conclude that reducing the amount of irrigation water applied to sugar beet has many advantages, provided that yields are not impaired. They note research from the USA suggesting that irrigation can be discontinued reasonably early in the growing season, provided that soil water reserves are sufficient, without significant reduction in sugar yield. In addition to (or in combination with) the use of improved irrigation methods, various techniques have been suggested for conserving soil moisture in sugar beet cultivation systems. It has also been shown that adopting systems of sugar beet cultivation that promote water conservation can contribute to enhanced sugar yields (Wiklicky, 1981). Aspects of modified tillage and mulching are discussed in Chapter 2. Other suggested methods include the use of subsurface asphalt moisture barriers (Gupta and Aggarwal, 1980).

BEET PROCESSING

Relatively large volumes of water can be consumed in the processing of sugar beet, in cleaning the roots, extraction of juice and other operations. For example, in the Ukraine, Nibit *et al.* (1994) reported that traditional processing of 1 t sugar beet required 20 m^3 water (see also figures below).

Measures for Reducing Water Consumption in Beet Processing

Partly as a consequence of root shape, mechanical harvesting of sugar beet results in a relatively high soil tare (soil removed at harvest, which has to be washed off before processing – see Chapter 6). Elliott and Weston (1993) note that, in addition to significant cost savings, the development of low-tare beet varieties would speed up harvesting, reduce soil degradation and reduce the water consumption of beet factories.

Recycling of the water used in beet processing reduces the total volume consumed and can contribute to reducing volumes of waste water effluent ultimately generated for disposal. Provided that entrainment of organic matter can be minimized by careful management of operations, barometric condenser cooling water is generally produced in large volumes with only low levels of contamination. Thus it can be ideal for recycling (e.g. for use in cooling towers, open or spray ponds), reducing total factory water usage.

Evidence of considerable success in reducing water consumption at sugar beet factories comes from various sources. Kuzminski *et al.* (1991) report that a Spanish sugar beet factory reduced its water consumption from 330 to 35 l/s between 1984 and 1990. Benhnini (1991) predicted that two sugar beet factories in Morocco would reduce their water consumption from 350 to 130 m^3/h and from 430 to 100 m^3/h. Nibit *et al.* (1994) report that sugar factories in the Ukraine introduced various recycling schemes decreasing water throughput from 300 to 50–60% on beet from the 1960s to the 1980s. Fornalek (1995) explains how the water requirement in Poland has gradually been decreased from 105 m^3/t sugar in 1950 to around 10 m^3/t in 1994. Also in relation to Polish sugar beet factories, Polec and Kempnerska-Omielczenko (1995) report declines in water consumption from averages of 2.8 m^3/t beet (or 22.5 m^3/t sugar) in 1989 to 1.1 (8.5 m^3/t) in 1993/94, thanks partly to extra closed circuits for cooling water. In some cases, at least, improved water management systems can 'pay for themselves': Kuzminski *et al.* (1991) report that the costs of a new system in a Spanish sugar beet factory were recovered from savings on discharge permit taxes and from profits obtained from better utilization of by-products.

Specific measures for reducing water usage in beet processing are examined by Benhnini (1991), Kuzminski *et al.* (1991), Nibit *et al.* (1994), Polec and Gozdek (1994), Fornalek (1995) and Volgyi (2002).

Klemes *et al.* (1999) argue that adoption of new methods in beet sugar processing, such as cooling crystallization of concentrated raw juice, as opposed to the traditional method of evaporating crystallization, has the potential to improve energy efficiency and to reduce atmospheric emissions, water consumption and polluting potential of effluents.

4

Impacts on Water Quality and Aquatic Ecosystems

Sugar production can have an impact on water quality and aquatic ecosystems through both cultivation and processing of sugar crops. In relation to cultivation, the main considerations arise from runoff and leaching, which can lead to pollution of groundwater (which may include sources of human drinking-water), surface water (including natural watercourses such as rivers and streams) and ultimately coastal environments. The main polluting agents here are nutrients (notably nitrates and phosphates derived from fertilizers, which can cause eutrophication), agrochemicals such as pesticides, and sediments arising from soil erosion. Irrigation can increase runoff and deep drainage, and impacts on water quality potentially arise from the use of waste or saline water for irrigation, which can also contribute to the salinization of soils (e.g. Ghassemi *et al.*, 1995; Feizi, 1998). Ometo *et al.* (2000) note that most published studies of land use impacts on aquatic systems focus on nitrogen dynamics in temperate American and European catchments, and less attention has been given to impacts in tropical areas of developing countries, where land management practices are often rather different, and where cane growing is concentrated. Johnson *et al.* (1997) conclude that downstream impacts of any form of agriculture are largely governed by the periodicity, volume and intensity of rainfall. Although based on observations in Australian cane growing areas, this probably holds true for most other (particularly tropical) regions. In addition, Johnson *et al.* (1997) note that it is the inputs associated with intensive agriculture in general (rather than management strategies for particular crops) that result in increased risks of pollution.

In relation to processing of sugar crops, the main consideration is pollution arising from the discharge of effluents from cane mills and beet factories. Given the nature of the materials being processed, it is not surprising that these effluents tend to be relatively rich in organic matter, including carbohydrates, when compared with those from other sources. Consequently, sugar processing effluents can represent pollutants with very high biological/chemical oxygen demands (BOD/COD). However, other potential pollutants occur in these effluents, including heavy metals, oil/grease and cleaning agents. In addition, characteristics such as pH and temperature of discharged effluent can influence environmental impacts. Various measures can be taken and forms of treatment used to reduce the quantity and polluting potential of sugar mill effluent. Recycling of processing waste water reduces consumption, thereby easing potential impacts of processing on water availability (see Chapter 3), as well as reducing the volume of effluent. Such measures are attractive, provided that large discharges of low concentration effluents are not simply replaced by smaller disharges of more concentrated effluents (Vigh, 1994).

The siting of sugar cultivation and processing activities affects potential impacts on water quality and aquatic ecosystems. As a relatively 'thirsty' crop, sugarcane in particular is grown where ground or surface water is readily available for exploitation, in water-rich catchments and areas which formerly supported natural wetlands (see Chapter 5). Such areas are also often located relatively near to coastal zones, exacerbating potential impacts on these (and marine) environments. Sugar crop processing is also a relatively heavy user of water, and so again will often be undertaken where the resource is readily available (and processing centres will often be sited close to cultivation areas to minimize costs of transporting harvested cane/beet from field to factory).

It may be difficult, however, to unambiguously assign water pollution impacts to sugar crop cultivation and processing in any given area. Although cane growing may dominate agriculture in some catchment areas, it is often part of a mosaic of potentially polluting activities. As noted elsewhere, the fact that beet is typically grown as part of a rotation can hinder the assignment of impacts to this crop, specifically, over periods greater than one growing season. Furthermore, although sugar crop processing may be the single (or dominant) activity at a given site, it is often undertaken in areas where there are other sources of effluents (other industrial operations, population centres generating human wastes, etc.).

Runoff

Water, derived from rainfall or irrigation, may drain from fields by running across the soil surface, depending on rates of input, topography and soil characteristics (notably traits such as permeability or porosity that influence infiltration rates). Hence surface drainage (runoff) is closely linked to soil factors, including compaction (which tends to increase runoff rates) and erosion (which tends to be promoted by runoff). The relationship between runoff and certain types of erosion is considered in detail by Monnier and Boiffin (1986). Surface drainage water can carry with it dissolved nutrients, soluble pesticide residues and soil sediments. The latter (arising from erosion) can contribute directly to a range of impacts downstream (e.g. see Arthington *et al.*, 1997), but can also carry with them insoluble chemical residues.

Leaching

Water that does not drain from fields across the soil surface can be assumed to infiltrate the soil (excepting that proportion that evaporates from the soil surface). During the growing season, much of the water that infiltrates the soil may be utilized by the crop (see above in relation to evapotranspiration), but a proportion may also drain from the field at the subsurface level. Rates of subsurface drainage will be enhanced when the crop is at an early stage of development, or absent (postharvest). Subsurface drainage water may carry soluble chemical residues with it (leaching), affecting the chemical composition of deeper soil strata or groundwaters, or flushing these agents out into waterways. Other environmental factors may influence the specific composition of leachates. Haynes and Hamilton (1999), for example, note that soil acidification promotes leaching of certain nutrients (such as Ca and Mg) and accumulation in the soil of others (notably Al). Leaching of nitrate can contribute to the eutrophication of freshwater and coastal ecosystems. Although most leached nitrate appears to originate from mineralized soil organic nitrogen, excessive application of inorganic fertilizers also contributes to the loss of nitrate from soils (Christensen, 2004). Effects of fertilizer and agrochemical inputs on runoff and drainage waters will be influenced significantly by local soil characteristics, as these affect the dynamics of processes such as leaching, as well as by climatic factors (rainfall) and management factors such as irrigation. The threat of water pollution also varies between different chemical residues, not just because of differences in their environmental effects, but because some are more soluble (and therefore mobile in solution) than others. For example, nitrate and sulphate tend to be relatively mobile, whilst phosphate and ammonium are relatively immobile (Draycott and Christenson, 2003). Consequently, nitrate tends to be relatively easily leached from soils, whereas phosphate might be expected to move more readily in association with sediments carried in runoff waters.

SUGARCANE CULTIVATION

Concern has been expressed over the impacts of sugarcane cultivation on the quality of natural water resources in many parts of the world, notably in Australia, where Arthington *et al.* (1997) note that three factors have been particularly detrimental: extensive vegetation clearing in the riparian zones of rivers and flood-plain wetlands (see Chapter 5); soil erosion and stream sedimentation; and contamination of water bodies with nutrients, pesticides and other discharges from diffuse sources. Although it is difficult to unambiguously identify the sugar industry as the source, evidence suggests that sugar cultivation has contributed to contamination of groundwater with nitrates and pesticide residues (Brodie *et al.*, 1984; Keating *et al.*, 1996, 1997; Weier *et al.*, 1996; Biggs *et al.*, 2000), although the levels of pollution involved have tended to be fairly low. None the less, Arthington *et al.* (1997) emphasize that the long-term effects of even trace quantities of pesticides in aquatic ecosystems are as yet unknown.

Considerable concern exists in Australia over the impacts of land use, including sugar cultivation, on downstream environments, not least the Great Barrier Reef (e.g. Arakel *et al.*, 1993; Christiansen and Hunt, 2000). Crossland *et al.* (1997) examined evidence that sugar production has had a negative impact on the Great Barrier Reef and adjacent marine environment. The principal difficulty is in distinguishing impacts of the industry from other land uses along the eastern coast of Australia. None the less, it is clear that the sugar industry has been a significant player in major infrastructural projects, including damming of the Burdekin, Tully and Barron Rivers, that have altered the pattern of freshwater flow into the marine environment. It is also apparent that sediment and nutrient loads have generally increased from diffuse sources in cane growing areas, and that major discharge events (i.e. following heavy rainfall) can carry these materials to offshore reefs as well as inshore environments. Also, drainage waters from acid sulphate soils under cane and other crops have had adverse impacts on the estuarine and marine environment, and pesticide residues have been detected in coastal sediments and marine organisms. Johnson *et al.* (1997) consider that, although soil erosion is a problem in many cane growing areas in northern Queensland, the precise rate and impact of sediment delivery to estuarine and marine environments remains poorly understood. Also, available evidence suggests that nutrient inputs to coastal waters arising from cane cultivation are small relative to natural fluxes, but, again, poor understanding of the dynamics of these ecosystems hinders confident prediction of long-term consequences. Although precise causes could not be identified, Gaus *et al.* (2001) reported elevated levels of higher polychlorinated dibenzo-*p*-dioxins (PCDD) in the coastal environment of Queensland. High octachloro-dibenzodioxin (OCDD) concentrations were associated mainly with sediments collected near the mouth of rivers that drain large catchments in the tropical and subtropical regions. High concentrations were found in samples from sugarcane drains collected from coastal regions, and lower concentrations in drain sediments from drier inland cotton growing areas.

Johnson *et al.* (1997) observe that the scale of the sugar industry in Australia results in a large potential for diffuse source pollution of ground and surface waters. However, a number of studies suggest that, for example, nutrient concentrations in streams in cane growing areas are only a cause for concern, or greater than those associated with other land uses, during peak flow events. For example, although not exclusively derived from cane fields, runoff waters in north Queensland catchments markedly increase stream concentrations of sediments and nutrients (mostly bound to sediments) during periods of greatest rainfall (Hunter *et al.*, 1996; Mitchell *et al.*, 1996). Recorded rates of soil erosion from Australian cane fields can be very high (e.g. see Prove *et al.*, 1995), but understanding of the ultimate fate of sediments in runoff waters is poor (Pailles and Moody, 1996). Arthington *et al.* (1997) suggest that emphasis on potential impacts on sensitive marine ecosystems, such as the Great Barrier Reef, may have led to an under-appreciation of effects in freshwater environments closer to the field. For example, Arakel *et al.* (1989) estimated a sediment deposition rate of 300,000 t/year (equivalent to cane field erosion rates of 150 t/ha/year) in the lower South Johnstone River, resulting in the formation of large sand spits and islands in the river mouth, but relatively little sedimentation in the estuary.

In Mauritius, with around 90% of its 90,000 ha of existing arable land under sugarcane, the sugar industry has traditionally been seen as a major source of pollution to surface and groundwater (Ramjeawon and Baguant, 1995; Ng Kee Kwong *et al.*, 1996). Umrit and Ng Kee Kwong (1999) refer to crystal clear watercourses becoming loaded with mud during and after heavy rainfall events in sugarcane cultivation areas in Mauritius, stimulating public fears over water pollution. Whilst this observation clearly demonstrates the off-field movement of sediments, studies suggest that movement of agrochemicals is less than might be anticipated. Although concerns remain that sugarcane cultivation degrades water quality and the wider environment, a study by Ng Kee Kwong *et al.* (1996) showed that concentrations of nitrate and herbicide residues in drinking-water were well below the maximum permissible levels. Such findings are supported by further research, discussed below, from Mauritius, where Ng Kee Kwong *et al.* (1998) note that most fertilizer and pesticide application occurs during the dry season, reducing the likelihood of pollution of ground and surface waters.

In Brazil, Ometo *et al.* (2000) demonstrated that the water chemistry and macroinvertebrate fauna of two streams in the Piracicaba River Basin were related to land use (including sugarcane cultivation) in their respective catchments. Silva *et al.* (2001) considered that sugarcane cultivation could be a contributory factor in the dynamics of dissolved nitrogen and phosphorus in the lower portion of the Paraiba do Sul River. Gomes *et al.* (2002) suggest that the disorganized spread of agricultural activities including cane cultivation has contributed to their impact on environmentally sensitive systems, including aquifers. Investigations in 1995–1998 revealed the presence of herbicide residues in water samples from a semi-artesian well in the State of São Paulo, where sugarcane is a major crop.

Tudor-Owen and Wyatt (1991) note that the cane producing areas of the South African sugar industry are situated mainly in the catchment areas of gently to steeply undulating land from which runoff flows into the major rivers of the Natal seaboard; it has been suggested that lack of adequate integrated soil conservation practices on these farms has resulted in soil erosion, leading to degradation of the rivers and estuaries.

In the USA, much concern has been expressed over the possible impacts of sugarcane cultivation on the Florida Everglades (see Box 5.1). Soicher and Peterson (1997) examined pollution of the coastal waters of West Maui, Hawaii, and established that, in addition to sewage effluent impacts, the principal agricultural activities in the area (sugarcane and pineapple cultivation) contributed to elevated loads of nutrients and sediments.

Runoff in Cane Cultivation Systems

Runoff phenomena tend to have been relatively under-studied in tropical agroecosystems (Ng Kee Kwong *et al.*, 2002), but there are a number of relevant studies from cane cultivation systems, relating to the movement of sediments and nutrients.

Soil erosion can facilitate the movement of nutrients from sugarcane fields. Studies from Australia suggest that 50% of N and 80% of P transported (in a flood event in the Herbert River catchment) were bound to sediments (Crossland *et al.*, 1997). In Louisiana (USA), soil erosion losses averaging around 17 t/ha/year resulted in annual nutrient losses from fields of around 18 kg N/ha, 14 kg P/ha and 104 kg K/ha. The amounts of nutrients lost from fields depended on application rates of inorganic fertilizers (Bengtson *et al.*, 1998). In Mauritius, Ng Kee Kwong *et al.* (2002) examined runoff from cane fields on slopes of 5–12%. There was a 6-week time lag between fertilizer application and the first runoff event, which contributed to the relatively small quantities of inorganic N and P transported from the fields. Data from five runoff events over a 2-year period indicated that less than 1 kg P/ha and 2–7 kg N/ha were lost from the cane fields. The transport of N and particularly P was intimately linked to that of sediment in the runoff water. Although insignificant from an agronomic viewpoint, the N and P transported were sufficient to raise concentrations in runoff water to levels considered to be of environmental concern.

Whilst noting that soil erosion is a problem in many cane growing areas in Australia, Johnson *et al.* (1997) observe that the precise rate of sediment delivery to estuarine and marine environments remains poorly understood. Although there have been reports that suggest cane cultivation results in significant sediment transfer to rivers and coastal areas, solid evidence is often scarce. However, clear demonstration of such impacts include studies for the Johnstone catchment (Pailles and Moody, 1996). Prove *et al.* (1995) suggested that, although adoption of no tillage reduced erosion rates under cane cultivation in Queensland, eroded sediments and a more mobile fraction of soil nutrients might be transported further from the field under this system of cultivation.

Lindau *et al.* (1997) examined runoff and surface water from cane fields on poorly drained soils adjacent to forested wetlands in Louisiana (USA). Fertilizer N draining into the wetlands from cane fields was estimated as only a small fraction of that applied, and nitrate and ammonium concentrations were low; around 3–4% of applied N was removed in runoff waters. In addition to sediment and nutrient transfer, runoff from cane fields has also been shown to transport potentially polluting levels of carbohydrates, including sugars. Bohl *et al.* (2002), for example, demonstrated that the BOD of irrigation runoff water from postharvest cane fields in Australia was well above state licensing limits for the discharge of sewage and industrial effluents. They concluded that the runoff water had the potential to cause serious environmental impacts, including fish kills, if it was allowed to reach local waterways.

Leaching in Cane Cultivation Systems

Ng Kee Kwong and Deville (1984, 1987) studied nutrient leaching under sugarcane in Mauritius. These studies used ^{15}N-labelled fertilizer (as $(NH_4)_2SO_2$ or $NaNO_3$), applied at a rate of 100 kg N/ha, on soils of different types (Oxisols and Inceptisols) in areas of differing rainfall. Vertical and lateral distribution of residual fertilizer N in the soil was not affected by type of fertilizer or timing of application. More than half of the residual N remained at the soil surface, and less than 30% moved laterally more than 0.3 m. The amount of N leached was affected more by the duration and intensity of soil drying preceding rain than by leachate volume. The cumulative N loss over 1 year was similar across sites, but leaching was greater from the drier (Inceptisol) soils. Soils with higher organic matter content displayed greater N leaching. Oxisols retained nitrate by absorption, reducing N leaching, but K and Ca were more readily leached than N. Ng Kee Kwong and Deville (1984, 1987) concluded that losses of cations might be a more acute problem than N leaching, which was not considered a cause for concern in cane growing soils in tropical environments like Mauritius. In Australia, Wei-Ping *et al.* (1993) studied fertilizer N dynamics in recently harvested, trash blanketed cane fields, and found no evidence of N leaching (although 20–30% of applied N was lost to NH_3 volatilization). Similarly, Chapman *et al.* (1994) found no significant loss of N by leaching following urea fertilizer application, but suspected appreciable losses of gaseous N due to denitrification.

In contrast to those studies which suggest that nutrient leaching under cane is not a major concern, other studies indicate that environmental problems can arise from this source. In Australia, Verburg *et al.* (1998) found that over-application of N fertilizer could substantially increase leaching from the root zone, and Keating *et al.* (1996) found that 2% of sampled boreholes in Queensland contained water with a nitrate concentration of > 50 mg/l. For comparison, Keating *et al.* (1997) note that 50 mg NO_3/l is the threshold set for infants under Australian drinking-water guidelines (other age groups have a threshold of 100 mg/l), and that general drinking-water guidelines laid down by the World Health Organization (WHO) and European Union (EU) set thresholds at 45–50 mg/l. Whilst Keating *et al.* (1996) suggest that some of the groundwater pollution that they report may derive from natural, geological sources, most came from alluvial aquifers close to sites associated with intensive horticulture and sugarcane cultivation. Studies in the USA have shown that leaching of fertilizer N under cane cultivation can be considerable. In Florida, El Wali *et al.* (1980) found losses of 6–24% of N applied as urea, depending on fertilizer application rate and irrigation level. Leaching of N was mostly as nitrate, but, when irrigation took place before the N hydrolysed from urea was completely nitrified, leaching of ammonium was also considerable. In Louisiana, Southwick *et al.* (1995) also demonstrated substantial leaching of nitrate, amounting to 15–60% of applied N.

In relation to leaching of pesticide residues, studies also show a mixed pattern of results, although in many cases levels of pollution are below those that would cause immediate concern. It is likely that the threat of long-term pesticide impacts has reduced in recent years, as less persistent agents have replaced older formulations. For example, Cavanagh *et al.* (1999) note that organochlorine pesticides were widely used in the Australian sugarcane industry from the early 1950s until the late 1980s, but have now been succeeded by less persistent chemicals. Whilst they found evidence of organochlorine residues in field soil samples, these authors found no detectable residues (< 5 pg/g) in sediments from inshore coastal regions of the Herbert and Burdekin Rivers. Reduced impacts associated with less persistent agents depend, of course, on new formulations being adopted by growers. Regulation of pesticide use and degrees of enforcement of regulations vary considerably between different countries.

Umrit and Ng Kee Kwong (1999) monitored off-farm transport of atrazine, diuron, hexazinone and acetochlor in runoff water at a 500 m^2 plot scale and across a 40 ha catchment at Valetta (Mauritius). Mean herbicide concentrations were low and did not exceed existing drinking-water standards. The total mass of herbicide lost by runoff from the 40 ha catchment over one growing season represented very low proportions of quantities applied (not more than 0.02% atrazine, 0.32% hexazinone, 0.07% diuron and 0.19% acetochlor). At plot scale, herbicide losses occurred mainly as sediment-bound residues, but, at catchment scale, 70–95% occurred as dissolved residues.

In the USA, Southwick *et al.* (1992, 1995) measured leaching of atrazine and metribuzin from soils under sugarcane in Louisiana. Leaching of both herbicides was greatest immediately following application, but decreased over a period of weeks. Total losses of 0.4–1.7% were found for metribuzin, but levels in drainage waters did not exceed health advisory levels for drinking-water. Losses of 0.4–2.0% were found for atrazine, the maximum levels of which in drainage water (82–403 mg/l) substantially exceeded the drinking-water health limit (3 mg/l), which was only reached 20–30 days after application. Bengtson *et al.* (1998) also examined atrazine and metribuzin leaching under sugarcane in Louisiana, and found that application method was the main factor determining rate of herbicide loss. In Brazil, Lanchote *et al.* (2000) measured residues of atrazine, simazine and ametryne in surface and groundwater, in an area where sugarcane is intensively grown and from which the water-table of an important aquifer is recharged. Ametryne residues were detected in a small number of surface water samples, but almost always at levels below internationally recommended environmental limits. In Australia, evidence to support pesticide contamination events arising from cane cultivation is scarce (Johnson *et al.*, 1997), although low levels of atrazine and heptachlor have been detected in the Burdekin Delta aquifer system (Brodie *et al.*, 1984; Keating *et al.*, 1996).

Acid sulphate soils

These particularly problematic soils undoubtedly occur in other cane growing countries, but have attracted particular attention in relation to cane cultivation in parts of Australia. They are typically found in association with mangroves and coastal wetlands, and undergo accelerated chemical changes when such areas are drained for agriculture (which may result in cultivation attempts being abandoned – Bowman *et al.*, 2000). The soils contain naturally elevated levels of pyritic minerals and become acid when these are leached or oxidized. Subsequently, drainage waters tend to have low pH and carry relatively high concentrations of heavy metals, including Al, Fe and Zn. Suspended iron oxide particulates may also carry elevated concentrations of arsenic. Polluting outflows tend to be concentrated during the wet season, when oxidation products that have accumulated in the soil during the dry season are flushed out (Bowman *et al.*, 2000). Drainage from acid sulphate soils consequently discharges acid, heavy metals and arsenic into downstream aquatic habitats such as estuaries (Bowman *et al.*, 2000; Keene *et al.*, 2003). A number of negative impacts on fish and other fauna have been reported as a consequence (Arthington *et al.*, 1997).

Impacts of Sugarcane Irrigation

In studies of sugarcane irrigation at Mackay, Australia, Chapman (1997) found that irrigation resulted in additional runoff and deep drainage, amounting to 29% of the irrigation water applied. In studies of cane irrigation systems in India, Inamdar *et al.* (1995, 1996a,b) found that amongst the advantages of a switch to drip irrigation was a reduction in soil erosion. Irrigation has also been shown to exacerbate the problem of soil salinization in cane growing systems (see Chapter 6). Saline soils tend to produce saline drainage waters (Meyer, 1997), and it would be reasonable to expect negative effects of these on freshwater ecosystems, although specific studies do not appear to be readily available. Similarly, no information appears to be available on the effects of salts (and possibly nutrients) flushed from soils in Guyana following flood fallowing (see Box 6.4).

Reduction of Cane Cultivation Impacts on Water Quality and Aquatic Ecosystems

As in other aspects of environmental impact, the adoption of good agricultural practice can do much to alleviate the effects of cane cultivation on water quality and aquatic ecosystems. In particular, this includes rational use of fertilizers and pesticides (see Chapter 2). As water quality impacts of agrochemicals are often mediated by soil factors, SASA (2002) notes that soil properties should also be managed, for example, to limit the effect of leachate from the soil on the wider environment. Decreased rates of runoff arising from soil conservation measures will also be expected to have benefits in reducing the pollution of surface waters. SASA (2002) also notes the importance of appropriate management of waterways (including establishment of bank vegetation), in combination with soil conservation measures, in regulating the transfer of runoff waters from fields into natural watercourses. Natural but rarely flowing (ephemeral) streams can provide ideal channels for controlled surface drainage. Bank vegetation should be managed by slashing or careful herbicide application (with removal of debris) rather than by activities such as hoeing, which disturb the soil; where soil has been deposited in drainage channels, it should be removed using hand tools, not with machinery. In addition, installation of storm water drains may be necessary to prevent uncontrolled runoff of heavy rainfall from natural habitats, fields, roads, buildings, etc.

Arthington *et al.* (1997) report that artificial wetlands have been tested for the removal of suspended solids and nutrients in irrigation drainage waters from cane fields in the Burdekin River irrigation area in Australia. Storm water treatment marshes also feature in strategies to reduce nutrient runoff pollution in the Everglades (Anderson and Rosendahl, 1997). Such technologies have also been investigated in relation to the treatment of cane mill effluents (see below).

SUGARCANE PROCESSING

Reports of water pollution by effluents from sugarcane processing come from a range of countries, although in many cases published studies report on impacts of pollution from a range of sources. Srivastava (1989) found that discharge of water from two sugar factories and a distillery into a stream without proper treatment in the Gorakhpur district in Nepal had rendered the water unfit for drinking, bathing or irrigation. Galindo *et al.* (2001) studied the Sali River in Tucuman (Argentina), and discovered various sources of ionic pollution derived from heavy human usage of the watercourse, as well as severe contamination by organic matter (mainly from sugarcane processing) in the lower course.

Thuresson, M. (2001) reported that water pollution problems have increasingly affected Lake Victoria in recent decades, and that sugar and allied industries were considered to be the main point sources of organic matter and nutrients in the Kenyan part of the drainage basin. A number of reports of pollution of surface and groundwater by cane processing effluents in India are summarized in Box 4.1.

In terms of biodiversity impacts, the effects of effluent discharge may be most readily apparent initially from changes in the plant community in affected aquatic habitats. In Cuba, Borhidi *et al.* (1986) studied the composition and stratification of aquatic vegetation in areas affected by various human activities, including the discharge of waste water from a sugar factory. They found evidence of oxygen deficiency, leading to dominance of aquatic plant communities by macrophytes, which (in some areas) resulted in thick mats of weeds that impeded the delivery capacity of canals and had an impact on sport fishing and tourism. Ali and Soltan (1996) studied the impact on submerged aquatic plants in drainage channels and the River Nile (Egypt) of effluents from two sugarcane mills, a paper/chipboard factory and a fertilizer plant. The main pollutant in the sugar factory effluent was organic matter, including

Box 4.1. Examples of reports of pollution by cane processing effluents in India.

Surface water

Baruah *et al.* (1993) found that sugar mill and distillery effluents resulted in deterioration of many aspects of water quality in the River Gelabil (Assam) for 8 km downstream of the discharge point. During periods when the factory was closed, pollution in the river system was negligible. Singh *et al.* (1998) found evidence of adverse impacts on water quality in the River Ramganga (between Moradabad and Bareilly), arising from discharges of sewage waste, and effluents arising from the sugar, rubber and paper industries. Rajendra Singh (2000) examined soil and water pollution levels in the Abu drainage area (Meerut, Uttar Pradesh), and identified indiscriminant discharge of effluents from a wide range of industries (including sugar mills). These resulted in serious effects on the local flora and fauna, and water was found to be unsuitable for human consumption, domestic use and irrigation purposes.

Groundwater

Ali and Ahmad (1993) studied the environmental chemistry of groundwater in parts of the Gandak Basin, where they found that fluoride levels were elevated in the vicinity of three sugar factories, although not sufficiently to render it unsuitable for drinking purposes. Singh *et al.* (1996) examined the effects of industrial effluents on groundwater quality in urban industrial units, including a sugar factory and distillery, in Sardarnagar (Gorakhpur, Uttar Pradesh) and found that both hand pumps and boreholes yielded poor water quality. At Sonai (Maharashtra), Pawar *et al.* (1998) found that effluents from the Mula sugar factory, released into a stream flowing through the area, had infiltrated the underlying aquifer. Relative to the discharge point, the resulting plume of polluted groundwater extended only a few metres upstream, but more than 400 m on either side of the waterway downstream. This resulted in a zone of polluted groundwater extending for more than 3.5 km^2. It was recommended that the base of sugar factory lagoons and the stream used for effluent discharge should be waterproofed for the protection of groundwater, which was the only source available locally for drinking and agricultural purposes.

carbohydrates. In drainage channels dominated by large growths of sewage fungus, submerged vegetation was absent, although some emergent vegetation survived. In the most polluted river sites, even up to 2 km downstream of discharge points, the flora was restricted to a single species (*Potamogeton pectinatus*), even though a more diverse submerged flora occurred elsewhere in the river. Kumar Arindam (1999) studied pollution effects of carbonaceous effluent from the Chakia sugar mill (India) on herbaceous plants along drainage channels. At stations 200 m apart along the channels, distinct plant communities were recorded, and differences could not be ascribed to natural environmental variability such as local climatic effects.

Sugar mill effluents can also have an impact on other aquatic taxa. Pearson and Penridge (1987) noted that relevant studies from tropical systems were few, but demonstrated that sugar mill effluent discharged into a stream in Queensland (Australia) substantially reduced the diversity of aquatic macroinvertebrates as a consequence of lowered levels of dissolved oxygen. Still further up the aquatic food chain, Lopez-Lopez *et al.* (2003) determined that the high level of pollution from sugar industry effluents in the De La Vega reservoir (Mexico) may impose a considerable stress on the native fish population of the Ameca basin.

Selective monitoring of biodiversity can provide a tool not only for assessing pollution (Pearson and Penridge, 1987), but for encouraging measures to reduce it. Henne *et al.* (2002) validated a family-level biotic index method for rapid assessment of organic pollution from untreated municipal sewage and sugarcane processing in a west-central Mexican river. The biotic index was highly correlated to dissolved oxygen and sensitive to different levels of pollution. Information from rapid assessment biomonitoring was used successfully by local natural resource managers to help bring about improvements in water resource management.

Indeed, many improvements have been made to effluent management systems at cane mills in recent years, as pollution impacts become less acceptable. By the mid-1970s, waste water from the Sainte Madeleine cane sugar factory was polluting the Cipero River (Trinidad) so badly that major expenditure on effluent treatment was undertaken (Millette, 1991). Arthington *et al.* (1997) consider that the extent of waterway pollution from cane mills in Australia has decreased since the 1960s (when low dissolved oxygen levels, increased temperatures, odour problems and fish kills were reported downstream of effluent discharge points). This is a consequence of improved effluent management in the industry. However, some problems have persisted, with oxygen depletion downstream of discharges, particularly during periods of low stream flow (Moss and Bennett, 1991).

Sources of Cane Mill Effluents

Cane washing

Payne (1991) notes that the waste water generated from cane cleaning presents an environmental problem, as the effluent is muddy and has a high BOD. UNEP (1982) suggests a BOD of 200–900 mg/l, depending on the washing system used.

Barometric condenser cooling water

UNEP (1982) considers that water from the cooling systems of barometric condensers represents one of the major sources of potentially environmentally damaging waste in a cane mill. Large volumes are produced and may easily become contaminated by sugar ('sucrose entrainment'), depending on the design and management of the equipment. However, provided that such entrainment and levels of contamination can be minimized, barometric condenser cooling water can be suitable for recycling (e.g. for use in cooling towers, open or spray ponds), reducing total factory water usage.

Other sources of waste water in cane mills

UNEP (1982) notes that mills can produce acid and caustic wastes, from the cleaning of equipment, although this tends to be in relatively small quantities and is generally insufficient to affect the pH of the combined waste flow. A significant quantity of contamination in waste water can also arise from accidental sugar and molasses spillages, and from poor maintenance of equipment (contributing oils and greases). Potential sources of pollutants also include chemical reagents used in the processing and testing of sugar products. For example, Wilson (1996) proposed an alternative method for raw sugar polarization, to reduce risks associated with the handling and disposal of the lead subacetate reagent of the International Commission for Uniform Methods of Sugar Analysis (ICUMSA) method.

Characteristics of Cane Mill Effluents

The extent to which effluents have been treated (see below) prior to discharge will have a substantial influence on their polluting potential. As a consequence of the materials and processes undertaken in the sugarcane mill, raw effluents tend to be rich in organic matter, including carbohydrates, and consequently high in COD/BOD, when compared with those from other sources (e.g. Ali and Soltan, 1996). As well as variations between sites, there may be substantial variation in effluent volume and composition from a single mill, throughout the season and even within the weekly cycle of processing operations (Wong Sak Hoi *et al.*, 1996).

A number of sources provide information on the characteristics of mill effluents, although it is not always clear whether these represent untreated or treated material; in a few cases, a comparison between the two is provided (e.g. Amitabh *et al.*, 1999). Example figures are given in Tables 4.1 and 4.2. For comparison, figures for toxicity of heavy metals are given in Table 4.3.

Reduction of Cane Processing Impacts on Water Quality

Various measures can be taken to reduce the quantity and polluting potential of sugar mill effluent. Some sources of waste water contamination, such as those arising from accidental sugar and molasses spillages, cleaning and poor maintenance of equipment, are best tackled using simple in-plant control measures, the costs of which are negligible compared with effluent treatment costs. A range of techniques are available for treating sugar mill effluents, from simple methods to settle out solid wastes, to various forms of aerobic and anaerobic treatment.

Cane washing

UNEP (1982) notes that minimizing the need for cane washing by reducing the extraneous material collected with the cane at harvest is desirable from an economic point of view as well as from an environmental perspective, because of the cost of the operation and because of the sucrose lost during cane washing. For example, Payne (1991) suggests that unburned cane, cut by hand in dry conditions, should not need cleaning prior to sugar extraction. UNEP (1982) notes that experiments have been conducted to investigate possible methods for dry cleaning of cane, using pneumatic processes.

Treating sugar mill effluent

Various methods are available for treatment of sugar mill effluents, some of which may be used in combination. Relatively simple methods, such as the use of settling ponds or lagoons, can be used for preliminary effluent treatments. Although these methods may be considered inefficient (Contreras Moya *et al.*, 1998), they can at least reduce pollutant levels far enough for release of effluents on to the land, as some regulatory systems allow for higher levels of pollutants in waste water discharged on to land than into waterways (Rao and Rao, 1992; Shukla, 1995). Consequently, irrigation with (partially treated) sugar mill waste waters may be an attractive proposition, compared with release into watercourses (Wong Sak Hoi *et al.*, 1996).

Table 4.1. Characteristics of cane mill effluents – (1) general characteristics.

	Source	pH	BOD (mg/l)	COD (mg/l)	Suspended solids (mg/l)	Total solids (mg/l)	Oil/grease (mg/l)	Temp (°C)	Notes
India	1a	6.8–8.4	667–1,660	890–2,236	504–936	792–2,043	Traces to 60	–	
Puerto Rico	1b	5.3–8.8	112–225	385–978	100–700	500–1,400	–	31–49	
Hawaii	1b	–	115–699	942–2,340	915–3,590	3,040–4,500	–	–	
Philippines	1c	5.3–7.9	130–1,220	–	50–1,880	240–5,440	Traces to 113	34–48	
USA (Louisiana)	1d	–	81–562	729–1,430	150–8,120	409	–	–	
Mauritius	2	–	–	75+	–	–	–	–	
India	3	4.2	–	5,200	1,102	1,670	74	39	untreated
Taiwan	4	5.2–7.9	28–164	–	–	–	–	–	? treated
India	5	5.8–7.4	500–1,000	1,000–2,000	300–800	550–2,000	20–40	–	untreated
Nepal	6	6.4–8.2	1.2–280	–	–	730–3,560	–	–	
Trinidad	7	4.7–11.4	2,000–15,000	–	176–2,580	–	–	24–53	
India	8	3.8–4.8	1,600–21,000	3,500–65,000	800–12,000	–	–	38–48	untreated
India	9	6.8	568	960	836	–	71	30	? treated
India	10	8.0	1,000	2,000	300	800	–	–	Pre-treatment
India	10	7.5	28	220	77	32	–	–	Post-treatment

Sources: (1a) Verna *et al.* (1978) (for 25 mills) in UNEP (1982); (1b) US EPA (1975) in UNEP (1982); (1c) G.A. Pecache (personal communication) in UNEP (1982); (1d) UNEP (1982); (2) Wong Sak Hoi *et al.* (1996); (3) Sinha (1993); (4) Lu and Chen (1991); (5) Rao and Rao (1992); (6) Srivastava (1989) (also gives figures for distillery effluent); (7) Millette (1991); (8) Baruah *et al.* (1993); (9) Hari Om *et al.* (1994) (cane mill and distillery combined); (10) Amitabh *et al.* (1999).

Table 4.2. Characteristics of cane mill effluents – (2) heavy metals.

	Source	Cadmium	Copper	Lead	Chromium	Zinc	Arsenic
Philippines (ppm)	1	0.0004	0.18	0.003	Undetectable	–	–
Taiwan (mg/l)	2	0	0.025	0	0	0–0.09	–
India (mg/l)	3	0–0.2	0.01–0.12	0.01–0.12	0–0.05	4.5–15	0.01–0.08

Sources: (1) Abotal and Cabigon (2001); (2) Lu and Chen (1991); (3) Baruah *et al.* (1993).

Table 4.3. Toxicity (mg/l) of heavy metals (after SASA, 2002).

	Copper	Zinc	Lead	Cadmium	Iron	Manganese	Aluminium
Toxicity in solution to plants	0.02	1.3	1.7	2.1	9.3	0.06	0.93
Toxicity in solution to fish	0.02	1.3	1.7	2.1	250	100	1.5
Drinking-water	1.5	15	0.1	0.01	1.0	0.5	–
Water for farm animals	0.5	25	0.1	0.05	–	–	5.0
Irrigation water	0.2	2.0	5.0	0.01	5	0.2	5.0

However, irrigation with effluents can have mixed effects (see Chapter 3).

More sophisticated chemical or biological (aerobic or anaerobic) treatments or combinations of treatments may be required when effluent is to be discharged into waterways. Avram-Waganoff (1990) recommended the purification of cane factory waste water in fully biological plants. Gunasekaran *et al.* (1999) note that, with appropriate biological treatment systems, BOD and COD of the treated water may be brought down below 5 ppm and 10 ppm, respectively. In some cases, partially treated effluent may be diluted in order to achieve pollution levels sufficiently low for discharge into waterways (preferably using uncontaminated waste waters). For example, Rao and Rao (1992) described a sugar factory in India where effluent was held in an anaerobic lagoon for 10 days (with daily organic loading 0.12 kg/m^3) and then treated in an anaerobic contact filter for 12 h. The BOD of this treated effluent was approximately 72 mg/l, but dilution with excess condensate and condenser outlet water lowered it to 30 mg/l, the permissible level for discharge into rivers.

Effluent treatment also has the potential to yield useful by-products other than irrigation water. Malmary *et al.* (2000) investigated the use of solvents to recover aconitic and lactic acids from dilute aqueous effluents of the sugarcane industry, in order to reduce environmental pollution and in view of the possible uses of pure solutes in the field of foods and pharmaceuticals.

Artificial wetlands have been investigated for their potential in removing pollutants from mill effluents, for example, as a tertiary form of treatment in Australia (Dawson *et al.*, 1995), where the method was used successfully to reduce BOD, nutrient levels and total suspended solids. Similar methods have been examined for the treatment of distillery effluent (see Chapter 8).

Recycling of cane processing water

Amongst processing operations, cane washing in particular does not require fresh water, but can be performed with water recycled from previous washings, or with other waste water, notably barometric condenser cooling water (UNEP, 1982). However, water quality is more critical in other operations. For example, boilers operating at relatively high pressures require especially high quality water (Payne, 1991). Appropriate monitoring and treatment may allow condensate water to be fed to boilers operating at above 4000 kPa. However, evaporator condensates in particular should not be used when boiler pressures reach 6000 kPa, because of the high likelihood of this water being contaminated by organic volatiles.

Considerable reductions in effluent volumes can be achieved using recycling of water at various stages in the processing of cane. Lu and Chen (1991), for example, reported on water recycling in sugar factories in Taiwan. Those producing raw sugar were able to reduce effluent outputs from nearly 15 to 1.5 t/t cane, and those producing white sugar reduced outputs from 17 to 3 t/t. Srimannarayana and Sudheer (2000) describe a 'zero pollution' system operated at a sugar factory in Andhra Pradesh (India), where recycled processing water is used as make-up water for spray ponds and (after suitable treatment) for a range of other purposes. After initial oil and grease removal, biological treatment in an anaerobic lagoon is followed by an anaerobic upflow filter and two oxidation/stabilization ponds. The water from these is clean enough to be added to the cooling water system or, outside the cane crushing season, used for irrigation.

SUGAR BEET CULTIVATION

Impacts of sugar beet cultivation on the surrounding environment can be difficult to assess unambiguously, as it is usually just one of a number of crops cultivated as part of a rotation at any particular site.

Most published studies examining the impact of sugar beet cultivation on water quality consider the effects of leaching of nitrogenous inputs derived from fertilizers, and it is such pollution in relation to drinking-water (from underground aquifers or rivers and streams) that has influenced much recent legislation in this area (Scott and Jaggard, 1993). Cooke and Scott (1993) note that the environmental acceptability of sugar beet as a crop is enhanced by the fact that it is a very effective scavenger of nitrogen fertilizer, leaving little in the soil at harvest to escape into groundwater. However, nitrate contamination of groundwater is also a function of the management of fertilizer application. In some cases, in Europe at least, there is evidence that typical levels of fertilizer (and pesticide) applications to beet crops have decreased (see Chapter 2). However, despite the considerable reduction in fertilizer inputs in the UK, Defra (2002) notes that concern remains over a lack of detailed knowledge of impacts such as eutrophication of ground and surface waters due to N and P inputs (and the impact of soil nitrates on buried archaeological features). Also, Businelli *et al.* (2001) modelled herbicide impact on groundwater quality in Umbria (central Italy), based on various chemical agents and weed control strategies, in crop rotations including sugar beet. They identified a significant risk of groundwater contamination at levels exceeding EU limits, even with weed control strategies of low pollution potential. A critical factor was the combination of crops in the rotation.

Runoff in Beet Cultivation Systems

Defra (2002) suggests that erosion of soil under beet cultivation in the UK can have potentially serious consequences, such as silting up of waterways. In general, runoff from eroding topsoils and from poorly managed application of organic fertilizers is the usual route of phosphate loss from beet fields, whilst nitrate tends to be more readily leached (Draycott and Christenson, 2003). Such transfer of phosphates can contribute to eutrophication of surface waters. Neeteson and Ehlert (1988) found that beet leaves left in the field after harvest substantially increased levels of residual soil P, potentially increasing the risk of pollution.

Leaching in Beet Cultivation Systems

Draycott and Christenson (2003) consider that leaching is the predominant means by which nitrogen is lost to the environment from most beet fields (i.e. relative to gaseous losses by volatilization and denitrification), particularly on free draining soils. Phosphates are only likely to be found at appreciable levels in leachates when soil phosphate concentrations are relatively high. In most temperate beet growing regions, the greatest risk of leaching is over winter, when the soil is wet and water is draining from it (Kolenbrander, 1978; Milosevic *et al.*, 1989; Allison *et al.*, 1996; Nievergelt, 2002). This effect can be amplified following a dry summer (when uptake of N by the previous crop has been restricted), and leaching can sometimes be a problem also during spring if conditions are wet (Draycott and Christenson, 2003). Leaching has also been shown to increase substantially in beet cultivation systems when heavy rain follows application of nitrogenous fertilizers. Last and Draycott (1975a,b) found that 40 kg N/ha was lost by leaching in wet years, whereas losses were neglible under normal conditions (rainfall of around 50 mm/month).

Some degree of nitrate leaching from beet growing soils is reported from a range of areas, including (for example) Lithuania (Sileika, 2000) and Poland (Borowiec and Zablocki, 1989). In Germany, Isermann (1989) concluded that intensive crop management in systems, including sugar beet cultivation, where inputs of organic and inorganic fertilizers were substantial was incompatible with the use of shallow wells for drinking-water supply. Draycott and Christenson (2003) review the evidence that nitrate leaching from beet fields constitutes a threat to the quality of groundwaters, noting that recommended limits on nitrate content of drinking-water are 50 mg/l in the EU and 45 mg/l in the USA. In a theoretical study, Johnston (1989) showed that rainfall in the average UK winter was sufficient to generate leachate from beet fields with nitrate levels above recommended levels. However, Draycott *et al.* (1997) measured the nitrate content of beet field leachates in the UK, and found that it was below recommended levels in most years where inorganic fertilizers were applied for optimum yield without excess. In either case, leachate from agricultural fields would be expected to be diluted by other sources of water entering aquifers, reducing impacts on groundwater quality. In the Netherlands, Neeteson and Ehlert (1988) found that there was little risk of nitrate leaching from beet fields receiving optimum levels of inorganic fertilizer, unless beet tops were left in the field after harvest and ploughed in, raising levels of residual N in the soil. Leaching of soil N derived from beet leaves is also reported by Thomsen and Christensen (1996).

Over-application of organic fertilizers (manures) can result in excessive soil N and increased risk of leaching. Isermann (1989) concluded that incorrect manure application may lead to greater nitrate leaching than with mineral N. Evidence of nitrate leaching where organic fertilizers are used to excess in sugar beet cultivation systems includes the work of Mathers and Stewart (1984) and Eck *et al.* (1990). Malzer and Graff (1995) reported greater leaching of nitrate in the second year than the first, following application of poultry manure, illustrating the lag that can be caused by delayed mineralization of material contained in some organic amendments.

In western Europe, where precipitation exceeds evapotranspiration for part of the year, there is evidence of K leaching from beet fields, particularly on freely draining soils (Draycott and Christenson, 2003). However, evidence of deleterious effects of K on the quality of natural waters is lacking (Syers, 1998).

Impacts of Sugar Beet Irrigation

Irrigation of sugar beet can increase the risk of nutrient leaching, particularly if it is excessive (Dunham, 1993), poorly scheduled (Bailey, 1990) and/or applied by particular methods (Draycott and Christenson, 2003). In an 11-year study of a rotation including sugar beet, Bizik (1989) found that irrigation had a marked influence on the downward movement of soil solution. By reducing irrigation from 50 to 30 mm, the risk of nitrate pollution of groundwater declined, while this change of method, in combination with reduced rates of fertilizer application, increased overall crop yields for the rotation substantially. In a sugar beet cultivation system in Montana (USA), Eckhoff and Bergman (2001) found greater nitrate concentrations in groundwater under a flood irrigation regime than under sprinkler irrigation, and demonstrated the presence of nitrates in flood irrigation runoff water.

However, it should be noted that the risk of nutrient pollution is determined by fertilizer application levels as well as by the effects of irrigation. Sharmasarkar *et al.* (2001b) note that mismanagement of nitrogenous fertilizers has caused serious nitrate contamination in many flood-irrigated regions of the western USA. In their own study of sugar beet cultivation in south-eastern Wyoming, these authors found that drip irrigation reduced the quantity of water leaching beyond the root zone, relative to flood irrigation. Drip irrigation resulted in greater residual soil nitrate, and the overall results indicated that beet cultivation could be sustained with lower water and fertilizer use by using drip irrigation. Where fertilizer application rates are well judged, it may be the case that irrigation can reduce the risk of leaching, by enhancing nutrient uptake by the crop. In a UK study, Groves and Bailey (1997) found that, in dry years on light soils, irrigation aided the uptake of N by sugar beet, reducing residual soil N by 31 kg N/ha (0–90 cm) compared with an unirrigated treatment (79 kg N/ha). The potential for N leaching during the ensuing winter was consequently more than halved. However, the additional N uptake associated with irrigation tended to concentrate in the beet leaves (tops), which are normally returned to the soil, and may make some contribution to future leaching risks (e.g. see Destain *et al.*, 1991; Brentrup *et al.*, 2001).

Irrigation with beet processing waste water can also result in water pollution risks, although the overall effects can be complex. Paulsen *et al.* (1997) examined the effects of long-term irrigation with beet processing waste water on the chemical composition of arable soils, leachates and groundwater. Drainage water showed raised concentrations of some alkali and alkaline earth metals, and oxygen-demanding substances (COD and BOD compounds) were also translocated down to the drainage level. In contrast, the nitrate content of leachates was lower than for non-irrigated soils. Limits on the use of waste water from sugar beet processing for irrigation purposes was considered appropriate on certain soils and in areas overlying deep groundwaters.

Reduction of Beet Cultivation Impacts on Water Quality and Aquatic Ecosystems

As noted in relation to cane cultivation, and with respect to other aspects of cultivation-derived environmental impacts, shifts in agricultural practice (including appropriate use of agrochemicals and soil conservation measures) can do much to alleviate the effects of beet cultivation on water quality and aquatic ecosystems (see Chapters 2 and 6). For example, Berg *et al.* (1997) showed that it was possible to reduce the leaching of nitrate to groundwater in cropping systems involving sugar beet by adopting an organic farming approach rather than a conventional or integrated strategy. In addition, specific measures have been investigated for reducing the particular risk of nitrate leaching that arises in many areas over the winter months prior to the sowing of beet. These include the incorporation of straw into the soil (Powlson *et al.*, 1985; Allison, 1989; Allison *et al.*, 1992), and the use of catch/cover crops (Allison and Armstrong, 1991; Allison *et al.*, 1993, 1998a,b; Duval, 2000) to remove residual nitrate from the soil after harvest of the preceding crop in the rotation. Catch crops can subsequently be used as green manures, being ploughed into the field prior to the sowing of beet.

In the low-lying agricultural plains of north-east Italy, controlled drainage systems, managed water-tables and wetlands (comprising *Phragmites australis*, *Typha latifolia* and *Carex elata*) have been found to be useful in controlling nitrate levels in drainage waters (Borin *et al.*, 2001).

BEET PROCESSING

Volumes of waste water produced in sugar beet processing can be considerable. For example, in the Ukraine, Nibit *et al.* (1994) reported that traditional processing of 1 t sugar beet generated up to 3 m^3 effluent. This waste water can have substantial polluting potential. For example, a series of studies relates the pollution of Danish coastal waters by effluent from sugar factories to bacterial pathogens and an ulcer syndrome in the cod, *Gadus morhua*. Fish living in polluted waters are exposed to considerably higher bacterial counts, reflected in a higher prevalence of disease (Larsen *et al.*, 1978; Larsen, 1982, 1985; Larsen and Jensen, 1982). The fish pathogen *Saprolegnia parasitica* was amongst microorganisms (which also included the sewage fungus *Leptomitus lacteus*) found by Zhan and Hu (1989) in reaches of the Songhuajiang River polluted by sugar-refining waste water.

Issues of effluent management and other sources of pollution in sugar beet processing in Poland, and their regulation, are summarized in Box 4.2.

Sources of Beet Factory Effluents

Beet washing

UNEP (1982) notes that a major difficulty with handling sugar beet is the large quantity

Box 4.2. Issues of effluent management and other sources of pollution in sugar beet processing in Poland.

A series of papers by Polec and co-workers (and others) examine aspects of the environmental impacts of sugar beet processing in Poland. Polec and Gozdek (1994) review progress in water economy and waste water treatment in the years 1893–1993; aspects of reduced water consumption reported by Polec and Kempnerska-Omielczenko (1995) can be found here in Chapter 3. Issues of atmospheric pollution reported by Wolski (1993, 1995), Polec and Kempnerska-Omielczenko (1995) and Tomaszewska and Polec (1997) are examined in Chapter 7.

In relation to effluent management, Marciniak (1989) reported that biological purifiers were installed at the Melno sugar factory in 1986, with the aim of eliminating the harmful effect of recycling water from and to Lake Skape, where pollution levels had exceeded acceptable limits for surface water for at least 8 years. In addition, Miniflox aerators were installed along the lake at depths of 4–9 m, to suppress putrefactive organisms. Comparisons of lake water quality in 1986 and 1988 indicated the success of these measures in reducing pollution effects. In particular, COD and BOD had decreased from 157–294 and 26–65 mg/dm^3 to 62–66 and 1.7–2.9 mg/dm^3, respectively, while dissolved oxygen had increased from 0–1.2 to 4.7–5.2 mg/dm^3. However, Polec and Kempnerska-Omielczenko (1995) found waste water treatment at Polish sugar beet factories to be widely unsatisfactory in 1993: 35 of the 78 factories used only mechanical treatment, while activated sludge treatment was preceded by closed anaerobic fermenters (with biogas recovery) at only three factories and by open fermenters at eight. The situation had improved somewhat in 1994, the corresponding figures being 28, five and 13, respectively. Indeed, the Koscian sugar factory received an award from the Polish Environment Minister in 1994, recognizing the good environmental quality of its processing, mainly in relation to quantity and quality of effluent discharged (Scigacz, 1995). Plichta and Nitzler (1999) assessed the performance of a waste water cleaning unit installed in 1989/90 at the Chelmza sugar beet factory, after 5 years of utilization. The quality of treated waste water from open fermentation tanks was comparable to that discharged from closed fermentation tanks. However, based on comparisons of a reservoir built in 1986 for the Hodonin sugar beet factory (Czech Republic) and a steel reactor built in 1987 at Wroclaw (Poland), Bielas (2001) concluded that closed systems had many advantages over open systems in the initial stages of effluent treatment. A detailed analysis of waste water cleaned in a biological treatment unit at the Koscian S.A. sugar beet factory was made in 1998 by an independent German company, to assess the efficiency of the biodegradation process. The results showed that, whilst during the first half of the processing season the unit performed well, the quality of waste water gradually deteriorated and was unsatisfactory in the last quarter of the season (Polec *et al.*, 1999).

Increasing regulation, partly associated with Polish accession to EU membership, is imposing controls on pollution from sugar beet factories. Marciniak (2000) and Polec (2000) summarized existing pollution levels and regulation of factors such as effluent content of nitrates, phosphorus, chlorine, sulphur, heavy metals and solids, in the context of existing Polish environmental regulations and those of the EU. Polec (2002) considered the impact on sugar beet factories of a new (from 1 January 2002) Polish environmental protection law, relevant to air and water pollution, waste disposal, noise and electromagnetic pollution. Although not all relevant control instruments were in place, beet sugar factories were required to obtain a number of permits from relevant authorities, based on the polluting potential of their operations.

of soil and trash ('tare'), particularly associated with mechanized harvesting and harvesting in wet weather, that is brought into the factory with the root crop. Despite attempts to reduce the quantity of solid waste delivered to the factory with the beet, a great deal of mud enters the system through the initial beet washing in the fluming process. Sugar may also be lost at this stage, particularly where hydraulic fluming (an otherwise effective and expedient means of transporting and cleaning beet) is used.

Barometric condenser cooling water

As in cane mills, beet factories can generate large volumes of waste water from the cooling systems of barometric condensers. Again, however, if contamination can be minimized, there are good prospects for recycling this water within the factory (UNEP, 1982).

Lime mud

UNEP (1982) notes that beet factory lime mud waste, principally from clarification operations, is generally recovered from vacuum filters at about 50% moisture content. Water is usually added, resulting in a (potentially highly polluting) slurry that can be easily pumped, generally to a holding pond in the first instance.

Other sources of waste water in sugar beet factories

As in cane mills, effluents from beet processing factories can become contaminated with acid and caustic wastes (from the cleaning of equipment), with organic material (from spillages) and by oils and greases (from poor maintenance of equipment).

Characteristics of Sugar Beet Factory Effluents

Much less information appears to be readily available (in English, at least) on the specific characteristics of sugar beet factory effluent than on effluent from cane processing. Example data are given in Table 4.4. Sugar beet processing generates relatively high levels of ammoniacal nitrogenous wastes, arising from amides in beet juice, causing decreases in pH during processing and atmospheric pollution, and contributing to the polluting potential of factory waste waters (Gryllus and Anyos, 1993; Morris and Herbert, 1997). As with cane mill effluents, polluting potential involves increased temperature as well as biochemical characteristics such as high oxygen demands (e.g. Morris and Herbert, 1997).

Reduction of Beet Processing Impacts on Water Quality

The need to reduce the pollution potential of sugar beet processing waste waters is recognized by a range of authors; incentives include the need to operate within legal standards (e.g. Kuzminski *et al.*, 1991; Marciniak, 2000). UNEP (1982) noted that in-plant control measures are important, including appropriate handing of beet prior to arrival at the factory, design of flume systems to facilitate dry handling techniques, dry methods for handling lime mud waste, and water recovery and recycling. As in sugarcane mills, it is likely that simple in-plant control

Table 4.4. Example of data on effluent from sugar beet factories (recorded in the Songhuajiang River, China, by Zhan and Hu, 1989).

	Temp. (°C)	pH	Dissolved oxygen (mg/l)	BOD	COD	NH_3-N (mg/l)
50 m upstream of the outlet	0	6.8	11.25	3.7	8.4	1.25
10 m beyond the outlet	5	5.9	3.14	135	143.6	3.96
300 m beyond the outlet	1	6.7	5.72	25.7	23.8	1.73

measures (e.g. prevention of spillages and of contamination of water with cleaning agents, oils and greases) will be more cost effective than removal of polluting materials from waste water at the treatment stage. Changes to processing methods, as well as treatment of effluents, can make an important contribution to pollution control in beet sugar factories. For example, Klemes *et al.* (1999) argued that adoption of new methods such as cooling crystallization of concentrated raw juice, as opposed to the traditional method of evaporating crystallization, has the potential to improve energy efficiency and to reduce atmospheric emissions, water consumption and the polluting potential of effluents.

Beet washing and fluming

The amount of extraneous material that needs to be washed from beet can be reduced by minimizing the collection of mud, leaves and trash at harvest. Shaking and screening the beet at the factory before processing can also reduce the amount of solid waste that would otherwise accumulate in waste waters. Recycling is an effective means of reducing quantities of flume water; spent flume water may be settled in holding ponds (possibly with the addition of lime to assist clarification) or fed through mechanical clarifiers, which demand less space at the factory site (UNEP, 1982). In reviewing plans for effluent disposal at sugar factories in the Ukraine, Nibit *et al.* (1994) noted that in the flume-wash circuit the amount of mud from vertical clarifiers was only 60–70% (w/w beet) whereas that from radial clarifiers was 100–120%. They also noted that it was possible to design an effluent-free, closed flume water circuit, where separated mud was sent to ground-level settling ponds whose decantate re-entered the circuit. In the USA, Fuentes *et al.* (2001) examined the potential benefits of using polymers, cationic coagulants and/or anionic flocculants for mud removal in a flume water clarifier, and identified an anionic polymer (KFLOC 4919) as the most effective compound. Plichta (2002) also reports on the use of flocculants and poly-electrolytes for thickening flume sludge, to reduce the volume for disposal and associated environmental impacts. Improved methods for dealing with beet washing are also discussed by Guyot *et al.* (2003).

Removal of solid wastes from effluent

UNEP (1982) notes that mechanical clarifiers and settling ponds (following coarse screening) are generally used to remove as much soil and other solid wastes as possible at beet processing sites. These need to be effectively operated and maintained, and waste retention times should be minimized to reduce the risk of fermentation and creation of odours. Clarifiers with retention times of 30 min to several hours can provide effective removal of solid wastes with minimal odour problems. Mechanical clarifiers can reduce settleable solids from 30–125 mg/l to 1.0 mg/l. Chlorination or pH control with lime may also be used to reduce odour production. Chemical flocculation at pH 10.5–11.5 will settle fine clay particles, and addition of lime not only suppresses fermentation, but serves to raise the pH to levels necessary for effective flocculation. Settling ponds are widely used in the beet sugar industry and perform a similar role to mechanical clarifiers in separating out solid wastes from effluent. Waste water retention times in ponds generally range from 24 to 48 h. Settling (or holding) ponds, as distinguished from waste stabilization lagoons, are used for solids removal and waste retention without discharge into surface waters. Long-term storage may be followed by water disposal by evaporation or filtration and discharge. Waste stabilization ponds (lagoons) are specifically designed and constructed to provide waste treatment for subsequent controlled land disposal, irrigation or discharge to surface waters.

Lime mud

UNEP (1982) notes that lime mud slurry is generally pumped to holding ponds, to separate solids from supernatant fluid waste.

Settled, solid lime mud waste is often allowed to solidify before it is recalcinated for reuse in the factory or removed for application to the land.

Treating sugar beet factory effluent

In addition to (or in combination with) methods to settle or filter out solid wastes, anaerobic and aerobic biological methods are widely used in the treatment of beet factory effluents, e.g. involving waste stabilization lagoons, activated sludge units, trickling filters and various types of biological reactor. Bugaenko (1996) gives a brief account of technologies used for waste water treatment in the beet sugar industry, with reference to biological treatments practised in the UK, Germany, the Netherlands, Austria, the Czech Republic, Slovakia and Poland, the types of reactor used, quality requirements for water entering anaerobic treatment systems and the appropriate location of effluent treatment stations. Murathan and Yildirim (2001) reported that an activated-sludge process was being used in the co-treatment of industrial effluents (including those from a cotton mill, locomotive factory and beet sugar factory) and residential waste water, to reduce organic loading on the Porsuk River in Turkey.

Kuzminski *et al.* (1991) reported that, in addition to substantial water use savings, the pollution load of effluent discharged from a beet sugar factory in Spain was significantly reduced (BOD decreased from 55 to 5 g/s), after modifications to processing methods. The most important changes were the conversion of the barometric condenser circuit from open to closed, with incorporation of cooling towers, and the installation of a high-rate anaerobic treatment unit for highly polluted waters; streams were segregated into those for recycling and those for treatment. This study also demonstrates how, in some cases at least, improved water management systems can 'pay for themselves': Kuzminski *et al.* (1991) report that the costs of the new system were recovered from savings on discharge permit taxes and from profits obtained from better utilization of by-products. In relation to by-products, Gryllus and Anyos (1993) describe how ammonia-rich condensates from one sugar beet factory were treated, reducing pollution potential and yielding 5.6 t/day $(NH_4)_2SO_4$ as a fertilizer by-product.

As in other areas of mitigating the environmental effects of sugar industry waste waters on the environment, reed beds have been investigated for their potential in the treatment of beet factory waste waters. In the UK, Morris (1996) and Morris and Herbert (1997) reported that reed beds were being tested as a low-energy alternative to existing waste water treatment systems (such as trickling filters, anaerobic digesters and activated-sludge units). Operation of vertical flow reed beds (two in series) during the beet processing season resulted in 87.3% reduction in COD, 87.7% removal of total suspended solids and a 79.5% reduction in ammoniacal nitrogen. Treating larger volumes of cooler, stored wastes out of season, the mean removal efficiencies were 73.9% for COD, 88.0% for total suspended solids and 93.4% for ammoniacal nitrogen.

5

Impacts on Terrestrial Biodiversity

Agriculture is arguably the predominant influence on the Earth's land surface (Tharme, 2003). Conversion to agriculture has led to substantial losses of many habitat types and associated biodiversity (with implications for ecosystem function) around the world. In many cropping systems, biodiversity has been reduced, although productivity has remained high (Anderson, 1994). The continued expansion of agricultural monocultures is seen as a particular driver in the loss of natural habitats, particularly in developing countries (Tinker, 1997). However, precise patterns of land use change (even in the relatively recent past) can be difficult to determine and their overall environmental impacts difficult to analyse (e.g. see examples given by Hartemink, 2003). None the less, there is an increasing pressure for the development of sustainable systems of land use. The sugar industry has a significant role to play in this, and biodiversity management is one aspect of environmental sustainability for which it will be held increasingly accountable (as in Australia – Woods, 2000).

The cultivation of sugar crops has led to loss of indigenous biodiversity in many parts of the world (as demonstrated, for example, by reductions in species richness associated with cane growing in Cuba – Vega *et al.*, 1999). Agricultural activities also threaten the sustainability of ecosystems, for example those of the sloping uplands of South-east Asia, where cane is grown (Garrity, 1993). Substantial areas have been cleared for cane cultivation, and in some places its cultivation continues to expand. As with other forms of agriculture, particular concern has been expressed for impacts of cane cultivation on wetland habitats and the resultant effects on associated ecosystems such as rivers and coastal zones (see Box 5.1). It is likely that only relatively small areas have been cleared specifically for sugar beet, which was adopted as a widely grown crop relatively recently and would often have been grown on areas already under some form of cultivation.

At a community or species level, most intensively cultivated agroecosystems are relatively lacking in biodiversity. Weeds and pests may provide important resources for other (non-pest) taxa. Hence, measures to control them may have knock-on effects for other taxa, either directly (e.g. through non-target effects of pesticides) or indirectly (e.g. through disruption of food chains). Of course, some of these taxa may be agronomically beneficial, as natural enemies of weeds and pests. The importance of the biodiversity of agricultural systems (agrobiodiversity), its evolution, function, management and environmental significance, is reviewed by Wood and Lenne (1999). One aspect of agrobiodiversity that is often overlooked is the suite of microorganisms associated with a crop. Most studies concentrate on particular pathogenic species, their epidemiology and control. More extensive work is to be encouraged, not least on soil microorganisms, which play such a critical role in ecosystem function, for example in the turnover of soil organic matter (Burke *et al.*, 2003). Knowledge of soil microorganisms also provides for more detailed environmental impact assessment. Although proxy measures (such as kinetic respiration analysis – Blagodatsky *et al.*, 2002) can be used to monitor soil microbial communities under environmental change, better characterization of the baseline community of microorganisms would be valuable (Colwell, 1992).

The conservation of crop genetic diversity is itself an important consideration. There is evidence that maintenance of diversity within the standing crop has agronomic benefits, and the conservation of cane and beet germplasm collections is vital. Such collections form the basis of breeding programmes, which allow for the development of new varieties. This is important from both an agronomic and an environmental perspective. The development of varieties with increased resistance to pests and diseases may allow for reduced pesticide inputs (often as part of broader integrated pest management strategies). Varieties which absorb and utilize nutrients more efficiently may facilitate the reduction of fertilizer application rates. The breeding of varieties more tolerant of drought may contribute to water conservation strategies. A combination of selective breeding objectives can produce varieties better suited to local growing conditions, which is effectively a contribution towards precision farming. Proponents of genetic modification are keen to stress the potential advantages that this technology provides in increasing the scope for breeding programmes, including the enhancement of varieties in ways that reduce the environmental impacts of their cultivation. None the less, there remains concern over the direct and indirect effects of growing transgenic sugar crops, particularly in relation to potential negative impacts on biodiversity. Whether through traditional selective breeding or genetic modification, however, it should be borne in mind that development of new varieties can be seen as treating the symptoms, not the causes, of many challenges to crop production. Where environmental constraints to production are exacerbated by environmental degradation, the development of new varieties may be less appropriate (and sustainable) than measures to prevent the decline in environmental quality (see Berding and Skinner, 1987; Garside *et al.*, 1997b).

SUGARCANE CULTIVATION

Ecosystem- and Habitat-scale Impacts

Land clearance for sugarcane cultivation

The greatest impact of sugarcane on biodiversity undoubtedly arises from the historical clearance of land for cultivation. This has had an impact on a wide range of habitat types in the tropics (and sometimes subtropics), including rainforest, tropical seasonal forest, thorn forest, semi-desert scrub and grass-dominated savannah (Bakker, 1999). In Puerto Rico alone, cane is grown in irrigated semi-arid valleys and plains, humid valleys and plains and humid uplands; for many years, a traditional use of labour in the off-season was the clearance of forests and perennial scrub, before decline in the industry led to reduced acreages (Alexander, 1985). Johnson *et al.* (1997) note that cane growing in Australia has led to the clearance of large areas of riverine rainforest and riparian habitats and the loss of some mangroves. The particular case of wetland habitats like these is outlined in Box 5.1. In Brazil, significant areas of cerrado habitat have been lost to sugarcane cultivation and other agricultural activities (Araujo *et al.*, 1999).

Land clearance impacts can have a long history. Inglese (1999) notes how 16th-century Spanish and Portuguese colonists began converting land in Latin America for large-scale cultivation of crops including sugarcane, fundamentally and permanently altering the nature of the landscape and the soil. Since the mid-19th century, when Europeans settled in the Herbert River catchment in Australia, land clearance has resulted in a substantial reduction in the area of *Melaleuca* spp., rainforest and *Eucalyptus*-dominated land cover (Johnson *et al.*, 2000). This has been accompanied by a marked increase in sugarcane plantations and a decline in landscape diversity, integrity

Box 5.1. The particular case of wetlands and riparian habitats.

Agriculture undoubtedly represents the main cause of loss of wetland habitats (Tharme, 2003). Half of the world's wetlands have been lost to drainage and conversion to agriculture (60–70% in Europe), and even protected wetland areas are subject to agricultural impacts. The conversion of natural vegetation to agricultural crops in major watersheds can have substantial impacts on the wider environment, including altered rates of rainfall infiltration, flooding and accelerated soil erosion, all of which modify aquatic, soil and riverine habitats, having an impact on biodiversity (e.g. McNeely, 2003). Natural peatlands also represent major carbon storage centres, and the conversion of these (and other habitats) to agriculture can result in substantial releases of carbon dioxide into the atmosphere (van Noordwijk *et al.*, 1997; Page *et al.*, 2002).

Low-lying and alluvial areas often support rich soils, and enjoy a good natural water supply, so it is not surprising that river valleys and wetland areas have attracted particular attention as agricultural activities have expanded. Whilst volcanic soils underpin cane cultivation in some areas, cane is often grown on alluvial soils (Hartemink, 2003). Clearance of land for cane cultivation has resulted in substantial loss of coastal wetlands in many areas of Australia (Johnson *et al.*, 1997), including flood-plain habitats like *Melaleuca* wetlands (Arthington *et al.*, 1997). A 60% reduction in such habitats occurred in the Johnston River catchment between 1951 and 1992 (Russell and Hales, 1996). The loss of wetland and river catchment habitats has also generated concern in other cane growing countries, including South Africa (Peel and Stalmans, 1999). Agriculture, including cane cultivation, not only threatens the biodiversity of natural wetlands and associated ecosystems, but can also threaten the traditional cultures and livelihoods of communities that rely upon them, for example, in the coastal wetlands of Brazil (Diegues, 1991) and Australia (Johnson *et al.*, 1997).

Riparian habitats in Australia

Large areas of natural riparian vegetation have been cleared in Australian cane growing regions, and revegetation of stream banks is often due to invasion by exotic weeds that suppress the regeneration of the original plant community (Johnson *et al.*, 1997). For example, around 45 km of stream bank vegetation was cleared in the Herbert River catchment, Queensland, between 1990 and 1995 (Perry, 1995). In New South Wales too, cane is often grown right up to the banks of streams, leaving no natural vegetation (Arthington *et al.*, 1997).

Natural riparian vegetation plays an important ecological role, providing habitat for wildlife and influencing water quality, stream morphology and ecosystem dynamics, whilst providing a buffer zone between agricultural systems and waterways (Arthington *et al.*, 1997). Riparian habitats provide a filter, reducing sediment loads and agrochemical concentrations in waters running off cane fields. Riparian vegetation also provides shading to waterways, which has a significant influence on the biological communities that develop and persist there. Bunn *et al.* (1997) found that lack of shading, resulting from the clearance of natural riparian vegetation, had led to choking of a lowland stream channel by a few, dominant aquatic and semi-aquatic plants, disrupting the dynamics of the stream ecosystem. Fish can be negatively affected by the clearance of riparian vegetation, as a consequence of loss of shade and consequent disruption of the food chain (Arthington *et al.*, 1997). The source of leaf litter entering streams is also important in driving the dynamics of the aquatic community, and there is evidence that material from sugarcane itself does not provide the necessary resources that would normally be derived from indigenous, bank-side vegetation (Bunn *et al.*, 1997). The failure of organic matter from cane leaf litter to be assimilated into the food web in the normal way may result in a pollution threat as great as that posed by runoff of nutrients from cane fields (Arthington *et al.*, 1997).

In addition to their wider ecological importance, well-maintained riparian habitats can provide benefits to agriculture. These include regulation of flooding following heavy rainfall, consolidation of stream banks (reducing erosion), provision of refuges for beneficial insects and even contributions to the control of rats and weeds (Arthington *et al.*, 1997).

The Florida Everglades

Throughout the 20th century, expansion of agriculture in Florida resulted in large areas of the northern Everglades being 'reclaimed', with the construction of an elaborate network of levees, canals and dams.

This activity led to the establishment of the Everglades Agricultural Area and of the Florida sugar industry, with nearly 200,000 ha now turned over to cane cultivation, representing about 25% of sugar production in the USA (Anderson and Rosendahl, 1997). By the end of the century, about half of the original wetland habitat had been drained, resulting in dramatic declines in biodiversity. For example, wading bird populations have fallen to about one-tenth of their former size (Schrope, 2001).

In addition to habitat loss, ecosystem impacts of agricultural development in the Everglades include major redistribution of water flows (e.g. Harwell *et al.*, 1996; Anderson and Rosendahl, 1997) and subsidence due to shrinkage, compaction and accelerated microbial decomposition of drained soils (leading to losses of 2.7–4.0 m of surface elevation in the last century, although at decreasing rates in recent years – Anderson and Rosendahl, 1997). The sugar industry has received much of the blame for such impacts, and particularly for the effects of nutrient pollution arising from fertilizer use. However, the industry has placed itself fourth in a list of environmental degradation sources, after water distribution, fire and invasion of exotic species (Anon., 1992). As with other major areas of sugar production adjacent to sensitive ecosystems (such as the Great Barrier Reef), it is very difficult to distinguish between impacts arising from cane cultivation and those derived from other land uses. Overall, the true environmental impact of sugarcane cultivation in the Everglades remains largely unquantified (Hartemink, 2003). None the less, research has been conducted to investigate the dynamics of (particularly N and P) pollution here, and to develop best management practices (BMPs) for cane cultivation to reduce impacts (Capone *et al.*, 1995; Izuno *et al.*, 1995; Thomas *et al.*, 1995; Anderson and Rosendahl, 1997, 1998; Hossain, 1998; Ivanoff *et al.*, 1998; Rice and Izuno, 1998; Stuck *et al.*, 2001; Rice *et al.*, 2002).

Policies that once favoured development are now being reversed, with increasing emphasis on the restoration of natural ecosystems and alternative forms of agriculture (Anderson and Rosendahl, 1997; Snyder *et al.*, 1999). However, arguments over the best way to proceed with proposed, major restoration projects reveal a lack of detailed understanding of the functional dynamics of the original ecosystem, and of the specific roles played by ongoing agricultural impacts (such as N and P pollution) in its degradation and that of adjacent ecosystems like Florida Bay and the Florida Keys (Schrope, 2001).

and ecosystem quality. Agriculture and associated environmental degradation can also have a complex history at any given site. For example, Conte (1999) discusses the history of the Mlalo Basin, an upland region in the west Usambara Mountains, Tanzania. During the Iron Age, alluvial soils, formed from sediments deposited by the Usambara Mountain streams, would have supported sorghum, millets and pulses. Agricultural intensification occurred between AD 800 and 1200, when farmers adopted other crops, including sugarcane, and introduced irrigation. By the 19th century, the Iron Age landscape had undergone several transformations, driven by cycles of deforestation, regeneration, agriculture and demographic change, and soils were exhausted and eroded.

Whilst the greatest land clearance for sugarcane cultivation is historical, land has continued to be cleared in some areas in recent years. The late 20th century saw a steady increase in global sugar production, partly driven by improved agronomic practices and yields, but also by increased areas under cultivation. In some cases, this reflected a switch from other crops, but, in other cases, new land was cleared. It was during this period, for example, that natural grassland and associated habitats were cleared in the Ramu Valley for the first commercial cultivation of sugarcane in Papua New Guinea (Chartres, 1981; Hartemink, 2003 – see Box 1.2). Power and de Araujo (1993) concluded that a programme initiated by the Federal Government in 1975, to use sugarcane as the raw material for fuel alcohol production, led to the deforestation of new areas in the State of Alagoas, Brazil, such that only 3% of the original rainforest cover remained. There are also reports of recent land clearance for agriculture, including sugarcane cultivation, in other parts of the world, for example in Sumatra (Gauthier, 1996), Thailand (Kobayashi, 1996; Fry, 1997) and Australia (where, increasingly, marginal areas and those of significant conservation value are being cleared of natural habitats – Johnson *et al.*, 1997; Ballantyne, 1998).

Whilst new areas continue to be cleared, there is an increasing recognition, in some parts of the world, of the need to protect natural habitats against the impacts of cane cultivation. SASA (2002) notes that, in South Africa, regulations apply to the cultivation of virgin land and that adjacent to certain natural habitats, and natural wetlands are protected from drainage and cultivation. Similarly, protection for riparian habitats is emerging in Australia (Arthington *et al.*, 1997).

In some areas, however, land has been taken out of intensive cane monoculture and converted to systems of agriculture that have greater potential sustainability and benefits to the local community (e.g. Mulkins, 2000) or allowed to revert to natural habitats. Unfortunately, the environmental damage that has already occurred can inhibit their return to a natural condition. Land clearance itself can promote soil salinization (Ghassemi *et al.*, 1995), and agricultural activities can lead to eutrophication of wetlands. Kent *et al.* (2000) note that eutrophic wetlands, such as those released from sugarcane cultivation in Puerto Rico and elsewhere, may be in a state of marshy, arrested succession because of a lack of forest species adapted for rapid reforestation of nutrient-rich habitat. SASA (2002) notes that remedial measures, such as removal of invasive species and closure of agricultural drains, may be required to promote the re-establishment of natural vegetation when wet agricultural land is abandoned and allowed to revert to natural habitat.

Impacts of ongoing cultivation practices

SASA (2002) recognizes that all natural (and cultural) assets, including uncultivated areas, of the farm form part of a natural resource (ecosystem) that can be utilized, but which needs to be appropriately managed, preferably through the development of a management plan. In South Africa, government programmes exist for the formal recognition of natural heritage sites and sites of conservation significance, both of which provide vehicles for the encouragement of environmental conservation by non-governmental organizations and landowners.

Fragments of natural habitats

SASA (2002) suggests that uncultivated areas of the cane farm should be mapped (according to a recognized habitat classification system), as part of the development of a management plan; natural habitats can be restored on areas of degraded land, providing wildlife corridors, a function also served by well-maintained watercourses. In a number of sugarcane cultivation systems, fragments of natural habitats persist within the agricultural landscape, and these can represent important refugia for indigenous biodiversity. Examples are given by Arthington *et al.* (1997) of the value of remnant wetland habitats (such as bulkuru sedge swamps) in the coastal zone of the Tully–Murray catchment in Australia, which provide nursery grounds for fish and support important bird populations. Martin and Catterall (2001) found that fragmented remnants of coastal heathland amid cane growing areas in New South Wales were also important for birds. Heathland fragments of 500 ha contained high densities of 'natural-vegetation-dependent' bird species, many of which were also found in heathland fragments down to 5 ha in area. However, Martin and Catterall (2001) suggest that the botanical character of smaller (tens of hectares) heathland fragments is unlikely to survive, given the degree of environmental change in the surrounding agricultural landscape, leading to longer-term declines in bird populations. Amador and Viana (2000) express similar concern for forest remnants in the plateau region of São Paulo, Brazil, and urge active restoration of such fragments of natural habitats. These authors report on experimental restoration work undertaken in patches of forest on a sugarcane plantation at Piracicaba. The importance to avian biodiversity of such forest fragments in the sugarcane growing landscape of São Paulo is demonstrated by bird surveys reported by Willis and Oniki (2002). Also in São Paulo, Araujo *et al.* (1999) showed that botanical diversity was similar in a 20 ha fragment of cerrado habitat which persisted in an agricultural area including sugarcane, and which had been designated a reserve in 1994, to that measured in other cerrado areas. Other biodiversity studies encompassing fragments of natural vegetation within cane growing areas include that of Moron *et al.* (1998), who studied the scarabaeid beetle fauna of Nayarit, Mexico, in sugarcane fields with remnants of tropical deciduous forest and oak forest.

Important questions over conservation mangement arise where relatively large areas of natural habitats survive in agricultural landscapes, including those that support sugarcane cultivation. For example, Sawarkar (2000) considers Terai grassland ecosystems in Uttar Pradesh, India. The conservation value of these ecosystems has been increasingly recognized and, of approximately 11,200 km^2 of Terai grassland, 19% is now included in Protected Areas. The Dudwa National Park (490 km^2), the Kishanpur Wildlife Sanctuary (204 km^2) and the Katerniaghat Wildlife Sanctuary (400 km^2) support 12 major vegetation communities and contain at least 24 plant, 12 mammal, 29 bird and five reptile species of conservation importance. The Dudwa National Park has a reintroduced population of *Rhinoceros unicornis*, and Katerniaghat is contiguous with the Royal Bardia National Park, Nepal. The Protected Areas are situated within a landscape of *Shorea robusta*-dominated forests, sugarcane and paddy-fields, scattered hamlets and small townships and thus have a large interface (and some conflicts) with a variety of human activities. Wildlife management plans typically address the Protected Areas alone, and forest work plans address only the managed forest. Sawarkar (2000) contends that an integrated landscape approach, encompassing all elements of land use in an holistic manner, is required if the ecological interests of the Terai grassland ecosystem are to be secured.

Diverse agriculture is a good thing

Intensification does not tend to promote diverse agroecosystems, which may persist where options for intensification are limited or where economic conditions favour alternative approaches. In a study in mixed agroecosystems (including sugarcane cultivation) in Veracruz, Mexico, Gallardo-Lopez *et al.* (2002) concluded that farmers with smaller farms, and with no access to irrigation, designed their agroecosystem structure and management in accordance with available resources and high subsistence usage, resulting in higher agricultural diversity. Although diverse agroecosystems (particularly those with fragments of natural habitats) are sometimes seen as providing for reservoirs of pest species, there is evidence that the opposite is true. Baliddawa (1985) reviewed literature on the effects of plant species diversity on crop pests, including those of sugarcane, and noted that populations of several pests were found to be depressed in situations where crop and/or weed diversity was relatively high. In some cases, such effects may be due to enhanced diversity and abundance of pest natural enemies in relatively diverse agroecosystems. For example, Salman *et al.* (1978) compared cane fields at two localities in Upper Egypt, and found much greater numbers of predacious arthropods (mostly spiders and coccinellid beetles) in the more diversified ecosystem. Garside *et al.* (1997b) note that there are potential agronomic advantages in maintaining diversity within the standing crop itself; domination of large cane growing areas by single varieties in Australia has led to serious problems with disease outbreaks. Indeed, Birch (1997) observes that genetic uniformity, even at the level of a single gene present in all varieties, can create problems if it leads to unforeseen disease susceptibility, citing such an example from maize cultivation in the USA in 1970 (see Zadoks and Schein, 1979).

SASA (2002) notes that, in addition to appropriate management of uncultivated land for conservation purposes, it may be possible for cane farmers to develop public recreational areas within the farm landscape. Although this has potential costs, in terms of provision (and maintenance) of facilities, it can also act as a tangible demonstration of the farmer's commitment to the conservation of natural resources, particularly if accompanied by the installation of signboards, maps, etc. Access may need to be restricted at certain times of year (for protection of fire-prone areas, for example), but provision of facilities for activities such as fishing, hiking and mountain biking can enhance the relationship between the farmer and the wider community, as well as demonstrating environmental credentials, particularly if this is done in collaboration with local interest and conservation groups. Although these specific examples apply to South African (and possible culturally similar) situations, they illustrate measures that can be taken in the management of the farm landscape which recognize the importance of the human dimension. Some studies indicate that local communities may wish to see natural habitats preserved within a cane growing landscape (e.g. Mallawaarachchi and Quiggin, 2001; Mallawaarachchi *et al.*, 2001).

Community- and Species-level Impacts of Ongoing Cultivation Practices

Papua New Guinea is considered to be the centre of origin of sugarcane, and is likely also to host a suite of species that have co-evolved with the ancestors of the crop plant. In this respect, Papua New Guinea is special in relation to cane-associated biodiversity, although it also means that the local sugar industry faces particular challenges from pests, diseases and weeds, many of which are indigenous (Hartemink and Kuniata, 1996; Kuniata *et al.*, 2001; Magarey *et al.*, 2002). In the majority of cane growing areas, where the plant is not indigenous, a complex of associated organisms develops (which will be considered pests, diseases and weeds, if they impair crop production). Species from adjacent habitats will also be expected to use cane fields, to varying extents.

Plants

Alexander (1985) suggests that the growth of many weed species in cane fields is likely to be suppressed by the highly competitive crop plant, particularly when the crop canopy is closed. Thus the cane field will be vulnerable to most weeds only when there are gaps in the mature canopy, and particularly during periods when the soil surface is exposed (between harvest and the establishment of a new mature crop). None the less, there is evidence that sugarcane is a relatively poor competitor against weeds, including in Papua New Guinea (see above) (Hartemink and Kuniata, 1996). Assessment of weed diversity in cane cultivation may be complicated by differences between species found in the seed bank and those that dominate after emergence (as observed in Sri Lanka by Witharama *et al.*, 1997).

The botanical significance of cane farming areas may be influenced by the presence of rare indigenous species or of invasive alien species that threaten other habitats. SASA (2002) urges that the various plant species that occur on a farm be recorded and measures taken to ensure the survival of rare or endangered species and to control invasive species. Examples of the latter which pose a serious threat to the environment in South Africa include triffid weed (*Chromolaena odorata*), tickberry (*Lantana camara*), Barbados gooseberry (*Pereskia aculeata*), bugweed (*Solanum mauritianum*), sesbania (*Sesbania punicea*), Mauritius thorn (*Caesalpinia decapetala*) and inkberry (*Cestrum laevigatum*).

The appropriate management of vegetation in the vicinity of the cane crop can promote beneficial insects, which may act to suppress pest numbers, as well as contributing to wider enhancement of biodiversity (see below).

Vertebrates

Avian biodiversity has been shown to be very low in cane fields. Petit *et al.* (1999) studied bird communities of 11 natural and modified habitat types, across a gradient from extensive forest to intensive agricultural land, in central Panama. They found that sugarcane plantations and introduced *Pinus caribaea* woodlands supported the fewest species compared to all other habitats, including rice fields, actively grazed pastures and residential areas. Natural wooded habitats generally supported the most species and individuals. Willis and Oniki (2002) also demonstrated the importance of natural woodland for birds in a landscape otherwise dominated by sugarcane cultivation in Brazil. Martin and Catterall (2001) found that cane fields supported very few bird species relative to surviving fragments of heathland (and residential areas) in coastal New South Wales, Australia.

Rats represent a particular vertebrate pest problem in many cane growing areas, and various control measures have been explored, some of which have had a negatively impact on indigenous biodiversity, notably historical introductions of the mongoose. Integrated strategies (as explored in India, for example – Srivastava, 1998) may provide more environmentally acceptable methods. Given that preharvest cane burning has been justified as a means of reducing the risk of snake bite where the crop is cut by hand, it seems likely that cane fields also provide useful habitat for reptiles in some areas. (Perhaps rat–snake–mongoose interactions would provide an interesting area for ecological study in some cane cultivation systems.) Whilst noting that forest remnants can act as reservoirs for pest vertebrates (including rats, but also pigs, monkeys, birds and deer) in a recently developed agricultural landscape in Sumatra, Gauthier (1996) found that areas where greater numbers (and diversity) of trees were actively planted as part of the agroecosystem were less affected by these pests.

SASA (2002) notes that wildlife (perhaps exemplified by vertebrate) populations provide important indicators of the effectiveness with which the farm ecosystem is managed. This can be formalized by regular censuses of such populations, in combination with specific measures to protect rare species from hunters and poachers. Ideally, the fauna of the cane farm (including, for example, nesting sites of important bird species) should be recorded as part of the development of a management plan. Farm workers should be encouraged to treat wildlife with respect, and (for example) should be discouraged from attacking and killing wildlife during cane harvesting operations.

Invertebrates

Studies of invertebrates in cane fields often focus on pest species and their natural enemies. The complex of invertebrates found feeding directly on sugarcane outside its native range will be made up primarily of indigenous species that have adapted to the introduced plant (Strong *et al.*, 1977) and other species imported (generally accidentally) from around the world. Charleston *et al.* (2003) note that in 1950 (around 100 years after its commercial introduction to South Africa) sugarcane growing in southern African countries had at least 33 species of indigenous insects feeding on it. Although a few more records have now been added and a small number of non-native species also feed on the plant, only two insect species (both indigenous) constitute major cane pests in South Africa. However, the length of time over which cane has been cultivated at a given locality is much less important in determining the number of arthropod herbivores associated with the crop than the size of area over which it is cultivated (Strong *et al.*, 1977).

The diversity of invertebrates in cane fields can be considerable. For example, in studies in China, Zhang (1989) collected 1035 spiders belonging to 58 species and 11 families, whilst in Brazil, Rinaldi *et al.* (2002) collected a total of 1291 spiders belonging to 73 species and 20 families. In studies of edaphic beetles in sugarcane in Cuba, Padiz and Garcia (1997) collected representatives of 38 species from 15 families, the greatest beetle density and biomass being concentrated in the first 10 cm of soil. In some cases, diversity of particular taxa may be greater in the crop than in surrounding vegetation (as found for cicadellid leafhoppers in Cuba – Hidalgo-Gato *et al.*, 1999), although this may apply principally to pest taxa and situations where adjacent habitats are highly disturbed.

Invertebrate diversity may be significantly enhanced by the presence of fragments of natural habitats in the sugarcane cultivation landscape. Such an effect probably contributed to the findings of Moron *et al.* (1998), who collected scarabaeid beetles representing 122 species and 11 subfamilies in sugarcane fields with remnants of tropical deciduous forest and oak forest in Mexico. In Hawaii, Topham and Beardsley (1975) found that populations of *Lixophaga sphenophori* (a cane weevil parasite) were enhanced where field margins contained plants which could act as nectar sources. Continuous elimination of such plants with herbicides resulted in a decrease in populations of the parasite, compromising its potential as a biological control agent against the weevil.

A significant proportion of the invertebrate diversity (of particular taxa) in cane fields may be accounted for by alien species, some of which may show invasive dynamics. Haynes *et al.* (2003) found that earthworm communities in agricultural soils (including those under sugarcane) in South Africa were dominated by exotic species accidentally introduced from Europe, India and West Africa. Cherry (2003) found that pitfall trap catches of arthropod ground predators in sugarcane fields in Florida were dominated by the imported red fire ant, *Solenopsis invicta*. This is a highly problematic invasive species, although its predatory zeal can be exploited in crop systems to counter potential pest outbreaks. For example, Ali *et al.* (1984) examined the influence of selected weedy and weed-free sugarcane habitats on diet composition and foraging activity of *S. invicta* in Louisiana, and concluded that *S. invicta* population levels could be enhanced through judicious vegetation management.

Specific cultivation practices can be a major factor influencing patterns of invertebrate agrobiodiveristy. Haynes *et al.* (2003) found that earthworm numbers were relatively low ($< 60/m^2$ vs. $230–310/m^2$ under permanent pasture), as were soil organic matter content and microbial biomass, in soils under conventional tillage sugarcane cultivation in South Africa. In studies of the pantropical earthworm *Pontoscolex corethrurus* in soils under cane in north-eastern Queensland, Spain *et al.* (1990) found that populations were substantially increased where harvest residues were retained as a surface mulch or mechanically incorporated into the topsoil, relative to situations where residues were burned. Srikanth *et al.* (1997) examined the impact of cultural practices, irrigation and postharvest trash burning on spider abundance in sugarcane at Coimbatore, Tamil Nadu, India. Exclusion of selected cultural practices (manual weed control, earthing up and three detrashing operations) significantly increased spider populations in the later stages of crop development, particularly the soil-associated *Hippasa greenalliae*. In comparisons of furrow, surface drip and subsurface drip irrigation, furrow irrigation disturbed the spiders most and subsurface drip least. Postharvest trash burning reduced spider numbers to 13.5% of pre-treatment levels (numbers recovered to 65% of pre-treatment levels within 3 weeks). In Florida, Cherry (2003) found that cane harvesting itself did not affect pitfall trap catches of arthropod ground predators, but that replanting reduced arthropod catches for 5–6 months. This suggests that, for most of its 3–5 year crop cycle, Florida sugarcane is a stable ecosystem at ground level for arthropod ground predators.

Microorganisms

Aoki and Salleh (2001) consider the diversity of fungal communities found on sugarcane and rice plants in Malaysia, and Mohawed *et al.* (1999) studied seasonal fluctuations of soil and root surface fungi of sugarcane in Upper Egypt. The latter study identified a total of 73 species and five varieties representing 33 genera, and monthly counts of these fungi fluctuated irregularly. Ingleby *et al.* (2000) found that the ectomycorrhizal inoculum potential (MIP) of soils from forest restoration sites in South Vietnam was much lower where sugarcane and rice had previously been cultivated than under wooded habitats. Declerck *et al.* (1998) report on experiments with arbuscular mycorrhizal (AM) fungi, isolated from the rhizosphere of bananas and sugarcane.

Burke *et al.* (2003) studied soil microorganisms found under natural habitats and tropical agriculture (including sugarcane cultivation), and found that soil type mainly determined the relative proportions of Gram-positive versus Gram-negative bacteria, whereas land use primarily determined the relative proportion of fungi, protozoa and actinomycetes versus other types of microorganisms. Significantly lower microbial biomass has been found in soils under long-term sugarcane cultivation than in soils at new land sites in Australia (McGarry *et al.*, 1996; Garside *et al.*, 1997a,b; Holt and Mayer, 1998). Dominy *et al.* (2001) found a decrease in both soil microbial biomass C and basal respiration under continuous sugarcane cultivation in KwaZulu-Natal, and associated this with a decline in soil organic matter.

Pankhurst *et al.* (2000) compared the diversity of bacterial populations in the rhizosphere of sugarcane growing in parts of the Burdekin (Queensland, Australia) that had been under continuous cane with minimal breaks for more than 20 years and those that had been taken out of cane and rotated to pasture and bare fallow for 3.5 years. Isolates from 25 different bacterial genera were identified. Bacterial diversity in soil under continuous cane cultivation was greater than that in the rhizosphere under rotation. Several genera, including *Acidovorax* and *Clavibacter* (which contain known plant pathogens), were present in the continuous cane rhizosphere, but not in soils under rotation. Conversely, soils under rotation yielded higher numbers of bacteria (notably *Pseudomonas* spp.) that were inhibitory to the growth of *Pachymetra chaunorhiza* and *Pythium graminicola* in a laboratory bioassay.

Impacts of transgenic sugarcane

Birch (1997) reviewed opportunities and limitations in relation to transgenic sugarcane, noting that the first field trials of such plants commenced in 1996, and that genetic modification had been used to confer on cane plants resistance to several major diseases, insect pests (e.g. see also Meyer *et al.*, 2000; Setamou *et al.*, 2002; Tomov *et al.*, 2003) and a herbicide. Further applications of the technology were predicted, providing environmental and consumer benefits. Smith (1997) considered opportunities for using genetic manipulation to improve the uptake and utilization of nutrients, predicting that such developments might reduce environmental impacts associated with high levels of inorganic fertilizer use in cane cultivation. Allsopp and Manners (1997) considered opportunities for transgenesis to control disease problems in cane, acknowledging that one concern over this technology relates to the use of genes from pathogenic viruses and the risk that they might recombine with other viruses (a phenomenon demonstrated in laboratory studies – Greene and Allison, 1994).

Sugarcane genetic resources

Some benefits of maintaining diversity within the cane crop itself are highlighted above. The maintenance of cane genetic resources in germplasm collections (Balakrishnan *et al.*, 2000; Tai and Miller, 2001, 2002; Balakrishnan and Nair, 2003) provides a vital facility for cane breeding programmes (Ramdoyal *et al.*, 2003). Germplasm collections have been supported by surveys of cane genetic resources, both in areas where the plant has been domesticated for some time (Kwon-Ndung *et al.*, 2000; Nair and Somarajan, 2003) and in Papua New Guinea, its centre of origin (Magarey *et al.*, 2003).

SUGAR BEET CULTIVATION

Ecosystem- and Habitat-scale Impacts

Land clearance for cultivation

It is likely that only very small areas have been cleared specifically for the cultivation of beet, relative to land clearance for cane, as the crop has only been cultivated relatively recently and in many cases has been grown on land that was already under some other form of cultivation. Consequently, published examples of such environmental impacts of sugar beet cultivation on natural habitats are relatively few. Sluyter (1998) provides one example, examining historical landscape changes in central Mexico. It had been argued that the agriculture of indigenous people was responsible for environmental degradation or that this accompanied the introduction of livestock by Spanish settlers in the 16th century. However, Sluyter (1998) contends that even high densities of livestock farmed between the 16th and 19th centuries were probably managed in such a way that over-grazing was avoided. Instead, it is suggested that agrarian reform, introduction of drainage and irrigation and the cultivation of sugar beet later in 19th century may have been responsible for greater levels of environmental degradation.

Impacts of ongoing cultivation

There is little available information on the impacts of ongoing beet cultivation on ecosystems and habitats, except where outlined in other sections of this review.

Diverse agriculture is a good thing

The general issue of diversity in land use in agricultural systems is discussed in more detail in relation to the cultivation of cane. In the UK, reduction in crop diversity is considered to have contributed to recent declines in the wider biodiversity of farm-land, notably bird populations (Defra, 2002). Sugar beet is typically grown as part of a crop rotation and is therefore already part of a relatively diversified system of agri-culture, which has environmental benefits over less diverse systems (e.g. Bramm, 1988). Where beet is included in the rotation, these benefits include reductions in agrochemical inputs during other phases of the rotation (by interrupting a potential build-up in pests/ diseases associated with other crops and by contributing organic matter to the soil in the form of root fragments and leaf material), as well as directly providing resources for farmland birds and other species (Defra, 2002). There is evidence that further increases to the diversity of crops (in space and time) would enhance sustainability in beet cultivation systems (Boinchan and Lykov, 1999).

Community- and Species-level Impacts of Ongoing Cultivation Practices

Many studies concentrate on the impacts of ongoing cultivation on particular broad taxonomic groups, but some consider a wider range of taxa. For example, Heijbroek and van de Bund's (1982) report on a long-term field experiment to examine the influence of crop rotation, insecticides and herbicides on soil arthropods, nematodes, soil fungi and weeds. Such multi-trophic studies have become more common in recent years. In particular, the use of pesticides in sugar beet cultivation can present risks to terrestrial and aquatic non-target species, and large variations in pesticide use have been found between farmers (e.g. de Snoo *et al.*, 1997). Reduced pesticide inputs in sugar beet cultivation systems can result in increased biodiversity across taxa (Esbjerg, 1998).

Plants

Weeds can be a serious agronomic concern in sugar beet crops. For example, El Antri (2001) reports that weeds, if uncontrolled, can reduce beet yields in Morocco by over 80%. Conversely, however, Covarelli and Onofri (1998) found sugar beet yields reduced by < 2.5% in central Italy where weeds were not removed. In the UK, Defra (2002) notes that efforts to control weeds in beet fields often tend to be most intense whilst the crop is becoming established, but (given the nature of the crop) control of broad-leaved weeds is relatively difficult. Consequently, beet fields tend to support more such weeds than cereal fields, and this is an important aspect of the biodiversity value of beet cultivation. Increased numbers of weeds result in larger numbers of invertebrates, which (along with weed seeds) provide an important resource for farmland birds (see below). Ishikawa and Takenaka (2002) recorded 24 weed species in sugar beet fields in Japan.

In general, it has been found that increasing intensity of agricultural practices leads to a reduced diversity of plants amongst the weed community. In agricultural systems including sugar beet cultivation in Spain and Morocco, Deil and Sundermeier (1992) concluded that traditional farming methods resulted in an intermingling of weeds, fallow species and subnitrophilous ruderal therophytes, whilst modern tilling techniques decreased both weed diversity (by 48 and 22 species/field in Morocco and Spain, respectively) and abundance. It was suggested that agropastoral phytocoenoses are disappearing in Spain. Barberi *et al.* (1995) also found a reduction in weed diversity with intensification of cultural practices in arable cultivation including sugar beet, particularly under high-input cropping systems. Von Bernhardt *et al.* (1991) found a high proportion of nitrophilous species amongst emergent weeds of maize and beet fields in the Osnabruck district (Germany), and attributed this to the use of N fertilizers. In this study, species belonging to the *Chenopodietea* and *Secalietea* were promoted by extensive cultivation (mostly hoeing), whilst intensive (cultural or chemical) weed control led to a reduction in species diversity. In relation to pesticide use in general, preliminary results from Esbjerg (1998) in Denmark suggest that, in a spring barley, winter wheat and sugar beet cultivation system, reduced dosages of herbicides and insecticides (1/4 and 1/2 compared with 1/1) rapidly resulted in significantly increased density and diversity of wild plants.

The seed bank is an important store of botanical diversity in arable cultivation systems, although von Bernhardt *et al.* (1991) found considerable differences between the species content of the seed bank and that of the weed cover of maize and beet fields in the Osnabruck district. Holub (1994) studied the vertical structure of the seed bank below wheat, sugar beet and lucerne at Tuchoraz (Czech Republic), and found that most seeds occurred in the top 30 cm of the soil, although under annual crops diversity and abundance of seeds were reduced in the uppermost 0.5 cm. In north-eastern Italy, Cantele and Zanin (1992) assessed the influence of crop rotation, irrigation and fertilizer use on the seed bank in the cultivated soil layer (0–35 cm). The average seed count was 7190/m^2, with peak values found under sugar beet (5482) and lucerne (8341) in the second year of a 6-year rotation. Different rotations did not greatly influence seed bank composition, but (generally) the shorter the rotation length the greater the augmentation of the seed bank. Irrigation did not affect the seed bank, while fertilizer applications enhanced the emergence of some species and suppressed that of others.

Crop edges and field margins can be important habitats for non-crop plants. In the Netherlands, de Snoo (1997) and de Snoo and van der Poll (1999) examined the arable flora on the edges of winter wheat, sugar beet and potato fields and on adjacent ditch banks. Edges were sprayed or unsprayed with herbicides and insecticides over 4 years. Leaving 3–6 m wide edges unsprayed increased the frequency and abundance of (mainly dicotyledonous) plants by factors of 4.8–12.1 and 1.5–2.7, respectively. Although the majority of the plants were common farmland species, there was a major enhancement of the floristic value of the unsprayed fields. In the sprayed centres of fields, frequency and abundance of farmland plants (and overall floristic value) were consistently lower than in sprayed and unsprayed edges. On adjacent ditch banks, potentially vulnerable to spray drift, only along the unsprayed winter wheat crop did the diversity and cover of dicotyledons (and floristic value) increase, with 65 species found compared to 50 on banks adjacent to sprayed edges, including a number that were found nowhere else in the study. The differences in ditch-bank vegetation associated with different crops were explained by differences in herbicides, dosages and spraying methods. However, leaving the crop edges unsprayed significantly decreased crop cover in sugar beet, but not other crops.

Vertebrates

As noted above, the weed community associated with beet fields provides a valuable resource for farmland birds in the UK. Defra (2002) reports that cereal stubbles from the preceding crop (particularly where these are not treated with herbicides) are another feature of beet cultivation systems that favours birds, for example, providing winter habitat for seed-feeding species such as finches and buntings. The open structure of the new beet crop in spring provides opportunities for ground nesting species such as stone-curlew, lapwing and skylark. Late-harvested beet fields also provide valuable winter habitat for birds, and postharvest fields (particularly where tops are retained) provide feeding grounds for many species, including pink-footed geese, swans, skylarks, golden plover, lapwing, pied wagtail and meadow pipit. Between a quarter and a third of the world's pink-footed goose population use postharvest beet fields in eastern England in this way.

Invertebrates

As noted above, the weed community and other aspects of the cultivation system in beet fields provide a valuable resource for farmland birds in the UK. This is partly a consequence of the invertebrate populations that develop in these situations (Defra, 2002). Peaks in invertebrate abundance are often strongly seasonal or resource dependent (e.g. aphids – Bennewicz *et al.*, 2001) and/or significantly influenced by climatic conditions and reproductive strategy (e.g. carabid beetles – Varis *et al.*, 1984). Spatial scale is also important, as particular taxa may be very localized in agricultural habitats (e.g. Kinnunen *et al.*, 2001).

Studies in sugar beet cultivation systems indicate that, at a landscape scale, invertebrate diversity is generally enhanced by environmental heterogeneity, including the presence of fragments of natural habitats, for example in Hungary (Ferenc *et al.*, 1998), the Netherlands (Booij and Noorlander, 1992) and the UK (Bedford and Usher, 1994). At the field scale, diversity and abundance of particular invertebrate taxa may vary considerably according to habitat and crop type. In Switzerland, Salveter (1998) found that 40% of hoverfly species occurred exclusively in a single habitat (either in forests, herb strips or apple orchards). Natural wooded habitats supported a particularly distinct fauna, an observation also made for carabid beetles and spiders in a UK agroecosystem (Bedford and Usher, 1994). However, for particular taxa, greatest abundance is not necessarily found in natural or semi-natural habitats, but may be associated with a particular crop. For example, in Hungary, Ferenc *et al.* (1998) found that abundance of carabid beetles was generally higher under winter wheat, pea and rape, but lower under sugar beet and in forest strips. In Belgium, Baguette and Hance (1997) found that crop type significantly influenced the diversity of carabid beetle assemblages in mixed arable fields including sugar beet. In Finland, Varis *et al.* (1984) also found that carabid numbers were affected by crop cover, with diversity greater under sugar beet and cabbage than under timothy grass. In the Netherlands, Booij and Noorlander (1992) related the influence of crop type on invertebrate diversity and abundance to crop growth characteristics. In a study of predatory carabids, staphylinids and spiders, these authors concluded that crops with greater cover early in the season, such as winter wheat and peas, were more favourable than late, open crops, such as onions or carrots.

The characteristics of field margins and adjacent land parcels are important influences on invertebrate diversity and populations in sugar beet cultivation systems. Invertebrate diversity can be greatest at the interface of natural and cultivated habitats, because this zone supports a mixture of the relatively distinct fauna associated with either habitat type (e.g. Bedford and Usher, 1994). In central Poland, Kaczorowski and Debek-Jankowska (1997) found that diversity and abundance of insects (particularly predatory carabids and hoverflies and parasitic Hymenoptera) was greater on the boundary of sugar beet fields than in adjacent fields. Closer scrutiny of field margins (different types of ditches, hedges and forest edges) revealed significant differences in diversity and abundance of hoverflies. Structural characteristics appeared to be important, as field margins with low vegetation (ditches) had higher numbers of hoverflies than other sites, as was the diversity of plant species present.

Levels of farm management intensity have been found to influence invertebrate diversity and abundance in sugar beet cultivation systems. In the UK, Buchs *et al.* (1997) found a general reduction of arthropod productivity with increasing intensity of crop management, but very different (and sometimes opposite) reactions at the species level, depending on life history characteristics. Certain pest species were enhanced by an increasing intensity of crop management, whilst (due to harvest and tillage procedures) some beneficial insects failed to build up stable populations in the crop, but did so in long-term set-aside (fallow) areas. Weiss *et al.* (1997) also found set-aside areas to be beneficial to spider diversity in an arable farming system including sugar beet. Booij and Noorlander (1992) studied predatory carabids, staphylinids and spiders under conventional, integrated and organic farming systems in the Netherlands, and found that abundance and diversity were clearly affected by farming system, although in most cases the type of crop appeared to be of greater importance at the field scale. In Lithuania, Saluchaite (2000) compared bio-organic, integrated and intensive cropping systems and found greater beetle diversity and abundance on organic sites than on intensive sites. The adoption of more environmentally sustainable systems of agriculture, such as integrated farming systems, has also been shown to benefit the general soil fauna of sugar beet fields (e.g. El-Titi and Landes, 1990).

The typical cultivation of sugar beet as part of a crop rotation rather than by continuous planting in the same plots also has beneficial effects on invertebrate diversity. Heijbroek and van de Bund (1982) found greater soil arthropod abundance and diversity in fields under a sugar beet/winter wheat rotation than in continuous beet. There was no difference in levels of damage caused by soil pests, but different species became predominant under rotation/continuous beet.

Specific farm management activities, such as application of manures and weed management practices, are also known to influence invertebrate biodiversity in sugar beet cultivation systems, as is the timing of cultivation practices. In some cases, application of inorganic fertilizers has been shown to have negative effects on insect diversity in sugar beet crops (e.g. Venegas and Aguilar, 1992). Curry and Purvis (1982) and Purvis and Curry (1984) investigated the influence of surface-applied farmyard manure and weeds on the distribution and abundance of selected arthropod taxa in sugar beet fields in the Irish Republic. Manuring enhanced numbers of soil Collembola, and resulted in a rapid but temporary increase in diversity of ground-dwelling arthropods and numbers of common predatory carabid beetles. Compared with plots treated with herbicides, weedy plots supported greater numbers of soil Collembola, enhanced diversity and abundance of other taxa (detritivores, weed-specific herbivores, predatory staphylinid beetles and parasitic Hymenoptera) and considerably smaller numbers of the pest aphid *Aphis fabae*. In crop rotations, Purvis *et al.* (2001) found that timing of cultivation (spring versus autumn), and microclimatic effects arising from the establishment of a crop canopy were notable influences on carabid beetle assemblages. In Belgium, Hance and Gregoire-Wibo (1987) found that manuring enhanced the diversity and abundance of carabid beetles. Cultivation and soil incorporation of green manures had an even greater positive effect. Spring-breeding carabids, with overwintering adults, were very sensitive to autumn ploughing and bare soil in winter.

Other studies also demonstrate the importance to invertebrate biodiversity of cultivation practices which have direct impacts on the soil. Baguette and Hance (1997) note that deep ploughing drastically changes soil structure, and is probably one of the most disturbing agricultural practices for soil fauna. In Belgium, they found that ploughing increased the abundance of the dominant carabid species (*Pterostichus melanarius*), while reduced or no tillage increased the abundance of other carabids. Soil compaction may particularly affect invertebrates in the upper strata of the soil, where numbers of certain invertebrates are greatest. Heisler (1994) considered the effects of soil compaction on springtails (Collembola) and predatory mites in the top 15 cm of soil in a German arable system including sugar beet. Species abundance was reduced on loaded plots, especially on the wheel track, where springtail density was just 30% (and mite density 60%) of that found in unloaded plots. Diversity of both groups was also negatively affected by mechanical loads.

Insecticide application clearly has the potential to negatively affect invertebrate biodiversity, although use of appropriate active ingredients, dosages and application methods can reduce immediate impacts. However, even in these situations, there may be long-term effects. The impact of insecticides on carabid populations in sugar beet cultivation systems in Belgium are amongst the issues discussed by Hance and Gregoire-Wibo (1987). In Germany, Epperlein and Schmidt (2001) found that imidacloprid seed dressings were effective against a range of pest insects, but did not have significant adverse effects on soil organisms including spiders (predominantly Linyphiidae), millipedes, carabids and staphylinids, whereas lindane spray applications did have a significant impact on non-target taxa. Baker *et al.* (2002) studied the effects of a range of insecticide seed treatments on beneficial invertebrates in sugar beet in the UK. They found no significant effects on numbers of earthworms, Acari or Collembola in soil cores or on numbers of carabids, staphylinids, spiders and Collembola in pitfall traps. However, although the trend was not statistically significant, insecticide-treated plots tended to have fewer organisms overall than untreated plots, possibly as a consequence of suppressed numbers of predatory arthropods in plots where their (pest) food source has been eliminated. It is suggested that long-term use of these insecticide treatments in a rotation could result in a reduction in biodiversity.

As in other systems, invertebrates can act as useful bioindicators of environmental impacts in beet growing systems. Diekkruger and Roske (1995) present a model (validated in crops including sugar beet) for the simulation of Collembola population dynamics under field conditions. This can be used to assist in assessing the environmental impact of agricultural practices on this group.

Microorganisms

A number of studies consider the soil microbial diversity under sugar beet crops, particularly in relation to pathogens and other organisms that have potential to inhibit them (Vesely, 1986) or to inhibit pests such as cyst nematodes (Schuster, 1997). There has also been work towards the genetic modification of these organisms, such as *Pseudomonas fluorescens*, a potential biological control agent against rhizomania. Whipps *et al.* (1998), for example, report on the survival, establishment, dissemination, gene transfer and impact on indigenous microbial populations of genetically modified *P. fluorescens*. Aspects of the potential environmental impacts of genetically modified (GM) microorganisms are examined by Colwell (1992), who notes that concerns over their release involve socio-economic, health-related and ethical, as well as environmental issues, and those relating to public trust in scientists.

Irrigation can affect the soil microbial community under sugar beet (see Chapter 3), as can other agronomic practices including agrochemical applications. In long-term studies on the use of pesticides on sugar beet, Mineyev *et al.* (1993) found adverse effects on soil microflora, resulting in an increase in spore bacteria, *Actinomyces* spp., *Penicillium* spp. and toxinogenic fungi under intensive cultivation.

Impacts of transgenic sugar beet

There is a rapidly expanding literature on the environmental impacts of transgenic sugar beet. It has been argued that crop plant transgenesis provides a potential mechanism by which to increase productivity and reduce the environmental impact of cultivation of this crop (e.g. Elliott *et al.*, 1996; Madsen and Sandoe, 2001; Pidgeon *et al.*, 2001). Risk assessment is an important factor in assessing the biosafety of transgenic sugar beet (e.g. de Vries, 1999). However, there are widespread concerns over the possible negative consequences of the cultivation of transgenic sugar beet, even where specific evidence of risks to human health or the environment is lacking. As Madsen and Sandoe (2001) note, the background to these concerns is complex, involving not just fear of specific food safety and environmental risks, but also reflecting an aversion amongst consumers to commercial monopolies in crop production. On-farm experiments to assess environmental risks associated with transgenic sugar beet and other crops, such as the recent UK farm-scale trials (see below), have themselves been controversial (e.g. Gura, 2001). Regulatory frameworks for the commercialization of transgenic crops, including sugar beet, and food products arising from them are also important considerations in this rapidly developing field, for example in Europe (Liegeois, 1997) and the USA (Harlander, 2002).

Specific environmental concerns over the cultivation of transgenic sugar beet (and other crops) include the risk of gene flow from transgenic plants to their wild relatives (e.g. Bartsch *et al.*, 2002, 2003; den Nijs *et al.*, 2004) as well as impacts on crop-associated biodiversity, including weeds, plant pathogens, soil organisms and pest species and their natural enemies (e.g. Lotz *et al.*, 2000). Impacts on these organisms could lead to wider indirect effects on species that rely on them, such as farmland birds and other wildlife (e.g. Watkinson *et al.*, 2000).

To date, work with transgenic sugar beet has centred on the investigation of genetically modified herbicide-tolerant (GMHT) cultivars (e.g. Brants and Harms, 1998; Villarias *et al.*, 2001). It has been argued that these have the potential to deliver environmental benefits through reduced application of herbicides (e.g. Madsen *et al.*, 1995; Madsen and Sandoe, 2001; Phipps and Park, 2002). Studies by Pidgeon *et al.* (2001) and Dewar *et al.* (2003) suggest that cultivation of GMHT sugar beet allows more controlled application (band spraying) of herbicides, accumulation of weeds between rows without yield reductions and positive knock-on effects for invertebrates and their predators (farmland birds and other wildlife). However, the precise pattern of crop management adopted with GMHT sugar beet is likely to be a critical factor in determining whether these provide benefits or otherwise for biodiversity. Watkinson *et al.* (2000) simulated the effects of GMHT sugar beet cultivation on weed (*Chenopodium album*) populations and consequences for seed-eating birds, especially the skylark (*Alauda arvensis*). They predicted that, depending on management practices, *C. album* populations could be reduced to low levels or practically eradicated, leading to severe reductions in food availability for birds.

In the UK, a set of farm-scale trials to assess the potential impacts on farmland biodiversity of GMHT beet, maize and oil-seed rape have recently been completed. These have been described as 'the largest scientific investigation of farm ecology the world has seen' (Coghlan, 2003a). Full reports of impacts on field and field margin plants and invertebrates have been published (Royal Society, 2003). In summary, these suggest that GMHT sugar beet has greater negative impacts on farmland plants and invertebrates than conventional beet. Not surprisingly, these results were largely dependent on the herbicide management regimes used in the respective crops. To some extent, more interesting findings from the studies relate to the impacts on biodiversity of particular arable crops and their management, whether GM or non-GM varieties (Coghlan, 2003a). In this context, a sugar beet crop appears to support more biodiversity than maize, oil-seed rape and probably wheat crops. The negative effects of GMHT beet found in the field-scale trials have been challenged; when a broader range of environmental impacts are taken into consideration using life cycle analysis, there appears to be evidence of environmental benefits (e.g. the work of Phipps and Bennett, described by Coghlan, 2003b). Such apparently contradictory findings illustrate the difficulties of assessing environmental impacts associated with complex activities like agricultural operations.

Sugar beet genetic resources

Wild forms of crop plants provide an important genetic resource for the improvement of their domesticated relatives. Kleijer (1993) notes that Europe is rich in wild plant species, of which 300 are ancestors of economically important plants, and the conservation of these genetic resources is an important consideration. There is some concern, for example, about potential negative effects of genes from cultivated beet crossing into populations of wild relatives (Bartsch *et al.*, 1999). The maintenance (for example, in gene banks) of a diverse pool of semi-domesticated forms (ecotypes and landraces) and of cultivated varieties (cultivars) is also important in the conservation of the biodiversity of agroecosystems. Heszky *et al.* (1999) consider such issues in relation to agriculture in Hungary, noting that the first decades of the 21st century may see reductions in the biodiversity of maize, rape, soybean and sugar beet in agroecosystems as multinational companies achieve a monopolistic position and promote a small number of genetically modified cultivars.

Germplasm collections provide beet breeders with the necessary genetic resources to modify and enhance varieties according to the needs of growers. Despite considerable residual genetic variation amongst cultivated beet, concern has been raised over the limited gene pool in cultivated varieties, highlighting the importance of genetic resources that reside in uncultivated forms of *Beta* (Asher *et al.*, 2001). An important area, for example, is varietal resistance to pests and diseases. Work on sugar beet, including a cross-European programme, scrutinizing the characteristics of cultivated and wild *Beta* germplasm, has identified lines with varying degrees of resistance to beet cyst nematode, *Cercospora* leaf spot, *Rhizoctonia* root rot, rhizomania, mildews (*Erisiphe* and *Peronospora*), beet yellowing viruses, beet mosaic virus, curly top virus and *Aphanomyces* and *Pythium* seedling diseases (Asher *et al.*, 2001). Similar properties have been identified by screening programmes in the USA (Doney, 1998). Genetic resources can also be drawn upon to enhance beet tolerance of abiotic stress (Richard-Molard and Cariolle, 2001), including water availability (of particular interest, for example, in areas like Iran – Sadeghian and Yavari, 2001). Other dimensions include plant developmental and seed characteristics (van Swaaij *et al.*, 2001; Zimmermann and Zeddies, 2001), yield/quality traits (Hoffmann and Marlander, 2001; Jansen and Burba, 2001) and characteristics to reduce soil tare at harvest (Olsen *et al.*, 2001). Integrated breeding programmes draw on all of the above to produce improved cultivars appropriate for local growing conditions (Gabellini *et al.*, 2001b). In Europe, the screening of germplasm (like that noted above for pest/disease resistance), and hence the development of breeding programmes, is assisted by a substantial database of *Beta* genetic resources (Germeier and Frese, 2001).

BEET PROCESSING

Over a period of 3 years, Dinter and Paarmann (1989) studied the arthropod community associated with a sludge disposal site for soils washed from sugar beet. Around 200 species were identified, mainly Brachycera (Sphaeroceridae, Ephydridae), Nematocera (Culicidae, Chironomidae, Scatopsidae), Coleoptera (Staphylinidae, Carabidae), Acarina (Eugamasidae) and Araneida (Linyphiidae, Lycosidae). Levels of arthropod activity and density of up to 9000 individuals/m^2/week were recorded. The results suggested that the structure and dynamics of the arthropod community were particularly influenced by soil quality and climatic factors.

6

Impacts on Soils

The direct environmental impacts on soils of sugar processing are relatively few, with the exception of localized impacts associated with the construction and operation of cane and beet processing facilities. However, indirect impacts may arise from the disposal of sugar processing wastes on land (see Chapter 8). The greatest range and intensity of impacts of sugar production on soils arise from the cultivation of sugar crops.

Soil is recognized as a fundamental resource in the cultivation of sugar beet (e.g. Morgan, 1986) and of sugarcane (e.g. SASA, 2002). SASA (2002) describes soil as a living, dynamic resource, made up of different-sized mineral particles (sand, silt and clay), organic matter and a diverse community of living organisms. They estimate that a healthy soil should contain 1000 kg/ha earthworms, 2700 kg/ha fungi, 1700 kg/ha bacteria, 150 kg/ha protozoa and 1000 kg/ha arthropods and other small animals. Conservation of soils, and maintenance of their physical, chemical and biological integrity is seen as vital for the sustainable cultivation of sugar crops (e.g. Morgan, 1986; SASA, 2002).

Many different types of soils are recognized, and systems of soil taxonomy are well established (Soil Survey Staff, 1975, 1999). Different soil types display different properties, including (for example) vulnerability to erosion and salinization. It is not within the remit of this document to summarize the range (and characteristics) of the many soil types associated with the cultivation of sugar crops globally. Summaries are given, for example, for South African cane growing soils by SASA (2002). However, it is worth stressing that baseline conditions vary substantially between different areas. Cane growing soils in Guyana are naturally acidic, for example, while the Barbados industry was built on alkaline soils (Blackburn, 1984). Alkaline soils are also a feature of cane growing in Pakistan, where lack of organic matter is the primary cause for concern (Nasir and Quereshi, 1999a,b; Arain *et al.*, 2000). Even within a given country, different soil quality issues may pertain in different areas, according to soil types and local management practices. For example, van Antwerpen and Meyer (1996a,b) suggest that in northern KwaZulu-Natal, acidification is a particular problem of dryland soils, whilst increased salinity and sodicity primarily afflict irrigated areas. In relation to soils, as much as any other subject area considered in this book, the environmental impacts reported reflect where published studies have been carried out, and extrapolation of general trends must be undertaken with caution. Irrespective of locality, however, knowledge of the types and characteristics of soils is very important for effective land use planning, be it at the landscape or the field scale. Soil characteristics also have a major influence on requirements for irrigation, drainage and nutritional management (e.g. fertilizer inputs).

Cultivation of sugar crops can contribute to soil degradation through negative impacts on soil quantity (by accelerating erosion and through removal of soil at harvest) and soil quality. Soil quality is a complex concept, involving a wide range of biological, chemical and physical variables. Haynes (1997) considers that soil quality can be broadly defined as the sustained capability of a soil to accept, store and recycle nutrients and water, maintain economic yields and maintain environmental quality. Negative impacts of cultivation systems on soils involve a range of factors, whose interactions create complex challenges for soil conservation. For example, soil compaction is undesirable in itself (restricting the rooting ability of crops and other plants, for example), but it can also promote erosion by increasing rates of surface water runoff. Tillage can ameliorate compaction (although it can also promote compaction in the longer term, under certain conditions), but it can also promote erosion by exposing soil aggregates to rainfall. The complexity of environmental impacts associated with soils is increased further by the close relationship between soil and water issues, for example in relation to erosion, movement of sediments, salinization and leaching.

The nature of the impacts on soils of cane and beet cultivation is similar in many respects, but also differs in a number of ways. In particular, the fact that cane is generally grown as a continuous monoculture, whilst beet is generally grown as part of a rotation, influences the types of impacts associated with either crop. It is also notable that harvesting of beet (as a root crop) results in particular soil loss problems, which are less pronounced in relation to cane.

Loss of soil fertility in agricultural systems is a source of major concern worldwide, notably in the tropics. Tropical countries are the main centres of the expanding global human population, and hence face particular challenges in terms of agricultural sustainability. Agriculture here must support increasing numbers of people, either directly through supply of food or through earnings associated with traded commodities. Particular concern has long been expressed over the effects on tropical soil fertility of cultivation of annual crops, but perennial crops (including sugarcane) have recently received increasing attention (e.g. Hartemink, 2003). Aspects of soil quality management, with an emphasis on temperate agriculture, have recently been reviewed in Schjonning *et al.* (2004).

Erosion

Soil erosion is recognized as a major problem, particularly in tropical agriculture, and general reviews are given by, for example, Lal (1990), Morgan (1995) and El-Swaify (1997). It is a matter of concern in a number of areas under sugarcane or beet cultivation. In agronomic terms, soil erosion is a major problem, resulting in the absolute loss of a fundamental resource or redistributing organic matter and nutrient-rich material at a landscape scale, along slopes for example (e.g. Schwertmann, 1986). Soil erosion also represents a substantial environmental threat, through land degradation and the washing of sediments (and associated nutrients and agrochemicals) into surface waters.

The physical loss of soil by erosion is influenced by a range of factors including water dynamics (rainfall and irrigation), wind, temperature, soil type, cultivation systems and topography. In particular, erosion is promoted by soil disturbance (a general feature of cultivated land), high-intensity rainfall (a feature of many cane growing areas) and increasing gradients of slope (e.g. Bakker, 1999). The erodibility of soil varies considerably with soil type, depending on the stability of soil aggregates (related to organic matter content) and the percentage of coarse primary particles resistant to erosion (e.g. Morgan, 1986). The presence of vegetation (including crops) reduces the risk of erosion, by consolidating the soil and protecting bare ground against direct exposure to water inputs. Hence, soils are vulnerable to erosion where natural vegetation has been cleared, or (on cultivated plots) during fallow periods, when the crop is in the earliest stages of establishment, and following harvest. One tool that has been used in an attempt to capture all of these factors in the analysis of soil erosion (in both cane and beet cultivating systems) is the universal soil loss equation (USLE) (see Box 6.1).

Erosion risk is greatest when the soil water infiltration rate is low, increasing runoff, which is also influenced by factors such

Box 6.1. The universal soil loss equation (USLE).

The universal soil loss equation (USLE) was developed and refined in the USA, as a tool for analysing rates of soil loss in relation to a range of factors (see Wischmeier and Smith, 1965, 1978; Wischmeier *et al.*, 1971). It takes the general form:

$$A = R \times K \times SL \times C \times P, \text{ where}$$

A = average soil loss (t/ha/year)
R = an 'erosivity' index, related to intensity of rainfall
K = an 'erodibility' index, related to soil characteristics
SL = a topography factor, based on slope gradient (S) and length (L) of land units
C = a crop management factor
P = a soil conservation practices factor

The applicability of the USLE in different situations has been a subject of debate, but it has been widely adopted (and adapted) as a basis for estimating potential soil losses due to erosion in areas where sugar crops are cultivated. For example, European studies which discuss the use of the USLE include Biagi (1986), Chisci (1986b), De Ploey (1986), Madsen *et al.* (1986) and Schwertmann (1986). Examples of the use of the USLE in cane cultivation systems include work in Australia (e.g. Sullivan and Sallaway, 1994) and South Africa (e.g. SASA, 2002).

as compaction and gradient of slope. The relationship between runoff and certain types of erosion is considered in detail by Monnier and Boiffin (1986). From an agronomic perspective, cultivation on flat or gently sloping land is generally preferable to that on steeper gradients, but cane and beet are grown on sloping land in some areas. Even on gentle slopes, erosion occurs as a gradual process (so-called 'interil' erosion) and may result in dramatic environmental degradation in the long term (Bakker, 1999). However, as the gradient of the slope increases, so does the rate of runoff and erosion. Wrigley (1985) estimates that velocity of runoff doubles for each fourfold increase in slope and that a doubling of runoff flow rate increases scouring capacity fourfold and carrying capacity 32-fold and allows particles 64 times larger to be carried in the flow.

Soil quality impacts

Soil quality is a complex concept, involving a wide range of biological, chemical and physical variables. Impacts of the cultivation of sugarcane or beet on soil quality include effects on physical properties (e.g. compaction) and biological properties (e.g. soil biodiversity). In addition, effects on chemical properties (e.g. changing nutrient levels, acidification and salinization) appear to be of particular concern in relation to cane cultivation in some areas. In addition to the complexity of soil properties that must be considered, further complexity exists in the range of crop management practices that can affect them. These include tillage, irrigation, mechanization and the application of soil amendments, including mulches and manures, as well as inorganic fertilizers and pesticides.

Compaction alters a number of the physical properties of soil. Bulk density and soil strength are increased (e.g. Martin, 1979; Soane *et al.*, 1982), while porosity, permeability and water infiltration rate are decreased (e.g. Hansen, 1982). Compaction can lead to surface sealing, which reduces infiltration rates and increases runoff and can thus exacerbate erosion problems (e.g. Morgan, 1986; Schwertmann, 1986; Hartemink, 2003). It can also negatively affect the soil mesofauna (e.g. Heisler, 1994) and inhibit the rooting ability of the crop. Loam-rich soils are more vulnerable to compaction than clays or sands, and compaction risk increases with soil moisture content. A degree of compaction can arise from the exposure of bare soil to the physical impact of heavy rainfall. However, it is the use of in-field transport and other heavy machinery (particularly when the soil is wet) which is associated with the most important soil compaction problems. In beet cultivation, the number of field operations (and therefore vehicle passes) used in preparation of the field and the fact that soils are often wet during harvesting contribute particularly to compaction risk.

Soil organic matter is a key attribute of soil quality, as (via mineralization) it is a source of nutrients such as N, P and S. It is also important in maintaining soil structural stability (e.g. Qongqo and van Antwerpen, 2000; Dominy *et al.*, 2001) and in influencing soil biological characteristics. As organic matter content affects soil physical properties, its reduction can render the soil more prone to structural breakdown by raindrop impact, resulting in blockage of pore spaces, reduced water infiltration rate, increased erosion risk and enhanced susceptibility to compaction (Sumner, 1997). Loss of organic matter is recognized as a problem associated with cultivated soils in general.

Acidification of soils appears to be a problem affecting cane growers in many areas. Haynes and Hamilton (1999) and Hartemink (2003) discuss the chemistry underlying acidification of cane growing soils, and conclude that the effect is mainly caused by the use of acidifying nitrogenous fertilizers such as urea and ammonium sulphate, coupled with nitrate leaching driven by heavy rainfall. Ammonium-based fertilizers have the greater acidifying potential (143 g H^+/kg N for ammonium sulphate versus 71 g H^+/kg N for urea), but a greater risk of volatilization of NH_3 is associated with urea fertilizers. Soil acidification promotes leaching of certain nutrients (such as Ca and Mg) and accumulation of others (notably Al).

Studies of nutrient balances in soils under sugar crops tend to focus on changes in levels of N, P and K, which are also the predominant components of inorganic fertilizers. However, other elements are also removed from the soil by cane and beet. Draycott and Christenson (2003) list 11 other elements that are known to be essential for sugar beet and four others that may be, and Wood *et al.* (1997) report that at least 14 elements are essential for normal cane growth and development. Failure to replenish soil reserves of any of these elements can give rise to soil nutrient imbalances and deficiencies, a situation exacerbated by recent trends for inorganic fertilizers to contain fewer impurities and trace elements (Wood *et al.*, 1997). However, there is also evidence of soil nutrient imbalances arising from the presence of such impurities. In some areas of Australia, Arthington *et al.* (1997) report that cadmium, probably derived from impurities in phosphatic fertilizers, is present in soils under long-term cultivation at levels seven times those found in uncultivated sites (although still within acceptable, background levels). The need to consider replacement of elements other than NPK in the cultivation of sugar crops may provide one argument for using processing wastes, which contain a wide range of nutrients, as soil amendments (see Chapter 8).

The global issue of soil and water salinization is reviewed by Ghassemi *et al.* (1995). In the context of sugar production, salinization of soils appears to be of greatest significance in certain cane growing areas. Haynes and Hamilton (1999) consider saline soils to be those where the concentration of soluble salts is sufficient to restrict plant growth (often taken as those where electrical conductivity of a saturation paste extract exceeds 4 dS/m – Sumner, 1997). They also note that saline soils also tend to be sodic (percentage of exchangeable cations present as Na (ESP) exceeds 15%).

A range of inorganic fertilizers, pesticides and other agrochemicals are used in the cultivation of sugar crops. The inappropriate management of these agents can result in impacts on soil quality, but also on air quality (e.g. through the release of nitrogenous gases and volatiles) and water quality (through leaching and runoff). The environmental fate of fertilizers is dependent on a range of factors, including soil type, climate and land use management practices (e.g. Knappe and Haferkorn, 2001). Herbicide applications can influence the environmental fate of fertilizer-derived nutrients (Sotiriou and Scheunert, 1994).

Inorganic fertilizers typically supply nitrogen, phosphorus and/or potassium in mineral form. The specific form in which they are applied can influence their environmental impact (e.g. Brentrup *et al.*, 2001). Environmental impacts typically arise because the nutrients applied with fertilizers are not matched against those taken up by the crop (Neeteson and Ehlert, 1988; Haynes and

Hamilton, 1999). Nutrients that are taken up by the crop may also pose environmental problems, as waste products of processing (Draycott *et al.*, 1997). A wide range of organic soil amendments are also used in the cultivation of sugar crops, which can also result in environmental problems (e.g. Isermann, 1989). Aspects of rational fertilizer and pesticide use are considered in more detail in Chapter 2.

SUGARCANE CULTIVATION

Soil degradation is recognized widely as a problem in the cultivation of sugarcane, particularly in relation to the effects of the intensive growing of cane as a continuous monoculture, which contributes to yield decline as well as environmental degradation (e.g. Haynes and Hamilton, 1999; Garside *et al.*, 2001; Meyer and van Antwerpen, 2001; Pankhurst *et al.*, 2003; Nixon and Simmonds, 2004). It is the introduction of such intensive agricultural practices that represents, in general, a particular threat to soils in tropical areas (Sparovek *et al.*, 1997; Sparovek and Schnug, 2001a). Past concerns over loss of fertility of tropical agricultural soils have focused particularly on those under annual crops. It has been argued that the characteristics of perennial crops, including their rate of return of organic material to the soil, ameliorate their impact on soil fertility. Such a case has been made for sugarcane. For example, Alexander (1985) considers that the cane plant's perennial habit results in

> an enormous underground commitment of decaying organic matter . . . that cannot be matched by temperate plants and seasonal-tropical species. For this reason it has essentially been possible to maintain cane cropping on the same field, literally for centuries, without apparent loss of soil productivity. There are sites in Puerto Rico that have improved in fertility through nearly four centuries of continuous sugarcane cropping.

Although cane yields have been maintained on some soils over very long periods of cultivation, a recent examination of the evidence by Hartemink (2003) concludes that the impact of sugarcane on tropical soil fertility, although less than that of annual crops, is greater than that of other perennials and plantation crops and is a significant source of concern. Specific aspects of the impact on soils of cane cultivation are examined below, and are illustrated by a case study from Papua New Guinea in Box 6.2. In at least one case, change of land use from sugarcane cultivation (to market gardening, in Martinique) appears to have resulted in increased soil degradation problems, owing to the adoption of increasingly intensive practices (Hartmann *et al.*, 1998).

Impact of Sugarcane Cultivation on Soil Quantity

Erosion

Sparovek *et al.* (1997) and Sparovek and Schnug (2001a) note that soil erosion is a particular threat to long-term productivity in tropical areas, because erosion rates here are usually greater than the rate of soil formation. In addition to environmental impacts, erosion in cane growing areas can have an impact on cane yields (e.g. McCulloch and Stranack, 1995) and may ultimately limit the sustainability of sugarcane cultivation (e.g. Glanville *et al.*, 1997).

Some soil types are more prone to erosion than others, which may be a consideration in cane cultivation systems (Ahmad, 1996; Hartemink, 2003). It has also been shown that inefficient or excessive irrigation of cane fields can increase runoff and/or erosion (Inamdar *et al.*, 1995, 1996a,b; Chapman, 1997; Hartemink, 2003). Vegetation cover is another important factor in regulating runoff and erosion. Consequently, cane fields are particuarly prone to erosion when the surface is most exposed, between harvest and the development of the next closed crop canopy, or during bare fallow. Rates of runoff from land under cane cultivation have been compared with those under certain other land uses. Leitch and Harbor (1999) found that conversion of cane fields to pasture reduced freshwater runoff into the near-coastal zone of

Box 6.2. The impact on soil of sugarcane cultivation in Papua New Guinea.

Commercial cultivation of sugarcane is a relatively recent activity in Papua New Guinea, having begun only in 1979 (see Box 1.2). Hartemink (2003) presents a case study of the impact of cane growing on soils, based on his work in Papua New Guinea (see also Hartemink and Kuniata, 1996). His findings are summarized below, and presented in diagrammatic form in Fig. 6.1. These findings are mostly based on data collected in the 1980s and 1990s, and so generally reflect changes that occurred over little more than 10 years of cane cultivation (a much shorter period than experienced by soils in many other cane growing areas). Although the effects of pests and diseases are probably the main constraints on cane yield in Papua New Guinea, decline in soil quality may become agronomically more significant in the future.

The Ramu Sugar Estate is the one area in Papua New Guinea given over to commercial sugarcane cultivation. It covers some 7000 ha, on alluvial soils (Fluvents and Vertisols) in the Ramu Valley, and the main crop is entirely rain-fed. Up to four ratoon crops are possible after harvest of the planted cane, after which the field is replanted (in some plots, cowpea *Vigna unguiculata* is sown, grown for a year and ploughed in prior to replanting of cane). Over the course of the cultivation period, there has been a shift away from preharvest burning, towards green cane harvesting and trash blanketing. This has been accompanied by a shift away from urea-based N fertilizers, towards ammonium-based preparations. Between 1991 and 1995, N was applied at an average rate of 90 kg/ha/year. P and K fertilizers have not been applied.

Soil erosion is a threat in some areas of the estate, and terraces have been installed to help manage the flow of surface water. There was evidence of compaction in cultivated fields, with soil bulk density increased to depths of 0.3–0.5 m. Compaction of topsoil resulted in a reduced water infiltration rate.

Evidence was found of soil acidification; the effect was greatest (a decrease of around 0.5 pH units) in topsoils, but remained statistically significant to depths of 0.6 m. The pH values were slightly lower within the cane rows than in the inter-rows. Overall, Hartemink (2003) suggested an annual decrease in pH of about 1%, with the initial decrease possibly due to an increase in the mineralization of organic matter. Thereafter, annual applications of ammonium-based fertilizers were identified as an important factor causing soil acidification. Another possible contributory factor was the shift away from preharvest burning, resulting in pH-increasing ashes no longer being returned to the soil.

Evidence was also found of a decrease in soil organic matter content. Soil organic C in cultivated soils declined between 1979 and 1996 by about 40% (from around 56 g/kg to around 30 g/kg). Soil organic C levels in topsoils were around 10 g/kg lower in the inter-rows (and about 8 g/kg lower within cane rows)

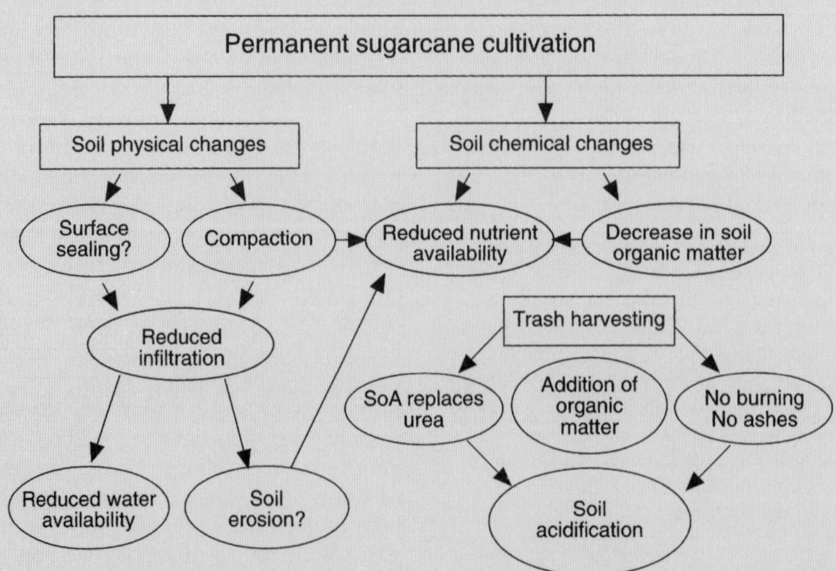

Fig. 6.1. Summary diagram showing major changes to soils under sugarcane cultivation in Papua New Guinea (after Hartemink, 2003).

than under natural grassland. There was some decrease in organic C content in inter-row subsoils, but that within rows was similar to the level found in natural grassland. The annual rate of decline in soil organic C was estimated at 1–3%. However, the shift towards green cane harvest and trash blanketing may result in enhancement of the soil organic matter content.

Total soil N was lower in the topsoil of inter-rows than in that of cane rows or natural grassland, but there was no difference between the three soil categories at subsoil depth. Available P was around 5 mg/kg less in the cultivated topsoil (inter-rows and within rows) than in that of natural grassland. Subsoils of inter-rows showed some decline in available P, but this was not apparent in the subsoil within rows. Exchangeable K levels were similar in the topsoils of rows and inter-rows, but below the levels found in uncultivated topsoils. At a range of depths, there was a striking increase in exchangeable Mg within the cane rows, and a possible elevation of exchangeable Ca levels.

Data were also collected on changes in the nutrient content of sugarcane leaves over the period of cultivation at Ramu. These broadly matched the trends for major nutrients seen in the topsoil data, with declines in N, P and K.

Overall, Hartemink (2003) concludes that a consistent decline in all soil chemical properties was found in both soil types (Fluvents and Vertisols) under cane cultivation in Papua New Guinea. Temporal data suggested that the decline was largest in the early stages of cultivation, and then began to level off.

The impacts of cane cultivation on soils reported here (and illustrated in Fig. 6.1) from Papua New Guinea largely reflect those recorded in other cane growing areas around the world, perhaps with the exception that levels of available P are generally shown to increase rather than decline.

the Holetown watershed in Barbados. In Thailand, Putthacharoen *et al.* (1998) found that (on an annual basis) soil losses due to erosion on fields with a 7% slope were highest under cassava grown for roots, followed by cassava grown for forage, sugarcane, mungbeans, sorghum, groundnuts, maize and pineapples. The relative value of sugarcane as a cover crop (e.g. da Silva *et al.*, 1986) is enhanced by the fact that the crop tends to remain in the ground for a number of years (due to ratooning), producing an extensive root system and (for much of the growing period) a closed canopy that protects the soil from the erosive effects of rain (e.g. Bakker, 1999). However, erosion in cane growing areas can still represent a very serious problem, leading not only to soil degradation, but to secondary impacts (for example) on waterways. Umrit and Ng Kee Kwong (1999) refer to crystal clear watercourses becoming loaded with mud during and after heavy rainfall events in sugarcane cultivation areas in Mauritius.

Cultivation on slopes

Cultivation on slopes is undesirable, as it tends to increase rates of runoff and erosion. It has been suggested that cane should not be grown on slopes greater than 8%, although slopes of 20–30% are used, for example, in parts of the Caribbean and South Africa (Bakker, 1999). The average slope of land cultivated with cane in South Africa has been estimated to be 20%, and associated problems have been noted by various authors (e.g. Landrey, 1978b; Gardiner and Cazalet, 1991; Tudor-Owen and Wyatt, 1991; SASA, 2002). In this context, it is worth noting that commercial sugarcane farmers in KwaZulu-Natal do appear to concentrate their soil conservation efforts on their most steeply sloping fields (Ferrer and Nieuwoudt, 1997). A number of socio-economic factors have contributed to the expansion of cane cultivation on to slopes in various parts of the world. For example, Gawander (1998a) and Seru (1998) report that until the 1940s sugarcane growing in Fiji was confined to flat land. Population growth, high demand, attractive prices and the land tenure system drove expansion into less fertile, highly weathered undulating terrain. Although a soil conservation programme was introduced in the 1950s to enhance sustainable development in the cane growing belt, this was compromised by further socio-economic changes in the 1970s. As a result, cane production on marginal land has led to higher production costs and lower productivity due to soil erosion. However, growers continue to sustain themselves on these farms mainly because of the high price of sugar due to preferential European Union (EU) markets to Africa, Caribbean and Pacific countries. This results in short-term gain for social sustainability and long-term misuse of land; a national land use plan would help to guide more sustainable practices.

Examples of figures for rate of soil erosion under sugarcane

Sparovek and Schnug (2001a), working in the Ceveiro watershed (Brazil), estimate a mean soil erosion rate of 15 t/ha/year for an area of mixed land use including sugarcane cultivation, and note that erosion was greatest on sugarcane areas, at 31 t/ha/year. Rates estimated in Australian sugarcane studies indicate soil losses of 42–227 t/ha/year on conventionally cultivated slopes of up to 8% (Sallaway, 1979, 1980) and 47–505 t/ha/year (average 148 t/ha/year) on conventionally cultivated slopes of 5–18% (Prove *et al.*, 1995). In the latter case, the considerable variation in erosion rate was largely explained by variation in rainfall. In Puerto Rico, Lugo-Lopez *et al.* (1981) found soil losses of over 15 t/ha/year from unmulched cane fields (half the rate of those found under coffee). In Louisiana (USA), soil erosion losses averaging around 17 t/ha/year were recorded under sugarcane by Bengtson *et al.* (1998). Based on experiences in South Africa, SASA (2002) estimates that, for every 1 mm/ha of soil erosion, losses equate to 10 t soil, 2 m^3 water and 0.02 t sucrose production. For a cane farm of 150 ha, where soil conservation is poor and 50 mm of erosion has occurred, this equates to 75,000 t soil, 15,000 m^3 water and 150 t sucrose. Rates of soil loss will depend on soil type and condition, according to which losses of between 4 and 23 t/ha/year may be acceptable. Shallow (< 400 mm), erodible soils are particularly in need of protection, with sandy, poorly structured soils being more vulnerable than well-structured soils with a relatively high clay content.

Because they depend on soil parent material and soil-forming conditions, precise generalizations for rates of soil regeneration cannot be made. However, broad estimates for soil formation rates of 25 mm every 300–1000 years have been suggested. In South Africa, soil regeneration probably occurs at around 10 t/ha/year, equivalent to 1 mm/ha/year (SASA, 2002).

Soil erosion risk under sugarcane varies with cropping methods, as demonstrated, for example, by Pohlan and Borgman (2000) in Central America, and noted in relation to various tillage systems in the coastal Burnett district of Australia (Sullivan and Sallaway, 1994). It may also vary according to farm layout: El-Swaify and Cooley (1980) noted that rates of soil erosion in Hawaiian watersheds where cane was grown increased as the proportion of land given over to roads increased. SASA (2002) also notes the particular erosion risks associated with roads, and recommends measures such as the grassing of secondary and tertiary roads and the siting of primary roads away from watercourses and wetlands, as means of reducing the potential negative environmental effects. Examples of the ways in which cropping methods can be adapted to reduce erosion risk are outlined below.

Measures to reduce erosion

A wide range of measures has been proposed and investigated for the reduction of runoff rates and soil erosion in sugarcane cultivation systems, particularly on sloping land. Aspects of modified tillage and mulching/trash retention are discussed in Chapter 2. Other measures include:

- Field layout (Landrey, 1978b).
- Terracing (Landrey, 1978b; Armas *et al.*, 1991; Gardiner and Cazalet, 1991; Ferrer and Nieuwoudt, 1998; Vanegas Chacon and Vos, 1999).
- Contour planting (Liao, 1972; Armas *et al.*, 1991; Gawander, 1998a; Vanegas Chacon and Vos, 1999).
- Strip tillage/planting (de Boer, 1997; Bakker, 1999; SASA, 2002).
- Hedgerows/vegetation 'live barriers' (Garrity, 1993; Mangisoni and Phiri, 1996; Gawander, 1998a,b).

- Drainage ditch design (Liao, 1979; Vanegas Chacon and Vos, 1999).
- Cover crops (Scandaliaris *et al.*, 2002).
- Enhanced vegetation of watercourses (Landrey, 1978a; Tudor-Owen and Wyatt, 1991).
- Soil amendments (Peng and Twu, 1980; Finegan, 1990).

The risk of erosion may increase with irrigation, and so measures may need to be taken specifically to counter this effect. Hartemink (2003) notes that appropriate soil erosion control measures (drains, bunds, ridges, strip cropping, etc.) are employed on most cane plantations. The integration of various soil conservation measures to minimize the risk of erosion from cane cultivation is outlined by SASA (2002), which provides recommendations for practices in relation to the cultivation of cane on new land of different slopes and soil erodibility classes, presented here in Table 6.1.

The value in erosion control of indigenous vegetation associated with waterways is noted by SASA (2002), who also recommend the active planting of grasses and wetland plants, the use of revets (bundles of cane tops) to line waterways until protective vegetation becomes established and (in extreme cases) structures such as rock baskets to consolidate unstable banks. With respect to ongoing maintenance of waterways, SASA (2002) suggests that vegetation should be managed by slashing or herbicide application (with removal of debris) rather than by activities such as hoeing which disturb the soil; where soil has been deposited in drainage channels, it should

Table 6.1. Recommendations for practices in relation to the cultivation of cane on new land of different slopes and soil erodibility classes (after SASA, 2002).

	Soil erodibility		
Land preparation/conservation practices	High (max. slope %)	Moderate (max. slope %)	Low (max. slope %)
Tillage: conventional Terraces: yes Harvest: burn + tops scattered	10	15	20
Tillage: conventional Terraces: yes Harvest: full trashing	15	20	25
Tillage: minimum Terraces: yes Harvest: burn + tops scattered	20	25	30
Tillage: minimum Terraces: yes Harvest: full trashing	25	30	30
Tillage: conventional Terraces: spillover Harvest: full trashing Other: strip planting	15	20	25
Tillage: minimum Terraces: spillover Harvest: full trashing Other: strip planting	No	No	30

Terraces imply water-carrying structures; full trashing implies leaving all residues after harvest; strip planting implies adjacent terrace banks planted with a 6-month difference in age; where a practice is not tried and tested in an area, it should not be used.

be removed using hand tools, not with machinery.

Sugarcane itself can be employed as a 'live barrier' or in border planting to reduce erosion (e.g. Mangisoni and Phiri, 1996; Hellin and Larrea, 1998), or as a soil conservation cover crop (Ismail, 1990). In the context of a large area under cane cultivation on land at risk from erosion, this is reflected in the practice of strip planting, or strip cropping (e.g. Bakker, 1999). Under this practice, strips (blocks, panels) of cane rows at different stages of development are established across a slope. Mature and maturing strips provide barriers against erosion from the relatively bare strips (those with cane at the earliest stages of development, or where harvesting has just occurred). SASA (2002) recommends that strip planting should be employed on all slopes of greater than 2%, except in certain areas, and that alternate strip planting must be practised on slopes greater than 12%.

Mathematical modelling can be used to help predict erosion risk, erosion impacts and the likely effectiveness of soil conservation measures under various conditions at a local scale (one of the most straightforward examples is the USLE, see Box 6.1). Increasingly, computer software is available to assist in the application of mathematical models. For example, on the north coast of KwaZulu-Natal, McCulloch and Stranack (1995) used the CANEGRO model to simulate sugarcane yields from various soil depths and to estimate losses in revenue resulting from erosion of the topsoil. In Queensland, Australia, Glanville *et al.* (1997) used the WEPP (Water Erosion Prediction Project) model to predict soil loss for five soils, for a range of cane furrow lengths and gradients, with different surface treatments. Remote sensing can be used to assess vegetation cover (e.g. at different stages in the sugarcane cultivation cycle) and hence erosion risk at a regional scale. Cavalli *et al.* (2000) tested such techniques in São Paulo, Brazil; the data obtained allow for better planning of the sugarcane crop in the region, avoiding possible environmental impact problems and maintaining sustainability for the crop.

In KwaZulu-Natal, Ferrer and Nieuwoudt (1997, 1998) found that cane growers' decisions over whether to adopt soil conservation measures, and over which methods to employ, were constrained by financial resources, management time and farmers' technical abilities. For such reasons (particularly in relation to small-scale agriculture), it is important that recommendations are tailored to farmers' needs and resources. For example, in Malawi, Mangisoni and Phiri (1996) considered that it may be easier to improve traditional soil and water conservation techniques than to introduce new methods. Hellin and Larrea (1998) found that farmers in Guinope, Honduras, were using live barriers to control erosion, as recommended by a non-governmental organization (NGO) programme in the 1980s. However, they tended not to use species promoted by the earlier programme (Napier grass, *Pennisetum purpureum* and king grass, *P. purpureum* × *Pennisetum typhoides* (*Pennisetum glaucum*)), but increasingly employed sugarcane and fruit trees. Whilst the grasses were more effective in retaining soil, farmers pointed out that they were invasive and that there was little demand for the amount of fodder produced. The species chosen by farmers were less effective in retaining soil, but contributed to the farm household in terms of domestic consumption and/or the sale of the products of the live barriers.

It is interesting to note that Honduran farmers rejected certain species of plant for soil consolidation on the basis of their invasiveness. The environmental dangers of invasive species are increasingly recognized, and should be considered when selecting plants for use in soil conservation projects; Tudor-Owen and Wyatt (1991), for example, specifically recommend indigenous tree species for stabilizing and rehabilitating watercourses and drainage lines.

Removal of soil with cane at harvest

UNEP (1982) estimates that harvested cane can contain up to 3–5% soil by weight; however, the amount of extraneous material (soil, trash) removed from the field depends on harvesting method and on soil and weather conditions during harvest (Payne, 1991). Generally speaking, increased mechanization of harvesting results in higher levels of extraneous material. The main factors determining whether cane is cut by hand or by machine are the cost of labour and the size of the area cultivated. Carefully hand-cut and loaded cane may contain just 1% extraneous material, but the rate of harvesting by this method is very low. The adoption of mechanical loading can increase this fraction to around 6%, and fully mechanized harvesting can increase it to 15% or above, even under favourable conditions. In the particular case of recumbent cane, where the growing cycle is such that the crop lodges at an early stage and stalks develop horizontally (as seen in Hawaii), the most straightforward harvesting methods can result in around 40% extraneous material in loads delivered to the factory, half of it soil (Payne, 1991).

Impact of Sugarcane Cultivation on Soil Quality

Soil quality declines under sugarcane cultivation have been widely reported and are associated with declines in yields in many parts of the world. As Garside *et al.* (1997a,b) note, a wide range of soil factors may be implicated in yield decline, the relative importance of each varying between locations, depending (for example) on soil type, environmental factors and patterns of crop management. However, very few studies examine closely the detailed relationship between cane yield and soil quality at a within-field scale (Hartemink, 2003). Even at larger scales, knowledge is limited, with much more known about impacts on chemical properties of soils than on physical and biological properties (Haynes and Hamilton, 1999; Hartemink, 2003).

Haynes and Hamilton (1999) provide a concise synthesis of world literature on the impact of sugarcane cultivation on soil quality. The main effects identified are:

- loss of soil organic matter;
- soil acidification;
- changes in soil nutrient levels;
- soil salinization and sodification;
- compaction of topsoil.

More recently, Hartemink (2003) has analysed the impacts of cane cultivation on soil fertility, with particular emphasis on chemical characteristics, including a detailed case study from Papua New Guinea. The above aspects of soil quality are explored further in the following sections, drawing on work cited by Haynes and Hamilton (1999) and Hartemink (2003), and additional material.

Organic matter

Bakker (1999) notes that the organic matter content of (even virgin) tropical soils tends to be relatively poor, partly because high soil temperatures promote the rapid mineralization of organic material. However, this organic matter content tends to decline further under cultivation. A decrease in soil organic matter when virgin land is brought under sugarcane cultivation is commonly observed (Haynes and Hamilton, 1999), as it is with other tropical crops (van Noordwijk *et al.*, 1997). Decreased organic matter content under cane is particularly consistently a feature of topsoils, with studies showing losses from those of Oxisols to be typically around 5–7% per year (Hartemink, 2003). Declines in soil organic matter under sugarcane have been specifically recorded in Australia (King *et al.*, 1953 – one of the first studies of its type; Wood, 1985), Brazil (Cerri and Andreux, 1990; Caron *et al.*, 1996), Cuba (Armas *et al.*, 1991), Fiji (Masilaca *et al.*, 1986; Gawander and Morrison, 1999), Indonesia (Sitompul *et al.*, 2000), Papua New Guinea (Hartemink and Kuniata, 1996; Hartemink, 1998), the Philippines (Alaban *et al.*, 1990), South Africa (van Antwerpen and Meyer, 1996a,b; Qongqo and van Antwerpen, 2000; Dominy *et al.*, 2001) and Swaziland (Henry and Ellis, 1996). However, it should be noted that some studies in Australia appear to show no overall decline in soil organic C after more than 20 years under cane, although various effects of cultivation (including changes in organic matter content) were identified in different soil types at different depths (Bramley *et al.*, 1996; Skjemstad *et al.*, 1999).

The level of decline in soil organic matter content in cane growing areas is illustrated by the following examples. Studies in Australia have shown levels of organic C under cane (after 20, 30 or more years of cultivation) to be less than half of that found in uncultivated soils (King *et al.*, 1953; Wood, 1985). Between 1979 and 1996, organic C content of soils under sugarcane cultivation in Papua New Guinea declined from around 5.5 to 3.2 g/kg (Hartemink, 1998). In the Philippines, Alaban *et al.* (1990) recorded a decrease in organic C from 13.3 to 9.9 g/kg over the course of 19 years of cane cultivation, accompanied by a decrease in pH. In Brazil, Cerri and Andreux (1990) found a decline in organic C to 46% of the level found in forest soils after 50 years of cane cultivation, and Caron *et al.* (1996) found topsoil organic C levels of 34 and 45 g/kg under primary forest (depending on soil type), compared with 16 and 30 g/kg, respectively, under cane after 20 years of cultivation. In this case, differences in soil organic C were also accompanied by a decrease in pH and extended to around 1 m depth. In KwaZulu-Natal, Dominy *et al.* (2001) estimated that organic C content was 46 g/kg under undisturbed vegetation, compared with 34 g/kg or 13 g/kg under continuous sugarcane cultivation, depending on soil type.

Biological indicators of soil quality (such as microbial biomass C, soil respiratory rate, soil enzyme activity and soil earthworm communities) are sensitive to changes in soil organic matter content, and can change markedly before any substantial changes in organic matter content itself are detected. Recent years have seen an increasing interest in the use of such biological indicators of soil quality (e.g. Pankhurst *et al.*, 1997; Sparling, 1997), and this area overlaps significantly with considerations of the biodiversity impacts of cultivation. However, relatively little is yet known in detail about changes in soil biological characteristics associated with sugarcane production (Hartemink, 2003). Significantly lower microbial biomass has been found in soils under long-term sugarcane cultivation than in soils at new land sites in Australia (McGarry *et al.*, 1996; Garside *et al.*, 1997a,b; Holt and Mayer, 1998). Dominy *et al.* (2001) found a decrease in both soil microbial biomass C and basal respiration under continuous sugarcane cultivation in KwaZulu-Natal, and associated this with a decline in soil organic matter.

Recent studies using computer modelling of soil organic matter dynamics have explored how levels of the original (pre-cultivation) organic material change in soils under sugarcane cultivation. Typical rates of turnover of pre-cultivation organic matter vary, for example, between 13 and 50 years depending on soil type (Burke *et al.*, 2003), although pre-cultivation organic C can remain detectable at some depths in cane soils for much longer periods (Vitorello *et al.*, 1989; Cerri and Andreux, 1990). New equilibrium levels of total soil organic matter are eventually reached, based on the gradual decay of pre-cultivation material and an even slower build-up of crop-derived material. This pattern is demonstrated for soils under sugarcane for a period of 50 years in Brazil (Cerri and Andreux, 1990; van Noordwijk *et al.*, 1997). These studies also show that pasture systems return greater amounts of organic C to the soil (7.5 t/ha/year) than cane (0.96 t/ha/year), resulting in a relatively rapid return of total soil organic C to pre-cultivation levels. The typical pattern of soil organic matter dynamics under cane appears to be a relatively rapid decline in pre-cultivation material, accompanied by a slow build-up of cane-derived material, leading to a new equilibrium level of organic C (below that seen in the pre-cultivation soils) after many years (Cerri and Andreux, 1990; van Noordwijk *et al.*, 1997; Sitompul *et al.*, 2000).

A number of the studies cited above emphasize the complexity of soil organic matter dynamics, including differences in patterns according to soil type, soil texture and particle size, depth at which measurements are taken and cultivation practices. For example, Haynes and Hamilton (1999) and Hartemink (2003) report evidence that organic matter can accumulate in subsurface horizons (> 30 cm depth) in soils under cane cultivation. This may be a result of topsoil being redistributed by deep tillage (e.g. Gawander and Morrison, 1999; Grange *et al.*, 2002) or as a consequence of turnover of root material from this reasonably deep-rooted crop. However, such effects are not entirely consistent across the industry. In a study in Australia, Skjemstad *et al.* (1999) found lower levels of organic C in subsoils of old (20–70 years) cane fields relative to virgin sites.

Soil acidification

Sugarcane tolerates a pH range of 5–8, and some studies have suggested an optimum pH of about 6, but some cane fields on very acid soils are known nevertheless to produce high yields (Blackburn, 1984; Coale and Schueneman, 1993). A decrease in soil pH when virgin land is brought under sugarcane cultivation is commonly observed, for most soil types, at a range of depths (Haynes and Hamilton, 1999; Hartemink, 2003). Such effects have been specifically recorded in Australia (Maclean, 1975; Wood, 1985; Garside *et al.*, 1997a,b), Brazil (Caron *et al.*, 1996), Fiji (Masilaca *et al.*, 1986), Florida (Coale, 1993), Papua New Guinea (Hartemink and Kuniata, 1996; Hartemink, 1998), the Philippines (Alaban *et al.*, 1990), Puerto Rico (Abruna-Rodriquez and Vicente-Chandler, 1967) and South Africa (Schroeder *et al.*, 1994; van Antwerpen and Meyer, 1996a,b; Qongqo and van Antwerpen, 2000).

The degree of soil acidification recorded under sugarcane is illustrated by results from South Africa, particularly the northern and southern coastal regions of KwaZulu-Natal, where Meyer *et al.* (1998) reported a decline in mean soil pH from 6.2 in 1980/81 to 5.6 in 1996/97, and an increase in the proportion of soil samples with pH < 5 (from 18% in 1980 to 43% in 1997). In Papua New Guinea, the pH of topsoils under sugarcane cultivation decreased from around 6.5 to 5.8 between 1979 and 1996 (Hartemink, 1998). In Fiji, declines in soil pH from 5.5 to 4.6 were recorded over the first 6 years of cane cultivation in some areas (Masilaca *et al.*, 1986), and, in the Philippines, Alaban *et al.* (1990) recorded a decrease in soil pH from 5.0 to 4.7 over 19 years of cane cultivation.

Sumner (1997) and Haynes and Hamilton (1999) discuss the chemistry underlying acidification of cane growing soils, and conclude that the effect is mainly caused by the use of acidifying nitrogenous fertilizers such as urea and ammonium sulphate, coupled with the nitrate leaching that occurs under the high rainfall conditions that often prevail in cane cultivation areas. Hartemink (2003) also identifies the use of ammoniacal fertilizers as a major contributory factor to soil acidification under sugarcane, owing to oxidation of ammonium to nitrate. Sumner (1997) notes that mineralization of organic matter can also contribute to soil acidification, through the oxidation of N to HNO_3 and S to H_2SO_4.

Sumner (1997) and Haynes and Hamilton (1999) also note that a range of other chemical changes typically accompany declining pH of soils under sugarcane, including an increase in exchange acidity, increase in exchangeable Al (which becomes soluble at pH < 5.5), decrease in exchangeable bases (Ca, Mg, K) and overall decrease in cation exchange capacity. Such effects have also been noted by van Antwerpen and Meyer (1996a) and Qongqo and van Antwerpen (2000).

The particular issue of acid sulphate soils is discussed in relation to water pollution in Chapter 4. Rayment (2002) suggests that a further hazard associated with this particular soil type is potential acid contamination of surface layers when spoil from drain construction is dumped.

Changes in soil nutrient levels

As well as directly influencing cane growth and development, soil nutrient levels can influence the impact of pests (Atkinson and Nuss, 1989) and diseases (Anderson and Dean, 1986), and have also been shown to play a role in the biological control of weeds (Kuniata, 1994; Kuniata and Korowi, 2001).

Haynes and Hamilton (1999) conclude that changes in soil nutrient levels under sugarcane depend primarily on whether fertilizer inputs exceed or are less than nutrients removed in harvested cane or lost by other means. It has been estimated that annual removals of nutrients in a 100 t/ha sugarcane crop are around 120 kg N/ha, 33 kg P/ha and 125 kg K/ha (de Geus, 1973; see also figures given by Srivastava, 1992; Hunsigi, 1993; Malavolta, 1994). Hartemink (2003) provides a detailed analysis of soil chemical changes under sugarcane cultivation, and concludes that the most typical general trends are decreases in total N and exchangeable cations (particularly K), and an increase in available P.

A range of processes (other than fertilizer additions and removal of nutrients by the crop) contribute to changes in soil nutrient levels under cane cultivaton. Levels of inputs from biological nitrogen fixation vary considerably, and many processes contribute to losses of soil nutrients (and may contribute to wider environmental impacts). Soil erosion, for example, can contribute to the loss of nutrients from cane fields. In Louisiana (USA), soil erosion losses averaging around 17 t/ha/year resulted in annual nutrient losses of around 18 kg N/ha, 14 kg P/ha and 104 kg K/ha (Bengtson *et al.*, 1998). Runoff and leaching are potentially important pathways for nutrient loss from cane fields (see Chapter 4). Nutrients may also be lost to the atmosphere, through denitrification and volatilization, and it has also been estimated (Valdivia, 1982) that preharvest burning may be responsible for as much as 30% of the annual N removal in a cane crop (see Chapter 7). Consequently, careful management of fertilizer inputs is important, not only for maintaining soil nutrient balance, but for controlling air and water pollution.

Sugarcane is regarded as a relatively heavy consumer of N (Malavolta, 1994), although its N requirements are similar to those of wheat and maize, despite its much greater biomass productivity (Keating *et al.*, 1997). However, a general N balance for the cane crop is difficult to construct, as a consequence of widely differing agronomic practices and growing conditions, and a lack of knowledge of certain processes (Ruschel and Vose, 1982). Under some circumstances, there is evidence of substantial potential inputs due to biological nitrogen fixation, equivalent to over 200 kg N/ha/year, providing 60–80% of total plant N (Boddey *et al.*, 1991; Urquiaga *et al.*, 1992). It has been argued that, in some situations, such biological nitrogen fixation allows cane to be grown continuously, with only low levels of inorganic fertilizer application, without serious depletion of soil N, despite substantial removals of N by the crop (Lima *et al.*, 1987). However, the dynamics of nitrogen fixation are complex and incompletely understood and are affected (for example) by applications of mineral N. Application rates for inorganic fertilizers and organic amendments vary substantially. A study of cane cultivation in Latin America and the Caribbean, for example, found inorganic fertilizer application rates ranging from 60 to 200 kg N/ha/year (Ruschel *et al.*, 1982), although a greater range of rates is undoubtedly used worldwide (see Chapter 2). In terms of N outputs, individual studies often fail to quantify N losses due to erosion, denitrification and leaching, further impeding the construction of complete N budgets for this crop (Hartemink, 2003). In addition, it has been shown that quantities of gaseous N can be lost from the plant itself. These results call into question many estimates of N recovery by sugarcane, which are generally of the order of 20–50% (Hartemink, 2003). For example, Ng Kee Kwong and Deville (1994b) showed that accounting for gaseous N losses from the plant increased recovery rate estimates from 13–18% to 20–40%. Such experiments also suggest that levels of denitrification and volatilization, which have generally been assumed to explain unaccounted-for N loss after other factors are taken into account, may have been greatly overestimated (Hartemink, 2003). With such complications in mind, Hartemink (2003) suggests that the N balance for cane cropping is generally negative, which is consistent with direct measurements of soil nutrient levels under cane, which generally show a decline in total N. Examples of such a decline include studies from Australia (King *et al.*, 1953 – total N in soils under cane cultivation for 22 years was less than half that found in uncultivated soils) and Swaziland (Henry and Ellis, 1996). Hartemink (2003) summarizes the results of a range of cane growing studies by suggesting that N losses from the topsoils of Oxisols, for example, are typically up to 10% per year.

In terms of changes in levels of exchangeable cations, effects of cane cultivation on K appear to be the most pronounced. Relatively high levels of K are typically removed from the soil by sugarcane, and depletion of soil K is a common phenomenon under this crop (Haynes and Hamilton, 1999). Summarizing the results of a range of studies, Hartemink (2003) suggests a typical K depletion rate of between 3 and 10% per year. Depletion of soil K under cane has been specifically reported from Fiji (Masilaca *et al.*, 1986), Papua New Guinea (Hartemink, 1998), the Philippines (Alaban *et al.*, 1990) and Swaziland (Henry and Ellis, 1996). Conversely, K levels are generally high in South African soils under cane, suggesting that there may be over-fertilization with this nutrient here (Meyer *et al.*, 1998). Levels of K are naturally elevated in the (illite-rich) soils of Pakistan, but rates of K removal by cane are a cause for concern and an influence on soil management strategies (Nasir and Quereshi, 1999a).

As well as K, levels of other exchangeable cations tend to decrease in soils under sugarcane (Hartemink, 2003). Haynes and Hamilton (1999) note that soil acidification promotes leaching of certain nutrients (such as Ca and Mg) and accumulation of others (notably Al). A decrease in topsoil Ca and Mg under cane, relative to uncultivated soils, was shown in Australia by Mclean (1975). Decreases in Ca and Mg and increases in Al have been reported for cane growing soils in Australia (Wood, 1985; Bramley *et al.*, 1996) and Costa Rica (Krebs *et al.*, 1974).

Studies show that levels of available P in soils under sugarcane generally increase, often (in the topsoils of Oxisols) at a rate of around 5–15% per year (Hartemink, 2003). Specific studies showing such an effect include work from Australia (Mclean, 1975), Fiji (Masilaca *et al.*, 1986) and Swaziland (Henry and Ellis, 1996). However, decreases in available P have been demonstrated for cane growing soils in Australia (Wood, 1985) and the Philippines (Alaban *et al.*, 1990). Hartemink (2003) concludes that P accumulation in cane growing soils most probably results from high application rates of P fertilizers.

Soil salinization and sodification

Salt-affected soils occur naturally and as a consequence of human land management activites (Ghassemi *et al.*, 1995; Sumner, 1997). Haynes and Hamilton (1999) consider saline soils to be those where the concentration of soluble salts is sufficient to restrict plant growth (often taken as those where the electrical conductivity (EC) of a saturation paste extract exceeds 400 mS/m – Sumner, 1997). They also note that saline soils also tend to be sodic (ESP exceeds 15%). In the context of cane cultivation in South Africa, SASA (2002) recommends that soil samples should be taken and assessed against the criteria outlined in Table 6.2.

Poorly managed irrigation is the main cause of human-induced soil salinization (Ghassemi *et al.*, 1995; Sumner, 1997; Haynes and Hamilton, 1999). As a consequence of poor drainage of irrigation waters, the water-table rises and salts dissolved in the groundwater reach the soil surface by capillary action, where they accumulate as water evaporates or is transpired by the crop. This process can be exacerbated by the use of saline waters for irrigation. Hence, prevention of soil salinity is largely a matter of appropriate management of soil water. Naturally shallow water-tables are generally associated with areas of low elevation in the landscape, and there is evidence that shallow water-tables are common throughout the sugar industry (Sweeney *et al.*, 2001a,b). Ghassemi *et al.* (1995) note that the risk of raising the water-table (and inducing salinization) by irrigation

Table 6.2. Criteria for assessment of soil salinity (based on samples from the upper 600 mm) in irrigated cane fields in South Africa. Soils suitable for irrigation are those with a sodium adsorption ratio (SAR) of < 15 and an EC of < 200 (after SASA, 2002).

EC (mS/m)	Rating	Effect on cane growth
0–200	Non-saline	None
200–400	Slightly saline	Slight
400–600	Moderately saline	Severe
> 600	Strongly saline	Very severe

is invariably underestimated – even where the natural water-table is 10–20 m below the surface, it can easily be raised to within 1–2 m of the surface, particularly where drainage is poor.

Excessive salt concentrations in soil under sugarcane have been reported from many areas, particularly where rainfall is relatively meagre and irrigation is practised (Haynes and Hamilton, 1999). Specific reports include those from Australia (Ham *et al.*, 1997; Nelson and Ham, 1998), Egypt (Nour *et al.*, 1989), India (Tiwari *et al.*, 1997), Iran (Sehgal *et al.*, 1980; Abbas Keshavarz, 1998; Sohrabi *et al.*, 1998), Jamaica (Shaw, 1982), South Africa (Maud, 1959; von der Meden, 1966; van Antwerpen and Meyer, 1996a,b), Swaziland (Workman *et al.*, 1986), the USA (Bernstein *et al.*, 1966) and Venezuela (Wagner *et al.*, 1995b). Salinity and sodicity of soils have been linked to serious cane yield declines in a number of these areas (Haynes and Hamilton, 1999). Ghassemi *et al.* (1995) give three examples of areas where irrigation of sugarcane has resulted in soil salinization. These are summarized in Box 6.3.

Remediation measures for saline soils include enhanced surface and subsurface drainage (see also Chapter 3), combined with accelerated leaching of salts by deliberate over-irrigation, and application of ameliorative agents such as gypsum or phosphogypsum (e.g. SASA, 2002). Where sodicity is a problem, addition of gypsum (to replace some of the exchangeable Na with Ca), combined with improved drainage to promote leaching of Na, may be effective; this also has the benefit of reducing the likelihood of clay dispersion and surface sealing (Sumner, 1997).

Accumulation of pesticides in the soil

Cultivation methods are likely to influence the levels of pesticide residues in the soil, either by affecting rates of runoff in water and sediments or by affecting rates of breakdown of pesticides *in situ*. For example, Srivastava *et al.* (1999) found that atrazine herbicide residues in soil under sugarcane cultivation decreased with increasing levels of N application and soil moisture regime. Umrit and Ng Kee Kwong (1999) found that, following applications of atrazine, diuron, hexazinone and acetochlor, there was a rapid dissipation of herbicide in the top 0–2.5 cm layer of the soil, and little herbicide was transported down the soil profile to below 30 cm depth. These authors found no evidence of on-farm build-up of herbicide residues. However, despite their widespread replacement by less persistent pesticides, Cavanagh *et al.* (1999) found easily detectable concentrations (0.01–45 ng/g) of organochlorine pesticide residues in soil samples from sugarcane fields in North Queensland, Australia.

Compaction

Many smallholders and subsistence farmers cultivate sugarcane, but most is grown on large-scale plantations with a high degree of mechanization, and heavy vehicles are often used for in-field operations such as tilling and harvesting (Hartemink, 2003). Particularly when the soil is wet, it is this that results in the most important compaction problems associated with cane cultivation, including breakdown of soil structure and direct damage to cane stools (e.g. Bakker, 1999; SASA, 2002). Compaction alters physical properties of soil, including bulk density and soil strength (increased), porosity and water infiltration rate (decreased). SASA (2002) provides a summary of critical values for bulk density and porosity for different soil types, shown here in Table 6.3.

SASA (2002) suggests that compaction can be identified by the following:

- Ponding on soils with a fine, sandy texture.
- Measurements with a penetrometer.
- Digging a pit, observing root growth and looking for evidence of a 'hardpan'.
- Smeared, shiny soil surface lacking structure.
- Lack of cracks, deep roots or earthworm holes.
- A platey soil structure.
- Grey, anaerobic topsoil.
- Mottling caused by impeded drainage.

Box 6.3. Irrigation of sugarcane as a cause of soil salinization

Ghassemi *et al.* (1995) give the following examples of cases where soil salinization has been directly or partially attributed to the irrigation of sugarcane.

Khuzestan, Iran

In ancient times, Khuzestan in south-west Iran was the land of sugarcane. However, around 700 years ago, cane cultivation ceased. Clearly, it is difficult to ascertain the precise cause, which may be ascribed to a combination of factors including an earthquake, invasion, siltation of irrigation canals and a breakdown in the elaborate institutions needed to operate the ancient irrigation system. However, it has also been suggested that salinization, resulting from a lack of knowledge of soil/water management, was a significant factor. Salinization as a consequence of inappropriately managed irrigation is believed to have contributed to the collapse of agriculture in other parts of the ancient world. Since the 1960s, sugarcane cultivation has returned to this area of Iran, although soils and groundwater are saline and remedial drainage programmes have been required. A measure of their success is the increase in cane yield since cultivation was reinstigated, from around 77 to 120 t/ha. More recently, Abbas Keshavarz (1998) has reported the problem of shallow water-tables and soil salinity, and measures to control it, as part of the 14,300 ha Amir Kabir Sugarcane Irrigation Project, in Ahwaz, Khuzestan.

Egypt

Despite having very limited rainfall, agriculture flourished in Egypt for thousands of years, based on surface irrigation methods tied to the annual flooding of the Nile River. This regular but infrequent inundation of the land did not result in salinization problems, because the local soils have a high capacity for natural drainage. However, systems for perennial irrigation were introduced in about 1820, to facilitate the cultivation of cotton and sugarcane. Subsequently, a number of major infrastructural projects (including the High Aswan Dam) and a network of canals have been completed to increase access to perennial irrigation. Seepage from these canals has resulted in a gradual but general rise in the water-table and accumulation of salt in the soil. Soil degradation and the need for improved drainage are now priority issues in many areas.

Kakrapar Irrigation Project, India

The Kakrapar Irrigation Project is one of the largest in Gujarat. The construction of a weir on the Tapi River at Kakrapar facilitated the introduction of increased irrigation to the area in the late 1950s, and the construction of the Ukai Reservoir in the early 1970s increased the area to which water was available on a perennial basis. The soils here are fertile, and can sustain increased crop production under irrigation, but are not freely draining. The increased availability of water led to a shift from traditional unirrigated crops to new irrigated crops, including sugarcane. By 1984, around 12% of the irrigation area was under cane. The shift in cultivation practices led to a water-table rise estimated at 0.3 m/year over a period of 20 years. Whilst 150,000 ha had a water-table deeper than 9 m in 1957, virtually no land in the area did by 1980. Salt-affected areas increased, and thousands of hectares became almost barren. Over-irrigation of sugarcane and the cultivation of cane in areas with soils least suitable for irrigation have been identified as significant factors in this process (see also Gajja *et al.*, 1994, 1997, 1998, 2000). Adherence to recommended cropping patterns (based on land irrigability classes), higher water charges and education of farmers are recognized as important means of recovering productivity in the Kakrapar area (Gajja *et al.*, 1997, 2001).

Table 6.3. Critical values for bulk density and porosity for different soil types (after SASA, 2002).

	Bulk density (kg/m^3)		Porosity (%)	
	Normal range	Critical value	Normal range	Critical value
Clay soil (> 35%)	1100–1300	1500	58–51	43
Sandy soil (< 20%)	1500–1700	1800	43–36	32

Haynes and Hamilton (1999) report that compaction problems have been reported from many areas where sugarcane is grown. For example, significant differences of 0.15–0.18 mg/m^3 in the bulk density of the surface 8 cm of soil have been shown for cultivated versus uncultivated land (Maclean, 1975; Wood, 1985). Further specific reports come from Australia (Braunack and Hurney, 1996; see also below in relation to inter-rows), Brazil (Centurion *et al.*, 2000), Colombia (Torres and Villegas, 1993), Cuba (Armas *et al.*, 1991), Fiji (Masilaca *et al.*, 1986), Guyana (Davis, 1997), Hawaii (Trouse and Humbert, 1961), India (Srivastava, 1984; Rao and Narasimham, 1988), Papua New Guinea (Hartemink, 1998), Thailand (Grange *et al.*, 2002), South Africa (Maud, 1960; Johnston and Wood, 1971; Swinford and Boevey, 1984; van Antwerpen and Meyer, 1996a,b) and Venezuela (Wagner *et al.*, 1995a,b).

Because of the pattern of vehicle movement through the cane crop, compaction can be most pronounced in the inter-rows (Haynes and Hamilton, 1999; Hartemink, 2003). Heavier mechanized traffic has the potential to cause greater inter-row compaction and also compaction (and stool damage) in the cane rows themselves, which is a greater threat to yields than inter-row compaction. Topsoil bulk density has been shown to be significantly greater in inter-row spaces than in cane rows or under natural vegetation. For example, in Australia, McGarry *et al.* (1996) found topsoil bulk densities of 1.85 mg/m^3 in the inter-row, 1.55 mg/m^3 within the row and 1.40 mg/m^3 on uncultivated land (see also Braunack, 1997).

Compaction is likely to be a particular problem under zero tillage, as the soil is not regularly loosened by tillage operations (Haynes and Hamilton, 1999; Grange *et al.*, 2002). However, this must be set against the reduced risk of erosion derived from zero tillage, particularly on sloping, high rainfall areas.

Surface sealing

Sumner (1997) notes that surface sealing and crust formation can occur on cane growing soils. This phenomenon results from the breakdown of soil aggregates, followed by the dispersion of fine clay particles which block pore spaces. This results in a relatively impermeable layer at (or close to) the soil surface, which can form a hardened crust on drying. Sodic soils are particularly vulnerable to sealing, and the loss of organic matter often associated with cultivation can also render (particularly low salt content) soils more susceptible. Sealing reduces water infiltration and increases runoff, increasing the risk of erosion and pollution of waterways. It also reduces the water available to the crop (i.e. by reducing the proportion that permeates the soil) and can inhibit seedling emergence.

Impacts of preharvest cane burning

There is evidence that sustained preharvest burning of sugarcane can contribute to a decline in soil quality. This applies to decline in the physical properties of the soil (as demonstrated in Brazil by Ceddia *et al.*, 1999) and soil microbial activity (as demonstrated in South Africa by Graham *et al.*, 2002). In Lucknow, India, Yadav *et al.* (1994) found that trash burning reduced soil organic carbon by 0.02%, available N by 15 kg/ha and available P by 16 kg/ha. Valdivia (1982) suggested that preharvest burning may be responsible for as much as 30% of the annual N removal by a cane crop. However, Hartemink (2003) suggests that preharvest burning may help to counter the risk of soil acidification, through the return of pH-increasing ashes to the soil, an idea supported by the results of Noble *et al.* (2003). Implications of a shift from cane burning to green cane harvesting and trash blanketing are outlined in Box 2.4.

Measures to Reduce Impacts of Sugarcane Cultivation on Soil Quality

Haynes and Hamilton (1999) conclude that sustainable systems of cane cultivation that maintain or improve soil quality are required to ensure the future of the sugar industry. A range of methods has been investigated in relation to reduction of the impacts of sugarcane cultivation on soil quality. Aspects of the rational use of agrochemicals, appropriate harvest operations, modified tillage and trash blanketing, all of which can make significant contributions to reducing or remediating impacts on soils, are discussed in Chapter 2. Other suggested methods include:

- Soil amendments (including cane processing wastes and animal manures) (Lugo-Lopez *et al.*, 1981; Ng Kee Kwong and Deville, 1988; Scandaliaris *et al.*, 1995; Haynes and Hamilton, 1999; Meyer and van Antwerpen, 2001; van Antwerpen *et al.*, 2003).
- Application of lime to counter acidification (Choudry and Corpuz, 1984; Meyer *et al.*, 1989; Meyer *et al.*, 1991; Turner *et al.*, 1992; Coale, 1993; Coale and Schueneman, 1993; Schroeder *et al.*, 1993; Kingston *et al.*, 1996; Sumner, 1997; Wood *et al.*, 1997; Haynes and Hamilton, 1999; Noble and Hurney, 2000).
- Crop rotation, fallowing, green manuring (Armas *et al.*, 1991; van Antwerpen and Meyer, 1996a,b; Pankhurst *et al.*, 2000, 2003; Dominy *et al.*, 2001; Garside *et al.*, 2003; van Antwerpen *et al.*, 2003; Nixon and Simmonds, 2004).
- Intercropping (Akhtar and Silva, 1999).

As the negative effects of cane cultivation on soils have often been related to the tendency to grow the crop as a continuous monoculture, the use of fallowing is of particular interest. Alexander (1985) suggested that the economics of tropical agriculture strongly mitigated against taking areas out of production. However, interest in fallowing cane fields (leaving soils bare, establishing pasture or growing a break crop or green manure) is not new (e.g. Chinloy and Hogg, 1968). In Guyana, an unusual situation exists in the routine flood fallowing of cane fields (see Box 6.4).

In Australia, a maximum of 6 months fallow (bare ground, weeds or sown legumes) has been common between ploughing out and replanting, and, in some areas of northern Queensland, rotation of sugarcane and bananas has been introduced (Garside *et al.*, 1997b). However, interest in fallowing has increased substantially in Australia in recent years, and it has been shown to enhance soil quality (van Antwerpen *et al.*, 2003) and yield (Pankhurst *et al.*, 2000; Nixon and Simmonds, 2004). Such effects have the potential to reduce the need for water and fertilizer applications (Garside *et al.*, 2003). Fallowing has also been found to reduce populations of detrimental soil organisms, while increasing numbers of beneficial organisms (Pankhurst *et al.*, 2000, 2003).

SUGAR BEET CULTIVATION

As in other respects, identification and characterization of the specific impacts of sugar beet cultivation on soils are complicated by the fact that beet is typically grown as part of a rotation of crops. There is evidence that the use of rotations, in itself, reduces soil impacts relative to monocropping, as well as ameliorating declines in crop yields. Zawislak and Tyburski (1992) and Zawislak and Rychcik (1997) present results of studies of continuous cropping of sugar beet (and other crops) versus rotation in Poland. Average yield reduction under continuous cropping (based on weighted means from numerous trials of various durations during the period 1957–1991) was 28% for sugar beet. Volkov (1986) stresses the importance of crop rotations both for the farm's short-term results and for maintaining long-term soil fertility. However, soils in continuous arable production (even with rotation of crops) remain at risk of degradation (e.g. Morgan, 1986).

Impact of Sugar Beet Cultivation on Soil Quantity

The main considerations in relation to beet cultivation impacts on soil quantity are erosion and removal of soil at harvest (the latter representing a greater problem with this crop than with cane). Soil under sugar beet cultivation can be prone to wind erosion as well as water erosion, according to local conditions.

Box 6.4. Flood fallow in Guyana.

The following summary of this unusual practice is based partly on an unpublished presentation given at a meeting to celebrate the 50th Anniversary of the Faculty of Agriculture and Natural Sciences, University of the West Indies, St Augustine, Trinidad (A.D. Dey and H.B. Davis, 1998, The effects of flood fallow on soil fertility maintenance under sugar cane monoculture).

Flood fallow is a practice dating back to the 19th century in Guyana and Surinam, and may have originated as a means of controlling insect pests such as *Castniomera licus* (Follett-Smith, 1934; cf. SAC, 2000). Current practice is to submerge cane fields with 30–45 cm fresh water for 6–12 months, following postharvest incorporation of crop residues into the soil. Apart from suppression of pests and weeds, the technique results in benefits to soil structure, with improvements seen in soil air–water relations and soil aggregate stability, and reduced bulk density. Such benefits are apparent for up to 4 years following flooding, and contribute to the alleviation of subsoil compaction (Davis, 1997). Soil nutrient status is also improved, from an agronomic perspective, through the liberation of substantial quantities of ammonium. This process is apparently independent of microbial activity, or reduction of nitrates, and has instead been ascribed to the release of ammonium bound to clay particles (e.g. Evans, 1962). It results in a reduced need for N fertilizer applications in the subsequent crop cycle (Dey and Davis, 1997). Other benefits include the flushing of salts from the soils, some of which are significantly saline.

This technique has proved suitable for use on around two-thirds of the cane growing soils in Guyana, specifically those based on high activity (swelling) clays. It is less applicable to the more porous soils, or those rich in fine sands and silts with a low percentage of swelling clays (where a fallow period under legumes may be more appropriate – Dey *et al.*, 1997). However, the precise mechanisms behind the physical and chemical effects of flood fallowing have not been clearly elucidated. It appears that flooding creates reducing conditions in the upper soil layers, and may promote anaerobic decomposition of organic matter, resulting in releases of gases such as N_2O, H_2S, NH_3 and CH_4. Gaseous releases contribute to the formation of cracks and crevices in the soil, allowing the reduction products of Fe_2O_3 to percolate downwards, contributing to the release of ammonium from lower soil layers. When the field is drained, oxidizing conditions return, and reconstitution and redistribution of iron contributes to the consolidation of improved soil structure.

Studies suggest that the agronomic benefits and increased yields arising from flood fallowing are sufficient to compensate for the time that land is taken out of production (McLean, 1982). From the environmental perspective, however, it is interesting to note that, whilst flood fallowing alleviates compaction and can reduce fertilizer (and presumably herbicide) inputs, it promotes the release of various gases from the soil and the removal of salts (and possibly nutrients) in drainage waters.

For example, Frede (1986) notes that, in Germany, wind erosion is mainly restricted to areas in the north, whilst water erosion occurs throughout the country, and areas of intensive sugar beet are amongst those most affected. Wind erosion causes damage to beet seedlings, and can therefore affect short-term yields as well as contributing to longer-term environmental degradation.

Chisci and Morgan (1986) note that soil erosion was not considered a major issue in Europe in the 1970s, as traditional agricultural systems were not seen as posing significant erosion risks. There was greater concern over the impact of farm machinery on soil structure (compaction), and issues related to soil organic matter content. However, concerns over erosion increased in the 1980s, particularly in relation to cultivation of sloping areas in the Mediterranean (e.g. Chisci, 1986a, and other papers in Chisci and Morgan, 1986), but also in the rolling, loamy landscapes of northern and western Europe (e.g. Monnier and Boiffin, 1986; Morgan, 1986, and other papers in Chisci and Morgan, 1986). The increase in erosion risk identified in these areas was related to recent changes in the pattern of land use, in combination with local topographical, geomorphological, pedological and climatic factors (e.g. Chisci, 1986a; Monnier and Boiffin, 1986). Key aspects of changing land use were altered cropping systems, notably the switch from pasture to arable, but also increasing crop specialization under arable (including switches from cereals to row crops such as sugar beet), increased mechanization and up-and-down cultivation of slopes (Chisci, 1986a; De Ploey, 1986; Eppink, 1986; Monnier and Boiffin, 1986; Morgan, 1986). Soil erosion in sugar beet cultivating areas can result in economic as well as ecological problems (e.g. in Germany – Heissenhuber and Schmidtlein, 1988), including beet yield declines (e.g. in the former USSR – Zaslawskij, 1982).

Wind erosion

Because beet fields are often ploughed in autumn and left bare over winter, and because the crop canopy takes a relatively long time to develop after sowing, wind erosion can be a serious problem, particularly on light soils (Henriksson and Hakansson, 1993). The seedbed (and seeds) can be blown away, and wind-borne soil particles can damage young seedlings, sometimes resulting in a need to completely resow the crop. Fornstrom and Boehnke (1976) estimated figures for annual soil loss due to wind erosion in sugar beet fields under conventional tillage in a 3-year experiment in Wyoming (USA) as 49, 19.5 and 13.1 tons/acre. Use of alternative tillage systems reduced these losses to 17, 5.5 and 13.1 tons/acre, respectively.

Water erosion

Cultivation on slopes enhances the risk of soil erosion. In Europe, this has been recognized as a problem in beet growing areas, particularly in the hilly areas of the Mediterranean (e.g. Chisci and Morgan, 1986, and papers therein). Even in the less hilly areas of northern and western Europe, where soil types increase erosion risk, increasing slope gradients exacerbate potential soil losses. For example, on loamy loess soils in the southern Netherlands, erosion risk becomes severe on slopes of > 8%, but remains a problem to some extent on slopes of > 2% (Eppink, 1986). These figures can be compared to those given by Tarariko *et al.* (1990) for the Obukhov district, Kiev region of Ukraine, where areas with slopes of > 7° were identified as being at particular risk of erosion, those of 3–7° were at moderate risk and those of < 3° were of least concern.

Exposure of bare soil increases erosion risk, whilst vegetation cover tends to reduce erosion. Sugar beet has been identified as a relatively ineffective cover crop in this respect, due to its slow canopy development during the first 2 months after sowing (Frielingshaus *et al.*, 1986, 1988; Schwertmann, 1986). Eppink (1986) also notes the significance of different crop cover on erosion risk in the southern Netherlands, suggesting that risk under sugar beet is intermediate to that under maize (higher) or cereals (lower). Crop combinations resulting in particular problems on the relatively erodible soils in parts of northern and western Europe have been identified as sugar beet/potatoes (Eppink, 1986) and sugar beet/cereals (Morgan, 1986).

In relation to rates of soil loss associated with erosion under arable systems including sugar beet, Eppink (1986) reports losses of 6.7 t/ha in an experimental area of the southern Netherlands between September 1983 and March 1984. De Ploey (1986) reports losses of a minimum of 3 t/ha/year on loess loams in Belgium, with 15 t/ha/year on recently cleared areas, and 10–100 t/ha/year on areas with steeper slopes. Under sugar beet specifically, Morgan (1986) suggests losses of 0.3–2.7 kg/m^2 in Belgium, and Graf *et al.* (1983) report losses of 1.57 t/ha in Wyoming (USA). Defra (2002) provides no figures for rates of water erosion under beet cultivation in the UK, but notes that soil losses of this kind are much smaller than those that occur with harvested beet.

Soil loss at harvest

The removal of soil from fields on mechanically harvested sugar beet can represent an important factor in soil degradation (e.g. Poesen *et al.*, 2001; Oztas *et al.*, 2002). Essentially, it is a consequence of the shape of the sugar beet root. Elliott and Weston (1993) estimated the proportion of soil in material delivered to the factory (soil tare) typically to be 10–30%, noting that the industry in the EU had to separate around 3 Mt of soil from harvested beet each year, at a cost of some £40 million. Poesen *et al.* (2001) considered soil losses at harvest for two commonly grown root crops in Belgium (chicory and sugar beet), and determined that soil loss was 9.1 t/ha/harvest for sugar beet. Assuming that root crops were grown every other year in the study area, mean annual soil losses would be 5 t/ha/year (0.33 mm/year). Based on these figures, the authors further estimated that, as these root crops had been grown in Belgium for at least 200 years, some 66 mm of soil losses could have accumulated. However, soil texture, soil moisture at harvest time and harvesting technique all influenced the amount of soil removed. In Turkey, Oztas *et al.* (2002) estimated that annual soil losses on harvested sugar beet were approximately 30,000 t in Erzurum and 1.2 Mt for the whole country, resulting in N, P and K losses worth approximately US$60,000 annually for the study area.

Defra (2002) found that removal of soil at harvest was the most significant impact on soils under beet cultivation in the UK, resulting in 350,000 t soil removed from fields each year. Although substantial, this represented the lowest soil tare (6.5%) in Europe (see Table 6.4), and all soil thus removed was returned to agricultural land or used in other applications. However, it was noted that returns to agricultural land had to be undertaken in such a way as to avoid undesirable impacts, such as excess nitrate leaching. Progress in reducing sugar beet soil tare has also been made elsewhere in Europe in recent years. Eeckhaut (2001), for example, estimates that soil tare was reduced from around 18.5% to 10.5% between 1980 and 2000 in Belgium.

Table 6.4. Soil tare rates for European beet cultivation (data from Defra, 2002).

Country	Soil tare	Country	Soil tare
France	20%	Portugal	8%
Hungary	15%	Austria	7%
Finland	13%	Denmark	7%
Belgium	12%	Spain	7%
Netherlands	11%	Sweden	7%
Italy	10%	UK	6.5%
Ireland	9%		

Measures to reduce impacts of beet cultivation on soil quantity

Wind erosion

Methods to combat wind erosion on land where sugar beet is cultivated include reduced tillage and winter cover crops (see Chapter 2), and the following:

- Establishment of cover intercrops (Hagen, 1974; Selman, 1976; Bastow *et al.*, 1978; Lumkes, 1981; Cherry, 1983).
- Exploitation of residues from the previous crop (Hagen, 1974; Simmons and Dotzenko, 1975).
- Application of farmyard manure (Hagen, 1974; Bakewell, 1980).
- Application of sugar factory waste lime (Hagen, 1974; Pickwell, 1974; Bastow *et al.*, 1978; Bakewell, 1980).
- Application of vinyl/bituminous emulsions (Hagen, 1974; Pickwell, 1974; Neururer, 1984).
- Creation of ridges in beet inter-rows (Styles, 1973; Hagen, 1974).
- Shelterbelts (Dzhodzhov and Georgiev, 1980).

In wind tunnel experiments, Knottnerus (1976b) studied a particularly wide range of measures, including methods of decreasing surface wind speeds with shelter belts, strip cropping, screens, straw trusses and rough surfaces, and of decreasing erodibility with manures, mulches, town refuse composts, crust-forming agents and artificial sticking agents. Measures to reduce wind erosion problems are also reviewed by Matthews (1983) and Henriksson and Hakansson (1993). Because wind erosion results in damage to sugar beet seedlings as well as soil loss, preventative methods can increase yields. For example, Styles (1973) saw root yields increased from 8–10 to 15–17 tons/acre when ridges were created between rows of beet to reduce wind erosion. Dzhodzhov and Georgiev (1980) found that a single-row shelterbelt of coppiced *Robinia pseudoacacia*

(9 m high, 2 m wide) significantly reduced wind erosion during a dust storm, and estimated that beet yield on the protected area was 32% greater than that obtained by resowing unprotected areas after the storm.

Water erosion/runoff

A range of measures is available for reducing soil erosion and runoff in sugar beet cropping systems. Aspects of modified tillage and the use of cover crops are discussed in Chapter 2. Other measures include:

- Contour cultivation (Zaslawskij, 1982; De Ploey, 1986; Tarariko *et al.*, 1990).
- Contour ditching (Chisci and Boschi, 1988).
- Terracing (Zaslawskij, 1982).
- Live barriers/contour strips (Zaslawskij, 1982; Schmidtlein *et al.*, 1987).
- Changes to crop rotation (Boschi and Chisci, 1978; Boschi *et al.*, 1985; Chisci and Boschi, 1988).
- Loosening of vehicle tracks (Schmidtlein *et al.*, 1987).
- Application of bituminous emulsion (Neururer, 1984).

Other papers which evaluate erosion control measures include Heissenhuber and Schmidtlein (1988) and Brunotte (1990). De Ploey (1986) recommends substantial changes in land use (reforestation and switches from arable to pasture) over significant areas in northern and western Europe, and also favours contour and reduced tillage, while dismissing the effectiveness of other potential measures, such as strip planting and the grassing of waterways.

Because slope gradient is such an important factor in determining erosion risk, various authors have commented on soil conservation measures and land uses appropriate for fields with different slopes. De Ploey (1986) discusses options for reducing erosion in the loamy loess areas of northern and western Europe, and suggests that (for comprehensive soil conservation) hundreds of thousands of hectares with slopes > 4–5% should be reforested, and contour tillage should be adopted on slopes of < 4%, in addition to other measures noted above. As part of a strategy to reduce soil erosion problems in the Obukhov district (Ukraine), Tarariko *et al.* (1990) divided cultivated areas into three groups, suggesting that those with gradients > 7° should be used for grass leys only, and those of 3–7° should not support row crops. Only areas with gradients of < 3° were considered suitable for a wide range of intensively grown crops. Tarariko *et al.* (1990) note that, whilst these and other soil conservation measures brought soil erosion under control, financing and the lack of a unified land service for coordination of the project were obstacles to its success.

A series of papers from Italy (Boschi and Chisci, 1978; Boschi *et al.*, 1985; Chisci and Boschi, 1988) examines the influence of crop type and other factors on soil erosion and runoff in sugar beet cultivation systems. Results indicate that the use of lucerne (*Medicago*) ley in the rotation has benefits. Erosion under lucerne was much less (3 t/ha/year versus 9 t/ha/year) than under arable crops (maize, sugar beet, wheat). Runoff was less than under arable crops, peak runoff discharge was retarded, runoff velocity was reduced and water infiltration was increased.

Soil loss at harvest

Elliott and Weston (1993) note that, in addition to significant (e.g. transport) cost savings, the development of low-tare beet varieties would speed up harvesting, reduce the problem of environmental degradation associated with soil losses from fields and reduce the water consumption of beet factories. Selective breeding for root shapes which carry less soil when harvested is one option for reducing soil tare. There are suggestions that new varieties could be developed that reduce soil tare by up to 50% whilst maintaining yield characteristics such as sugar content and juice purity (e.g. Olsen *et al.*, 2001).

Impact of Sugar Beet Cultivation on Soil Quality

Amongst the factors contributing to declining soil quality under arable systems including sugar beet in Europe are loss of organic matter content and compaction (Chisci and Morgan, 1986) and impacts associated with the use of fertilizers and pesticides. However, Winner (1993) suggests that the growing of beet as part of a rotation generally leaves the soil in relatively good condition for the following cereal crop.

Organic matter

Relative to the tropical soils in which cane is grown, beet growing soils in temperate regions typically display rather slow rates of mineralization of organic matter, as a consequence of lower temperatures. None the less, Draycott and Christenson (2003) note that most beet growing soils in Europe contain little organic matter (no more than 1–2.5%), having been under some form of cultivation for hundreds of years. Morgan (1986) reports a loss of organic matter content from soils under arable systems including sugar beet in northern Europe (possibly related to the removal of grass leys from the farming system). As organic matter content affects soil physical properties, this has made the soil more prone to structural breakdown by raindrop impact, increasing erosion risk.

Soil pH

Processes that promote soil acidification (such as the use of ammonium fertilizers and winter leaching of Ca) do occur in agricultural systems where beet is grown (Brentrup *et al.*, 2001; Draycott and Christenson, 2003). However, beet is very sensitive to soil acidity, and application of lime (often that recovered from beet processing) is commonly used to ensure a close to neutral soil pH. In order to obtain an appropriate pH, beet growing soils should be tested well in advance of sowing, and more than one lime application (followed by ploughing to ensure thorough incorporation with the soil) may be required (Draycott and Christenson, 2003).

Changes in soil nutrient levels

Again, because beet is typically only one crop in a wider rotation, soil nutrient dynamics in beet fields are complex. Among other factors associated with beet growing specifically, erosion can contribute to nutrient losses (Oztas *et al.*, 2002), and the practice of leaving beet tops in the field after harvest can substantially increase levels of residual N and P in the soil (Neeteson and Ehlert, 1988). In some areas (e.g. parts of the USA) where timing and quantity of rainfall do not promote leaching, nitrate has been shown to accumulate in soils to such an extent that beet quality factors are impaired (Draycott and Christenson, 2003). Nitrate levels can also be elevated in the lower horizons of beet growing soils where leaching is occurring if inorganic or organic fertilizer N inputs are sufficient (Isermann, 1989; Milosevic *et al.*, 1989). However, such patterns are contrary to the usual soil N dynamics under sugar beet, provided that inputs are kept to reasonable levels. Owing to its deep, spreading root system, the plant is a very efficient scavenger of soil nutrients. For maximum production, a beet crop generally requires twice the amount of N that is available naturally in most arable soils, which is around 100 kg/ha for the growing season, so fertilizer inputs can be important for maintaining soil N levels. In contrast, soil P concentrations tend to have increased markedly in beet growing soils after long-term cultivation. This has been recognized and has led to a reduction in P fertilizer inputs in many areas (Draycott and Christenson, 2003).

Soil salinization and sodification

Like other members of the plant family *Chenopodiaceae*, sugar beet is a halophyte, and therefore relatively tolerant of saline conditions in the main gowth stage (Elliott and Weston, 1993; Jarkovsky, 1997). It is able to utilize soil Na, which it can partially substitute for K, and may remove sufficient from some soils (e.g. in western Europe) for Na additions in fertilizers to be required

(Draycott and Christenson, 2003). Whilst sugarcane growth can be severely affected by salinity of 4–6 dS/m (SASA, 2002), beet yields are unaffected by soil salinity up to 7 dS/m (Draycott and Christenson, 2003). Dunham (1993) suggests that, amongst common field crops, only cotton and barley are more salt tolerant.

None the less, saline and sodic soils can be a problem in relatively arid beet growing areas, for example, in parts of Albania (Garo, 1994), Greece (Floras and Sgouras, 2002) and the USA (Kaffka *et al.*, 2002). As with saline and sodic-cane growing soils, remedial measures include addition of gypsum and improved drainage to leach excess salts. Tregaskis and Prathapar (1994) review the effects of salinity on sugar beet (and other crops) within the Murray Basin, Australia. Salt stress does increase the level of impurities in beet juice, and beet is very sensitive to salinity in the germination stage, 6–12 dS/m reducing emergence by half, with only tomatoes and onions showing equivalent sensitivity (Dunham, 1993). However, as sugar beet is typically grown in rotation with other crops, their salt tolerances will be equally important in determining fitness of land for agriculture, including beet.

Compaction

Morgan (1986) reports that compaction has occurred in soils under arable systems including sugar beet in northern Europe, particularly as a consequence of the use of heavy machinery. This has led to reduced infiltration rates and increased water erosion (particularly along tractor wheelings, where runoff rates have increased). Compaction is also noted as a problem associated with cultivation systems including sugar beet by Sommer and Lindstrom (1987), Sommer (1989), Brunotte (1990), Sommer *et al.* (1995) and Birkas (2001). Whilst acknowledging improvements made in the UK sugar industry, Defra (2002) considered that soil compaction remained an 'at risk' aspect of beet cultivation.

Tillage systems influence soil physical characteristics (e.g. Gemtos and Cavalaris, 2001), and Henriksson and Hakansson (1993) note that the various field operations conducted for seedbed preparation and sowing of sugar beet can contribute to soil compaction problems. Up to nine operations may be carried out, often in a random traffic pattern, leading to a total track area (for tractors and other vehicles) of three times the area of the field. Degree of compaction varies according to soil type (and particularly soil moisture), the distribution and number of passes by vehicle wheels, the loads on those wheels and wheel characteristics such as tyre pressures. Morgan (1986) notes that the first pass of a vehicle generally has the greatest impact, although repeated passes lead to longer-term problems at greater depths. Cooke and Scott (1993) note that seedbed preparation techniques for beet cultivation methods in eastern Europe, which may involve two or three ploughings and subsequent operations with ineffective implements, have often resulted in over-compacted soils. The work of Pidgeon and Soane (1978) also emphasizes the importance of tillage in determining risks of compaction, showing compaction to a depth of 300 mm when harvester wheels passed over soils previously deep ploughed, to 180 mm following mouldboard ploughing, 150 mm with chisel ploughing, and 60 mm with no tillage.

In addition to seedbed preparation, Henriksson and Hakansson (1993) note that sugar beet harvesting imposes unusually high risks of compaction. Harvesting is often carried out in relatively wet conditions (when compaction risk is greatest) using relatively heavy vehicles. For small-grain cereal production, annual traffic intensity has been estimated at 150–200 t km/ha, but in sugar beet production this load can be inflicted during harvest alone, with total annual traffic loads being twice as high. Axle loads of greater than 6 t can cause compaction deeper than 40 cm, which can persist for decades (and may become permanent). This can impair productivity, and it has been estimated that (when operating heavy machinery on wet soils) compaction costs can exceed the costs of machinery and labour. Tobias *et al.* (2001) studied compaction of wet and dry soil under excavators and sugar beet harvesters, estimating that caterpillar track loads of excavators varied from 13 to 19 t and wheel loads of harvesters from 6 to 11 t. On dry soil, excavators did not cause significant plastic deformation at 30 cm depth, but beet harvesters on wet soil led to soil sinkage of 1–2 cm, even at 60 cm depth. Increased wheel load in subsequent passes led to greater subsidence than during the first pass.

Impacts of pesticide inputs

There is some evidence from beet cultivating systems of the persistence of pesticides in soils (e.g. Fuhr, 1986; Wevers, 1997). In terms of environmental accumulation of pesticides, persistent agents are the most problematic. Fedorov (1999) reports on persistent organic chemicals in the former Soviet Union. A number of persistent organic chemicals (POCs) were found across the landscape, most notably dioxins, dioxin-like compounds, DDT and miscellaneous pesticides.

Impacts of irrigation

Irrigation of sugar beet crops can affect soil quality. In part this is related to the removal of nutrients by leaching (see Chapter 4), but irrigation can also affect soil physical properties such as infiltration rate, total porosity and bulk density (Abedi-Koupai *et al.*, 2001), although such effects are not necessarily negative (Artigao *et al.*, 2002). In addition, irrigation can affect the microorganism populations of sugar beet soils (Piotrowski *et al.*, 1996), with implications for disease agents, wider microorganismal biodiversity and soil health. Irrigation with beet-processing waste water can also influence soil quality. Paulsen *et al.* (1997) found that long-term irrigation with beet-processing waste water had a liming effect on soils. Drainage water showed raised concentrations of some alkali and alkaline earth metals, and chemical/biological oxygen demanding substances (COD and BOD compounds) were also translocated down to the drainage level.

Measures to reduce impacts of sugar beet cultivation on soil quality

A range of measures has been considered in relation to reducing impacts on soil quality of cultivation systems including sugar beet. Aspects of the rational use of agrochemicals, modified tillage, mulching and appropriate harvest operations are discussed in Chapter 2. Draycott and Christenson (2003) also note the development of a range of synthetic soil conditioners which may have benefits in improving the structure of beet growing soils.

7

Atmospheric Impacts

Carbon dioxide is emitted from agricultural soils as a consequence of the breakdown of organic matter, but is also fixed by the photosynthetic activity of crop plants. A few studies have attempted to address aspects of the carbon dioxide dynamics of sugar production, particularly from cane. This is seen as an important issue, given the high-profile status of carbon dioxide as a greenhouse gas. However, complete carbon dioxide budgets for sugar production are complex, involving emissions from processing operations (such as the burning of bagasse) as well as the balance between soil emissions, other sources related to cultivation (from preharvest cane burning, vehicles and machinery, for example) and the carbon dioxide assimilated by the crop.

Other emissions from agricultural soils include those of nitrogenous compounds, which are exacerbated by the use of fertilizers. Nitrous oxide may be released through microbially mediated nitrification of ammonium or denitrification of nitrate. In relatively wet environments, the latter is the predominant process (Keating *et al.*, 1997), although very wet (waterlogged) conditions reduce N_2O emissions by encouraging conversion of N_2O to N_2 (Smith and Arah, 1990). Nitrous oxide is 350 times more potent than carbon dioxide as a greenhouse gas, and 90% of global emissions of this gas are thought to be derived from soils (Byrnes, 1990). Seventy per cent of global nitrous oxide emissions have been linked specifically to agricultural land (Mosier, 1994). In addition, ammonia may be released through volatilization, contributing to acidification and unwanted enrichment of nutrient-limited natural ecosystems when returned from the atmosphere by wet or dry deposition (Christensen, 2004).

In addition to emissions from soils, cultivation of sugarcane can result in air pollution where the crop is burnt prior to harvesting, releasing carbon dioxide, nitrous oxide and (in the smouldering phase) methane. Cane burning also has a range of knock-on effects, notably negative impacts on soils. These, and air pollution impacts, can be eliminated by the adoption of green cane harvesting and trash blanketing, which has a range of other beneficial effects (see Box 2.4).

Aerial application of agrochemicals also has the potential to generate air pollution and associated health risks. However, published data from cane and beet cultivation systems are apparently lacking.

The processing of sugar crops results in a range of air pollution effects, principally the release of gases and particulates (particularly from power plants) and unpleasant odours and gases (including H_2S) from effluents and other waste materials. Forms of anaerobic fermentation are often used to treat effluents rich in organic material such as cellulose, giving rise to methane and carbon dioxide.

SUGARCANE CULTIVATION

Aerial pollution from cane cultivation derives mainly from two sources – emissions from soils and preharvest cane burning. Comprehensive assessments of the impact of cane cultivation on atmospheric composition are rare, but Weier (1998) attempted such an assessment for Australia, focusing on greenhouse gas dynamics. The results are summarized here in Box 7.1.

Emissions from Soils under Sugarcane Cultivation

Concern over nitrogenous emissions associated with fertilizer use have been expressed in a number of cane growing areas, including Mauritius (Ng Kee Kwong *et al.*, 1996, 1999a) and Australia (e.g. Keating *et al.*, 1997). Hartemink (2003) notes that many studies of the N balance in cane cultivation show an appreciable fraction of fertilizer N unaccounted for, and it has often been assumed that gaseous emissions through denitrification (releasing N_2O and N_2) and volatilization (releasing NH_3) are the main components of this. However, it has been shown that significant quantities of gaseous N can be lost from the plant itself (Ng Kee Kwong and Deville, 1994b). Such experiments suggest that previously assumed levels of denitrification and volatilization may have been greatly overestimated (Hartemink, 2003). Weier (1998) also urges caution in the interpretation of indirect estimates of nitrogenous emissions from cane cultivation, which have often suggested that denitrification (or N_2O emissions specifically) might account for 25–60% of N losses from the system (e.g. Balasubramanian and Kanehiro, 1976; Chapman *et al.*, 1994; Vallis *et al.*, 1996b).

Weier *et al.* (1996) examined the potential for biological denitrification of fertilizer N in soils under sugarcane in Australia. In glasshouse studies, denitrification losses accounted for 13–39% of N applied, and most of the loss ocurred as N_2. In the field, denitrification

Box 7.1. The greenhouse gas dynamics of sugarcane cultivation in Australia.

The following are the results of an assessment by Weier (1998), who urges caution over possible inaccuracies and lack of precision in estimates of this kind, based on extrapolation from smaller-scale studies which are themselves limited by the reliability of techniques available to researchers. Estimates were made for the total area under cane cultivation in Australia, in relation to carbon dioxide (CO_2), nitrous oxide (N_2O) and methane (CH_4) data from 1994.

CO_2 emissions arose from uncropped cane fields, either left bare (1058 kt C/year) or trash blanketed (1331 kt C/year). Mean emission rates from trash blanketed fields were nearly double those from bare soils (23.4 versus 13.0 kg/ha), probably as a result of differences in soil moisture and temperature. A further 5.2 Mt C/year was released as CO_2 through preharvest cane burning, giving a total of 7.6 Mt C/year from cane cultivation (around 1.5% of Australia's total annual CO_2 emissions, and about half of the emissions derived from cultivated land). The assimilation of CO_2 by sugarcane was estimated to be 13.4 Mt C/year (suggesting a net removal of 5.8 Mt C/year from the atmosphere), but around 50% of this would be expected to be released again by the growing crop as CO_2 through respiration. Unfortunately, the assessment does not go on to estimate the quantity of the remaining sequestered carbon that was subsequently released as CO_2 in the burning of bagasse or other processing and by-product utilization operations.

Denitrification in soils following inorganic fertilizer application led to N_2O emissions of 4.4 kt N/year from cane fields. In addition, preharvest burning contributed 6.3 kt N/year as N_2O. Hence, total N_2O emissions from cane cultivation were 10.7 kt N/year, around 1% of Australia's total annual emissions of this potent greenhouse gas (and around 18% of that derived from croplands).

The smouldering phase of preharvest burning resulted in CH_4 emissions of 6.7 kt C/year. This is a very small proportion of Australia's total annual CH_4 emissions, which were estimated at 6.5 Mt C, to which domestic livestock made the greatest single contribution. However, cane cultivation was also found to contribute to the removal of CH_4 from the atmosphere, through trash blanketing (which could result in the CH_4 equivalent to 15–105 kt C/year being absorbed) and CH_4 oxidation in soils.

losses of around 1–20% of applied N were estimated, with considerable differences according to soil type. It was concluded that denitrification was an important process in the loss of fertilizer N from fine-textured soils, with N_2O making up 45–78% of the gaseous product (depending on soil water content and nitrate concentration). Weier (1996) found greater N_2O emissions where nitrate-based (as opposed to urea-based) fertilizers were used, and somewhat increased rates of N_2O emissions from trash blanketed versus bare soils where urea and water were applied. Hartemink (2003) observes that the Australian research summarized here did not specifically take into account the gaseous N losses from the cane plant itself, noted above from the work of Ng Kee Kwong and Deville (1994b).

Other studies of denitrification emissions include that by Ng Kee Kwong *et al.* (1999a) in Mauritius, which estimated N_2O releases of 16–20 kg N/ha over one growing season. Soil emissions of N_2O and N_2 were monitored at four sites, with or without fertilizer application at 140 kg N/ha. Despite between-site variabilities, emissions invariably showed a base level rate of 30–50 g N/ha/day, with a pulse of two to three times this level during the rainy season (December to April). When fertilizer was applied, large emission pulses (300 g N/ha/day) were recorded at one site, while at the other three sites applications hardly affected emission rates. The former was also the only site where gaseous losses of fertilizer N were of agronomic significance: 16% of the 140 kg N/ha applied escaped to the atmosphere over the 1996/97 growing season, compared to less than 3% at the other study sites. This was also the only site where N_2 production was significant. Here, 80% of the soil pore space was water-filled for prolonged periods and, although emission rates could not be predicted by any single soil variable, soil wetness was the most important factor controlling both the size of the flux of N_2O and N_2 and the $N_2O : N_2$ ratio.

In relation to volatilization, Hartemink (2003) notes that studies in Australia have shown relatively high NH_3 emission rates associated with trash blanketed cane fields receiving urea fertilizer in relatively dry conditions (e.g. Freney *et al.*, 1994). Trash, like other plant material, contains the enzyme urease, which breaks down urea to CO_2 and NH_3. In relatively dry conditions (with small amounts of moisture provided by dew, condensation and light rain), slow but steady NH_3 emissions were found to represent some 20–40% of the N applied in urea fertilizer. Heavy rainfall tended to wash urea from the trash, reducing NH_3 losses to 17% of applied N. Trivelin *et al.* (2002) quantified differences in the fate of N from urea fertilizers in Brazil, according to method of application. Under ratoon cane, surface application led to a 50% loss of N, from volatilization in the soil or via the plant, but this was reduced to 19% (all lost via the plant) when urea was applied at a depth of 15 cm.

The particular practice, in Guyana, of flood fallowing appears to promote the release of gases such as N_2O, H_2S, NH_3 and CH_4 from cane growing soils (see Box 6.4).

Preharvest Cane Burning

In Australia, preharvest burning was adopted from about the 1930s, to combat the spread of Weil's disease amongst cane cutters (Garside *et al.*, 1997b). In other cane cultivation areas it may be considered a 'traditional' form of harvest management, often (anecdotally) justified as a means of clearing snakes from the field before hand cutting. Whether or not preharvest burning is practised may be most significantly influenced by the cost of labour and size of area cultivated (larger estates may be more inclined to burn); there is also a more established history of burning in countries such as Brazil, Australia and Hawaii than, for example, in Mauritius, India, the Philippines, Thailand and Taiwan (Payne, 1991). Clearly, preharvest burning will not be appropriate where the intention is to maximize the collection of cane biomass for use as a fuel material; up to half of the fibre in the crop may be found in tops and trash (Payne, 1991). Preharvest burning is now declining in many cane growing areas, as the potential benefits of green cane harvesting and trash blanketing are increasingly appreciated (see Box 2.4).

UNEP (1982) noted that dust and particulate emissions may result from the burning of cane prior to harvest, but considered this to be a problem only for short periods, and one which should not produce deleterious effects if carried out under proper controls. However, other authors consider preharvest burning to be severely polluting (e.g. Szmrecsanyi, 1994; Ripoli *et al.*, 2000). Kirchhoff *et al.* (1991) found substantially elevated levels of carbon monoxide and ozone in the atmosphere around cane fields in São Paulo (Brazil) at the time of preharvest burning. Consistent with figures presented by Weier (1998) (see Box 7.1), Park *et al.* (2003) reported that preharvest burning was the greatest source of greenhouse gas emissions in Australian cane cultivation. Concern has also been expressed over the direct impacts of cane burning on human health, although the extent of such impacts has been questioned (e.g. Whalen, 1989). The practice certainly causes nuisance to local communities, such as ash fall-out (e.g. Cock *et al.*, 1999). In Brazil, dos Santos *et al.* (2002) analysed selected atmospheric pollutants in and around the city of Campos dos Goytacazes, and concluded that vehicular exhausts and the burning of cane trash/bagasse were the two major sources of pollution in the area. Sugarcane burning was not the main source of toxic compounds, such as polycyclic aromatic hydrocarbons (PAHs, like benzo(*a*)pyrene), which were derived principally from vehicle exhausts. However, it did result in elevated levels of other compounds, including laevoglucosan, particularly in areas surrounded by cane plantations. There is also evidence that cane burning can significantly influence rainwater quality (e.g. Lara *et al.*, 2001), and Patrick *et al.* (1994) found that it was a minor (2%) contributor to atmospheric mercury pollution in the Everglades.

There are also indications that cane burning can contribute to a decline in soil quality (see Chapter 6) and potentially reduce sugar recovery (Boonthum *et al.*, 1997) and quality (Narayan, 1999 – who observed that the extent of cane burning in Fiji had increased from 19 to 62% between 1968 and 1997, contributing to massive inefficiency in the industry).

Reducing atmospheric impacts of sugarcane cultivation

Nitrogenous emissions, in particular, from soils under sugarcane cultivation, can be reduced by improving fertilizer management practices (Weier, 1998). Control of soil nitrate levels may be particularly beneficial in reducing N_2O emissions due to denitrification. General aspects of rational fertilizer use are discussed in Chapter 2.

The impacts of preharvest burning can be eliminated by a switch to green cane harvesting and trash blanketing (see Box 2.4). Although this has a wide range of environmental benefits, it can contribute to enhanced emissions of NH_3 under certain circumstances (see above). If cane is burnt prior to harvest, a number of straightforward measures can be taken to reduce negative environmental impacts. Measures should be taken to ensure that the fire does not spread uncontrollably (e.g. by ensuring that suitable fire breaks are in place and by carefully considering wind direction before ignition). Wind speed and direction should also be checked to ensure that smoke and smuts are not blown into sensitive areas (such as residential or industrial areas, roads, parks and the routes of power lines) from further afield (SASA, 2002).

In some areas, specific regulations exist that limit cane burning activities. For example, SASA (2002) notes various legislative instruments which apply to cane burning in South Africa (The Forest Act (1984), the Road Traffic Act No. 29 (1996) and the Atmospheric Pollution Prevention Act No. 45 (1965)), and Rozeff (1998b) discusses regulation of cane burning in the USA. As well as formal legislation, voluntary codes of practice have been developed to limit the negative impacts of cane burning, as in South Africa (SASA, 2002).

CANE PROCESSING

Amitabh *et al.* (1999) report that, along with waste water, emissions from boiler stacks are the main sources of environmental pollution from cane sugar factories. Bagasse is routinely used to fire boilers (see Chapter 8), but supplementary fuels are commonly used for boiler start-up, or when bagasse supply is limited (Payne, 1991). Wood and oil are frequently used as supplementary fuels, but coal may also be used in some areas, and particularly where co-generation (see Box 8.2) is an objective. In general, UNEP (1982) noted that mills should be sited downwind of populated centres, to minimize nuisance from gaseous emissions. As awareness of the environmental and health threats posed by air pollution has increased, regulation of emissions from sugar mills has become more rigorous, and this trend is likely to continue (McBurney and McBurney, 1997; Kroes and Dixon, 1998; Lora and Jativa, 1999).

Table 7.1. Examples of particulate emission standards for bagasse boilers.

Jurisdiction	Standard (mg/Nm^3)
Australia (1991)	450
Australia (1990, for new boilers) [a]	250
South Africa (1991)	450
South Africa (1996, for new boilers)	120
Hawaii (1991)	870
India (1992) (inclined grate boilers)	250
India (1992) (spreader-stocker boilers)	800
Mauritius (1999)	400
Malaysia (1999)	400
World Bank (1997) (recommendations for project funding)	100–150

[a]Joyce and Dixon (1999); all other data from Lora and Jativa (1999).

Particulates

The main concern over particulate emissions from cane processing relates to those released in the burning of bagasse to fuel boilers. Other particulate (dust) pollution may arise as a consequence of the handling of bagasse (e.g. Payne, 1991), lime and coal (where this is used as a fuel source) and as a result of heavy transport traffic (UNEP, 1982). Lora and Jativa (1999) suggest that untreated exhaust gases from typical, commercial, bagasse-fired boilers carry particulate emissions of around 3000–5000 mg/Nm^3, although higher concentrations are quoted by Chang and Lee (1989, 1991), at 6000–10,000 mg/Nm^3, and 8000–12,000 mg/Nm^3 respectively. Examples of emission standards for particulates from bagasse boilers are given in Table 7.1 for comparison.

Hindy (1990) reported that the sugar factory (situated close to residential areas) was the only major source of air pollution in Abu Qurqas (Egypt). Particulate samples were collected from the tallest building in the area (approximately 20 m high and 50 m downwind of the factory), in a period when the factory operated at full capacity and no rain fell. Particle size distribution and heavy metal content of the bulked sample were measured. Mass deposition rate, for particulates and metals, increased with decreasing particle size, and 75% of particulates were < 20 μm. Metals consistently detected were (in $mg/m^2/day$): Fe (74), Pb (15), Zn (12), Cu (5), Cr (4), Mn (2.5), Ni (1.2) and Cd (0.9). While Fe and Mn may have originated from soils via bagasse, the others may have been derived from fuel oil. The Pb deposition rate exceeded those recorded in urban areas and industrial centres.

Odours

Odour problems associated with sugar mill effluents are reported by Yang *et al.* (1991) in Hawaii and by Sinha (1993) in India. Odours may also arise from drying operations and storage areas where deterioration of cane residues and other organic matter occurs (UNEP, 1982).

Gases

The burning of bagasse to fuel boilers results in CO_2 emissions, but these should be viewed in the context of the overall carbon balance of the sugarcane bioenergy system, and compared with emissions derived from non-renewable fossil fuels (Beeharry, 2001).

Boilers often also emit NO_x and SO_x (e.g. Chang and Lee, 1991), although releases may be within statutory limits (Kroes and Dixon, 1998). UNEP (1982) suggests that SO_2 may be a particular concern where oil is used as a fuel source. In contrast, one of the environmental benefits of burning bagasse in cane factory boilers is the relatively low sulphur content of this fuel material (Payne, 1991). Although their concentrations may be expected to be relatively small (Shukla, 1995), boiler fire emissions such as NO_x and PAHs need to be considered in environmental assessments (Lora and Jativa, 1999). In addition to potential emissions from boilers, sulphitation may be used in the processing of cane sugar, potentially resulting in the release of SO_2.

Other Issues

Chang and Lee (1991) note that high (90%) relative humidity in boiler exhaust gas can cause an opaque white plume, which could infringe pollution control standards. Heat exchangers can be used to avoid this, using the waste heat of incoming flue gas to reheat the exhaust gas, decreasing relative humidity and opacity.

Reducing Air Pollution arising from Sugarcane Processing

Because the nature of the combustion process itself influences emissions, the basic design and operation of cane mill boilers may be important in controlling the release of atmospheric pollutants. Aspects of boiler design in this context are considered by Silva Lora and Olivares Gomez (1995) and Basu and Talukdar (1997).

Particulates

Drying of bagasse prior to its use as boiler fuel can increase burning efficiency (see Chapter 8) and reduce emissions. Paz *et al.* (2001) found that bagasse-drying resulted in a substantial reduction in exhaust gas particulates, from 4500 mg/m^3 to < 300 mg/m^3. Boiler flue gases can themselves be used in the bagasse drying process, further improving cane factory efficiency whilst enhancing environmental protection (Avram-Waganoff, 1990). This, for example, is the basis of the first of three main types of bagasse driers described by Paturau (1989), the cylindrical rotating drier; the other designs are the pneumatic drier and the Exergy drier. The issue of bagasse drying using boiler flue gases is considered in detail by Payne (1991).

Lora and Jativa (1999) conclude that regulatory standards on particulate emissions of < 120 mg/Nm^3 cannot be attained without the use of control devices. They compare those used in the cane sugar industry to reduce boiler particulate emissions: (multi)cyclones; wet scrubbers (perforated tray, spray tower or venturi scrubber); and electrostatic precipitators, as well as a new device – the core separator. They consider that the spray tower is the most widely used device, although this (along with other wet scrubbers) has the disadvantage of generating waste water effluents that require treatment. However, suitable treatment can allow the water to be recycled back to the scrubber (e.g. Chang and Lee, 1991). Particulate emission control devices used in the cane sugar industry are also reviewed by Chang and Lee (1989, 1991), Marie-Jeanne (1993), Saechu (1995), McBurney and McBurney (1997) and Joyce and Dixon (1999).

Calero *et al.* (2001) analysed the dispersion of particulates from a Colombian cane mill, and concluded that modifying the combustion process and adopting emission control systems could reduce particulate emissions by around 98%.

Odours

UNEP (1982) suggests that much can be done to reduce odours from treatment systems by screening to remove organic solids (reducing the load on anaerobic or aerobic holding ponds and lagoons). In general, well-designed lagoons and avoidance of overloading can reduce the threat of odour problems, as can the adoption of alternative treatment systems.

Gases

Emission control devices for boiler flues are discussed above. Liu and Ho (1989) reported that sulphitation of cane juice under vacuum, rather than under pressure, reduced air pollution due to leakage of SO_2. Jadhav *et al.* (1990) reported that removal of unwanted colour using hydrogen peroxide in place of SO_2 resulted in a higher quality sugar product, reduced air pollution and required no new equipment.

SUGAR BEET CULTIVATION

Emissions from Soils under Sugar Beet Cultivation

There is an increasing interest in greenhouse gas dynamics associated with agriculture (including beet cultivation) in Europe, partly inspired by the Kyoto Protocol (e.g. in Austria – Dersch and Bohm, 2001). However, detailed analyses of the greenhouse gas dynamics of beet cultivation are not readily available. Fuchs *et al.* (1995) estimated CO_2 emissions from fields of beet in Germany to be greater than those supporting winter wheat and winter barley crops. Draycott and Christenson (2003) estimated nitrogenous emissions from soils under sugar beet cultivation to be relatively small in most cases, suggesting an average loss of 10 kg N/ha (see Fig. 2.5). Risks of nitrogenous emissions through denitrification were predicted to be greatest where large amounts of organic manure are used and on soils with poor drainage characteristics. Stoeven *et al.* (2002) examined N_2O emissions from soils under crop rotations including sugar beet in Germany, but found no relationship with crop species. Draycott and Christenson (2003) suggest that the risk of NH_3 volatilization is likely to be greatest where fertilizers are applied in the form of urea or organic manures, particularly when application is to the soil surface without incorporation into the soil (see also Jacobs, 1995; Chambers and Smith, 1999). Sotiriou and Scheunert (1994) found that N applied to soil as urea (under a rotation including sugar beet) was more likely to be lost through volatilization and denitrification than through leaching. This loss was greater from sandy loam than from loamy sand, and was slightly increased by incorporation of straw into the soil, but decreased by application of monolinuron herbicide.

SUGAR BEET PROCESSING

As in relation to cane, the principal air pollutants arising from the processing of sugar beet are emissions from power plants, other gases and odours.

Air Pollution from Power Plants

Power units (boilers) used to provide energy for beet factories can represent sources of particulates and polluting gases including oxides of carbon, nitrogen and sulphur, invariably as a result of the burning of fossil fuels (e.g. Evertz, 1991; Urbaniec, 1996; Zajac, 2000). UNEP (1982) suggests that SO_2 emissions may be a particular concern where oil is used as a fuel source, although Brejski (1998) suggests that, with effective systems, SO_2 emissions can be decreased by switching from coal-fired boilers to natural gas or oil. Polec and Kempnerska-Omielczenko (1995) reviewed air pollution from Polish sugar beet factories in 1993/94. Few had installations for flue gas desulphurization, and it was estimated that limits set for SO_2 and NO_x emissions were infringed in 70% and 13% of cases, respectively.

Odours and Gases from Processing, Effluents and Waste Water Treatment

Odours at beet sugar factories, ranging from musty smells to hazardous gases, are a nuisance to the factory personnel and to the community (Helge and Larson, 1993). Gryllus and Anyos (1993) note that NH_3 is a potential atmospheric pollutant from beet processing, generated from amides in beet juice. A range of odour problems and gaseous emissions are associated with factory wastes and their treatment, including releases of H_2S and methane, notably from holding ponds and lagoons (UNEP, 1982; Nibit *et al.*, 1994; Sarmento and Robbins, 2001).

Wolski (1993, 1995) studied emissions from an open fermentation chamber for effluent treatment in a Polish sugar beet factory, and found that the anaerobic digestion process generated 75–80% methane, 19–24% CO_2 and 1% others (ammonia, H_2S, mercaptans and water vapour). The methane concentration inside the fermenter was 3 g/m^3 (0.42% v/v), and was estimated to become zero at 140 m from the fermenter. Air sampled 30 cm from the sump contained 15.8 mg/m^3 ammonia, the concentration of which was estimated to reach zero approximately 271 m from the fermenter. The highest H_2S concentration, detected above the water being stored before fermentation, was 0.008 mg/m^3, well below the permitted limit of 0.03 around the fermenter.

Tomaszewska and Polec (1997) studied air quality (including CH_4, SO_2, H_2S, NH_3 and CO concentrations and abundance of microorganisms) associated with the waste management infrastructure of a sugar beet factory. The most important sources of gaseous pollution were reservoirs for accumulating waste waters, flume mud and carbonatation mud, near which H_2S concentrations were up to 400 μg/m^3 (compared with a regulatory limit of 30 μg/m^3), and NH_3 concentrations reached 2500 μg/m^3 (compared with a limit of 400 μg/m^3). Near fans for cooling condenser water, NH_3 concentrations were as high as 9–10 mg/m^3. Regulatory limits for CO (5 mg/m^3) and SO_2 (600 μg/m^3) were not exceeded. However, levels of SO_2 near the reservoir receiving flume mud and the closed fermentation chamber of the biological treatment system equalled or exceeded proposed new limits of 400 μg/m^3. Regulatory limits for total bacterial count were often exceeded, particularly near the closed fermentation chamber, the reservoir receiving carbonatation mud and the spray cooler of condenser water.

Other Sources of Atmospheric Pollution

Dust may represent a hazard in sugar beet factories. Polec and Kempnerska-Omielczenko (1995) reviewed air pollution from Polish sugar beet factories in 1993/94, and found that most factories had inefficient equipment for dust removal, such that limits set for the emission of dust were infringed in 63% of cases. Increasing regulation has sought to limit such sources of pollution (see Box 4.2); for example, Marciniak (2000) noted that all boilers at the Melno sugar beet factory had been equipped with dust filters.

Reducing Air Pollution arising from Sugar Beet Processing

Incentives for reducing air pollution from beet processing include the need to operate within legal limits for emissions (e.g. Urbaniec and Suchecki, 1994; Urbaniec, 1996; Tomaszewska and Polec, 1997; Marciniak, 2000; Zajac, 2000). A range of measures can be taken to reduce particular aspects of air pollution from beet processing (see below). Integrated systems may be adopted, such as that proposed by Friedemann (1992), where deammoniation and acidification of diffuser feed water are combined with removal of noxious gases (SO_2, NO, NO_2) and smuts from flue gas. Changes in processing methods may also have the potential to reduce air pollution and other environmental impacts. Klemes *et al.* (1999) argue that cooling crystallization (versus the traditional method of evaporating crystallization) has the potential to improve energy efficiency and reduce atmospheric emissions, such as ammonia and CO_2, water consumption and the polluting potential of effluents. Dust in sugar beet factories can be managed using appropriate ventilation systems (e.g. Evfimenko, 2001).

Equipment such as separators can be used to reduce particulate emissions from beet factory boiler flues (e.g. Zajac, 2000), and a range of equipment and techniques is available to reduce gaseous emissions (e.g. Evertz, 1991; Zajac, 2000). Urbaniec and Suchecki (1994) report on methods of desulphurizing boiler flue gases used in Polish sugar beet factories. Wet, semi-dry and dry methods typically yielded efficiencies of 90%, 70–90% and 40–50%, respectively.

A range of odour control measures are available for use in the management of sugar beet factory wastes. Removal of solid wastes from effluents can help to reduce odour problems in subsequent handling and treatment of these wastes. UNEP (1982) notes that mechanical clarifiers and settling ponds (following coarse screening) are generally used to remove as much soil and other solid wastes as possible. These need to be effectively operated and maintained, and waste retention times should be minimized to reduce the risk of fermentation and creation of odours. Clarifiers with retention times of 30 min to several hours can provide effective removal of solid wastes with minimal odour problems. The problems of fermentation and odours associated with handling of beet factory lime mud can be reduced by the use of shallow holding ponds and/or aeration. Alternatively, waste mud from beet washing can be dewatered using presses (such as the belt press, e.g. Kallstrom *et al.*, 2001), which can reduce air pollution associated with mud ponds and the costs of mud transport. Additives can also be used to reduce odour problems. UNEP (1982) notes that odour problems associated with holding ponds, where flume water is settled and clarified, can generally be controlled by the addition of lime and maintenance of pH above 10. Enzymes and organic scavengers can be used for the control of sulphides such as H_2S (e.g. Sarmento and Robbins, 2001). In the USA alone, the beet sugar industry annually spends at least $4.5 million on minimizing odour problems, using odour-masking agents, pH control reagents, antifoams, settling aids, bioaugmentation and other measures (Helge and Larson, 1993). Management of beet factory odour problems may also involve improved communications with local communities affected by the nuisance (e.g. Smith, 2001).

IMPACTS OF AIR POLLUTION ON CULTIVATION OF SUGAR CROPS

In addition to impacts of the sugar industry on air quality, the potential impact of air pollution and changing atmospheric conditions on cultivation of sugar crops has attracted attention from some authors. The following is by no means a comprehensive review of published information in this area, but gives some indication of the scope of available material.

Sugarcane

Climatic phenomena are an important consideration in the cultivation of sugarcane. Major phenomena like El Niño can have substantial impacts, sometimes with significant social and economic consequences (Marcus, 1992; Mafla Cifuentes, 1997; Singels and Bezuidenhout, 1999; Vos *et al.*, 1999; Palwarty *et al.*, 2001). Studies on specific aspects of the impacts of air pollution on the cultivation of sugarcane include assessments of the effects of SO_2 pollutants (Chen, 1984; Lin *et al.*, 1984) and enhanced concentrations of CO_2. Increasing CO_2 concentrations are anticipated to have direct effects on the development of crop plants such as sugarcane, possibly leading to increased growth and yield (Singh *et al.*, 1991), although weed species may show even greater enhanced development (Ziska and Bunce, 1997). However, the influence of increasing CO_2 concentrations also involves indirect effects through changing climate, which may result in major perturbations in natural and agricultural ecosystems, including those supporting sugarcane crops (Singh *et al.*, 1991; Rosenzweig and Hillel, 1993).

Sugar Beet

Studies on sugar beet include assessments of the effects of a range of atmospheric pollutants, such as NO, NO_2, SO_2, HF and O_3, and of air pollution in general (Bell, 1984; Fuhrer *et al.*, 1989), as well as more specific studies on the impacts of SO_2 (Maly, 1986), O_3 (Olszyk *et al.*, 1988; Naf, 1991) and O_3–herbicide interactions (Dixon *et al.*, 1995, 1996). Increased atmospheric concentration of CO_2 may enhance the growth and productivity of sugar beet (Dahlman *et al.*, 1985; Demmers-Derks *et al.*, 1998). However, wider impacts on beet cultivation associated with climate change under enhanced CO_2 concentrations may be substantially more far-reaching and harder to predict (El-Maayar *et al.*, 1997; Demmers-Derks *et al.*, 1998; Lorenzoni *et al.*, 2000). Draycott and Christenson (2003) observe that declining levels of sulphurous atmospheric pollutants in the UK have actually contributed to an increased need for S additions in fertilizers to provide for beet nutrition. Jaggard *et al.* (1998) suggest that predicted climate change appears likely to increase the severity of both drought and disease stresses on beet in the UK.

8

Use and Impacts of By-products

The disposal of large volumes of waste materials can be an expensive and environmentally threatening operation. However, if alternative uses can be found, disposal costs can be avoided and added economic value can be obtained from the conversion of wastes into by-products. The extensive utilization of wastes from the processing of sugar crops (particularly cane) as by-products is a notable, positive environmental aspect of the industry. However, environmental hazards can also arise from the handling and further processing of these by-products.

Particularly notable uses of sugar industry by-products, discussed in more detail below, include:

- Burning of bagasse as fuel to power cane processing and provide surplus electricity.
- Fermentation and distillation of sugar-rich wastes, notably molasses (many millions of tonnes of which, from cane and beet, are traded internationally each year).
- Use of various wastes as soil amendments, either by direct application or after composting (e.g. James and Hasibuan, 2002; papers by Madejon and co-workers).
- Use of wastes in livestock feed (e.g. Leng and Preston, 1986; Preston, 1993; Hernandez and Babbar, 2001), where recent advances in animal nutrition have greatly improved understanding of such by-product utilization.

Originally explored with some enthusiasm in the 1970s, there is also interest in the contributions that sugar crops and their by-products can make in the field of renewable energy sources, notably ethanol and biomass. Renewed attention has been paid to such strategies, in response to recent, increasing concerns over greenhouse gas emissions and the development of the Kyoto Protocol (e.g. Askew and Holmes, 2002; Atkinson, 2004).

The economics of by-product manufacture and utilization are heavily influenced by external factors (such as the availability of alternative materials). However, developments in various technologies (including genetic engineering, e.g. Cooke and Scott, 1993) promise increased opportunities for economically rewarding recovery and utilization of sugar industry by-products and alternative outputs. Also, integrated systems can enhance efficiency: it may be helpful, where possible, to combine on the same site (and under the same management) the manufacture of sugar with the production of one or more by-products (Paturau, 1989).

SUGARCANE

Sugarcane has well-explored potential for yielding useful by-products. The sugar industry in Cuba, for example, has at least 30 utilizable outputs other than sugar itself (Almazan, 1994). Indeed, it has sometimes been suggested that sugarcane has such rich potential for other uses that it should be grown for the specific purpose of obtaining materials other than sugar. Two notable examples, both initially driven to a large extent by the global oil crisis in the 1970s, relate to cultivation of energy cane (a holistic approach, recognizing the energy generating potential of many cane materials; see Box 8.1) and the cultivation of cane for the production of fuel alcohol (see below). Whilst most cane cultivation currently remains geared mainly towards sugar production, it is worth considering that future energy crises, perhaps again related to the cost and availability of fossil fuels, might catalyse changes in the purpose, dynamics and impacts of cane cultivation.

The use of sugar processing wastes as soil amendments is a particularly attractive proposition in some areas of cane cultivation, given the expense of inorganic fertilizers and shortage of organic manures (Ng Kee Kwong and Paul, 1997; Nasir and Quereshi, 1999b; Wynne and Meyer, 2002). It also has a particular role to play in the organic cultivation of sugar crops (Hallmark *et al.*, 1998).

Allen *et al.* (1997) predict an increasing emphasis on the utilization of by-products in the cane sugar industry, for example, in Australia, as the value of sugar as a primary product declines in real terms. One important development may be the more widespread use of machinery designed to separate sucrose from other materials (principally bagasse) in the initial stages of cane processing, to increase the efficiency of operations and by-product recovery. Other key developments may include the development of technologies to enhance the recovery of minor (but relatively high value) impurities and their derivatives, such as aconitic acid, polysaccharides, glycolic acid and octacosanol.

Paturau (1989) reviewed the industrial utilization of cane sugar by-products, estimating that cane sugar production at 66 Mt yielded approximately 66 Mt dry weight of bagasse, 17 Mt of molasses, 5 Mt dry weight of filter muds and 215 Mt of cane tops and trash. In Australia, Allen *et al.* (1997) estimated that each 100 t of cane generated 14.3 t raw sugar, 27.2 t bagasse, 5.2 t filter cake, 2.6 t molasses and 50.7 t waste water.

Bagasse

Bagasse is the fibrous waste left after the crushing of cane stalks and extraction of juice. UNEP (1982) notes that bagasse is generally used as a fuel in cane mills, except where alternative fuels are cheaply available, although a range of other uses have also been explored. The particular value of bagasse as a fuel for co-generation (the production of electricity and thermal energy, as steam) is a subject of great interest in the cane sugar industry (e.g. Payne, 1991 – see Box 8.2). The fuel potential of cane-derived lignocellulose, either directly as a material for burning or as the basis of other fuel production methods, underpins much of the energy cane concept (Alexander, 1985 – see Box 8.1). From these perspectives, bagasse (or the complete lignocellulose resource represented by bagasse, tops and trash) can be seen less as a waste and more as a primary product of cane cultivation.

Paturau (1989) describes bagasse as being composed principally of cellulose (from which paper and similar products can be obtained), pentosans (90% xylan, from which furfural and its derivatives can be obtained) and lignin (from which plastics can be obtained). These compounds make up the two main physical components of bagasse: fibre and pith. Some non-combustion uses may require that the pith component is removed, adding to the costs of handling bagasse prior to its utilization.

Immediately after juice extraction, bagasse typically contains 40–50% moisture and 1–3% sugar (Payne, 1991). Where bagasse is not used rapidly, as a fuel source at the cane factory, storage becomes an issue. Owing to the presence of sugar, moisture and micro-organisms, fermentation quickly occurs in stored bagasse, producing alcohol (followed by acetic and other acids), heat and carbon dioxide. Fermentation results also in a breakdown of the fibrous constituents of the raw material. The microbiology of stored bagasse is discussed in more detail by Lacey (1980). Fermentation can be suppressed by drying (e.g. by storage of baled bagasse in aerated stacks) or chemical treatment (Paturau, 1989). However, a degree of fermentation is often

Box 8.1. The energy cane alternative.

> It is sometimes difficult for the sugar technologists who have devoted their entire careers to sugarcane to recognise that sugar production is not the only thing that this plant does well. It is equally difficult to believe that sugar production is not even the thing that it does best. (Alexander, 1985)

The oil crises of the 1970s stimulated an increased interest in the potential of biomass as an alternative source of fuel. In this context, sugarcane was seen as a particularly promising source of biomass, given its efficiency in converting solar energy into organic matter. With particular reference to the Puerto Rican sugar industry, Alexander (1985) considers in detail the potential for refocusing cane cultivation and processing, away from an exclusive interest in the manufacture of sugar, on to biomass production – the energy cane alternative. The underlying concepts, as outlined by Alexander, are summarized below. The value of cane residues (bagasse) as fuel in co-generation (the production of electricity and thermal energy, as steam) as part of the conventional sugar industry are explored by Payne (1991, see Box 8.2).

Alexander (1985) presents plant biomass as a multifaceted commodity, acknowledging its values as a source of oxygen, food, fibre and energy and its less tangible contributions to human well-being through its aesthetic qualities and its role as a 'friend of man'. Such a perspective resonates with more recent promotion of the broad range of values and benefits associated with botanical biodiversity.

As a source of energy, in particular, herbaceous land crops and managed forests are likely to be most effective in providing the principal commodities – fermentable solids and lignocellulose. Sugarcane and related tropical grasses have particular potential, based on their combined anatomical, physiological and agronomic features, which provide for continuous growth and efficient harnessing of solar energy (see Box 1.3). These underpin a potential for high biomass yields (in tropical countries, at least) and greater efficiency of biomass production than could be achieved with woody species. From a biomass perspective, sugarcane is reinvented as 'energy cane', a source of fermentable solids (cane juice, sugars, syrups, molasses) and lignocellulose (tops, trash, bagasse). In this context, the whole plant (not just millable cane) becomes a harvestable commodity. Cultivation for maximum growth may also allow for greater efficiency of nutrient use (including reduced leaching, even at greatly increased levels of fertilizer inputs) and of water use, through enhanced development of the root system of the plant.

As presented by Alexander, the energy cane concept is holistic and complements the natural tendency of the cane plant to maximize its exceptional growth potential (see Box 1.3), rather than its yield of recoverable sugar. A new blend of food and energy planting is proposed, with cane explicitly recognized as a multiple-product commodity. Greater emphasis is placed on lignocellulose yield, for the manufacture of fuel, fuel products and industrial feedstocks, together with expanded opportunities for the exploitation of fermentable solids (e.g. in the production of fuel alcohol, see below). Essentially, those materials that are otherwise seen as wastes and by-products of cane sugar manufacture all become primary products of cane processing, with cane sugar itself becoming a by-product. Consequently, the crop is managed for lignocellulose as well as sugar production, and the whole cane is harvested. Cane factories shift from being 'sugar mills' to being 'biomass dewatering mills' – along with high yields, removal of excess moisture and size reduction/compaction of derived materials are the main constraints in energy cane production.

The energy cane concept is seen as providing a wide range of benefits, largely based on a new 'grand alliance'. Here, the grower's objectives are brought into line with the biological imperatives of the plant (maximizing production of biomass rather than recoverable sucrose), and there is a shift from a qualitative perspective to a quantitative one.

Although particularly inspired by the oil crises of the 1970s and broader limitations of fossil fuels, the energy cane concept also recognizes (and seeks to serve) a wider need for diversification within the sugar industry. A need driven by a complex of factors, including increasing sugar production costs, shifts in consumer tastes and preferences (competition from alternative sweeteners), changes in agricultural labour patterns (increased mechanization, urbanization), increased environmental awareness (and regulation), stagnation of productivity and yield declines. However, unlike more radical changes to the industry, the energy cane concept allows for diversification whilst retaining the essential, familiar features and infrastructures associated with the cultivation and processing of sugarcane.

The enthusiastic promotion of the energy cane concept may not be fashionable currently, but the challenges that led to its development (a crisis in fossil fuel supply and socio-economic pressures on the sugar industry) have not been consigned for ever to history. Nor has the interest in biomass fuels been entirely forgotten. Perhaps future developments will lead to a reassessment of the potential of energy cane and a reinvention of the cane sugar industry along the lines advocated by Alexander and others in the 1970s–1980s.

acceptable, and large quantities of bagasse may simply be bulked in storage sheds or silos, or in the open, where the piles of material can reach 'monumental' size (Payne, 1991). Where bagasse is piled up in the open, the wind may scatter the surface material until the pile stabilizes. Fire may also be a hazard in bagasse stores, and Allen *et al.* (1997) note that storm water runoff from stockpiled bagasse can cause severe environmental problems.

Use of bagasse as a fuel source

The burning of bagasse to fire boilers in cane factories is widespread and reduces the reliance of the industry on fossil fuels. As well as providing a means of generating steam and electricity (co-generation – see Box 8.2) for use on the factory site, the quantities of bagasse produced in cane processing also allow for the production of surplus electricity, which can be sold. Paturau (1989) estimated that it was possible for a cane factory to produce around 60 kWh of surplus electricity per tonne of cane processed, translating into 36 billion kWh/year of surplus electricity potential globally, based on an annual total of 600 Mt of cane processed. In some cases, this surplus electricity can make a significant contribution to meeting national energy needs, particularly in small countries with a strong cane sugar industry. For example, Payne (1991) estimated that, in the mid-1980s, the 'national grid' of Hawaii was able to obtain around 10% of its electricity (400 million kWh/year) from the surplus produced by 13 cane factories, whilst Mauritius obtained around 25% of its electricity (115 million kWh/year) from 15 factories. The Kyoto Protocol, in particular, has inspired increased interest in such contributions from the sugar industry for meeting national requirements for electricity, for example, in Australia (Dixon and Burbridge, 2000; Downing *et al.*, 2002), Thailand (Wakamura, 2003) and Zimbabwe (Mbohwa and Fukuda, 2003).

The quantities of bagasse consumed and energy obtained (and hence of potential surplus bagasse and energy) will depend to a large extent on the efficiency of furnaces and boilers used in cane factories, as well as on the energy efficiency of cane processing operations. Some relevant aspects of power plant design are summarized briefly in Box 8.2. Further details of improved efficiencies in the use of bagasse directly as fuel in cane factories and of methods for the generation of electricity for export from the factory are given by Paturau (1989) and Payne (1991).

The calorific value of bagasse can be improved by reducing its moisture content (e.g. using boiler flue gases), a process which can also reduce atmospheric pollution by particulates on burning (e.g. Paz *et al.*, 2001 – see Chapter 7). Bagasse immediately recovered from crushing and juice extraction has a moisture content of around 50% and a net calorific value of about 7500 kJ/kg (Paturau, 1989). This compares with around 42,000 kJ/kg for fuel oil, 28,000 kJ/kg for bituminous coal, 50,000 kJ/kg for natural gas and 15,000 kJ/kg for wood. Drying of bagasse to 25% moisture increases its net calorific value to around 12,500 kJ/kg and bagasse of 0% moisture can yield around 17,500 kJ/kg. Wang *et al.* (1990) found bagasse drying from 50 to 35–40% moisture content increased net calorific value by 12–17%, heat transmitted to steam by 9–13% and bagasse surplus by 1–2%. However, although drying (and increased calorific value) can result in some fuel economy, Paturau (1989) suggests that substantial improvements should not be expected, particularly if the existing boiler is well designed and has a good thermal efficiency when using bagasse at 50% moisture.

In addition to its direct utilization to fire cane factory boilers, bagasse can be further processed to yield fuel materials. Paturau (1989) highlights the following approaches:

- Densification – formation of small briquettes by depithing, drying and compression. Such processing is not cost free, but the resulting briquettes (or pellets) can be relatively easily stored, transported and utilized in a range of situations.
- Charcoal production, by:
  - the Java process – bales of bagasse are carbonized (as in the production of charcoal from wood); charcoal

Box 8.2. Co-generation in the cane sugar industry.

Co-generation refers to the combined production of electricity and thermal energy (steam) by appropriately designed power plants. This subject is considered in some detail in relation to the cane sugar industry by Payne (1991), from which source the summary below is largely derived.

Co-generation has been a long-standing feature of cane sugar factories, which have been well placed to exploit the technique. This is principally a consequence of the ready supply of a biomass fuel (bagasse) and the high level of usage of low pressure (process) steam in sugar recovery operations. The latter largely accounts for the higher levels of thermal efficiency seen in cane sugar factories, relative to utility power plants. However, higher pressure steam and high temperatures are generally required for generation of electricity, and so furnace and boiler design is important in efficient co-generation. The types of turbines/turbogenerators used influence the reliability of electricity supply for export from the factory, and therefore represent another important consideration.

There is a high level of direct generation of steam from the burning of bagasse, relative to other fuels, as a consequence of the relatively high moisture content of the material. Energy can be recovered from the steam in flue gases through the use of sensible heat recovery equipment, such as air heaters, economizers and bagasse driers.

Furnace/boiler design

Although bagasse is invariably the principal fuel source, in most cases cane factory furnaces and boilers will also need to be able to handle supplementary fuels, such as wood, oil, coal and municipal waste. As well as greatly influencing the efficiency of energy production, furnace and boiler design and types of fuel used also affect the potential levels and characteristics of atmospheric emissions from flues. In cane factories, a spreader-stoker is often employed, which sprays a mixture of bagasse and air into the furnace. A travelling grate allows for continuous discharge of ash. Boiler specifications (e.g. in terms of output steam pressure) will depend to some extent on the types of turbines/turbogenerators employed. Efficient co-generation generally requires a boiler pressure of around 6000 kPa (with steam at around 460°C), although lower pressure boilers (around 3000–4000 kPa) may be suitable for relatively small-scale operations. Based on the net calorific value of bagasse fuel (at 40–50% moisture content), boiler efficiency in a cane factory would be expected to be around 85%.

Turbines/turbogenerators

With no use for low pressure (process) steam, utility power plants tend to operate high pressure boilers in combination with condensing turbogenerators. Back pressure/mechanical drive turbines are suitable for use in cane factories which generate process steam and power only for their own consumption, or with excess electricity exported from the factory as an interruptible supply. Although this method of co-generation is efficient in terms of electric energy heat rate, electricity production is bound to fluctuate with the factory's demand for process steam. If a more stable or increased supply of (export) electricity is required, condensing turbogenerators are more appropriate, although their heat rate efficiencies may be lower. Back pressure turbines and condensing turbogenerators may be used in combination, to meet the various energy needs (including electricity exports) of the cane factory.

Process steam

Process steam is generally utilized at 65–140 kPa and is often superheated. It may be recovered from the exhaust of back pressure/mechanical drive turbines and/or the steam extraction points of condensing turbogenerators. A lower energy form of process steam (vapour) is that produced by the evaporation of cane juice. The use of steam at different pressures and temperatures for various operations (including the generation of electricity) in the cane factory can result in complex steam cycles. Efficient management of these, as well as safe operation of equipment, requires reliable instrumentation and effective monitoring and control systems. Efficient use of process steam is a key element in effective co-generation, as it maximizes the steam available for the generation of electricity.

thus obtained is mixed with molasses and formed into briquettes, which are carbonized by baking. Inputs of 270 kg bagasse and 100 kg molasses yield around 100 kg charcoal. Resulting briquettes can be used as a substitute for wood charcoal, but their relatively high content of potassium salts may limit their use in some specialized processes;
    - destructive distillation of bagasse – although there are methodological difficulties, destructive distillation can be used to yield charcoal (around 45% of product), combustible gases (35% of product, which can be fed back into cane factory furnaces) and small quantities of tar, acetic acid and methyl alcohol. By this method, around 10 t bagasse might produce 1 t charcoal, and molasses could again be used as a binder in the production of briquettes.
- Biogas/methane production – bacterial anaerobic fermentation (in a digester) of cellulose-rich organic material can yield biogas (sludge gas), a 35 : 65 mixture of methane and carbon dioxide. Biogas has a calorific value of about 22,000 kJ/kg (27,500 kJ/m^3). The anaerobic fermentation process is sensitive to operating conditions, and is inhibited by the presence (as in bagasse) of lignin. It is therefore more suited to the treatment of wastes such as vinasse, but this can be augmented with bagasse-derived material, such as surplus pith. Theoretically, 1 kg cellulose can yield 415 l methane.
- Producer gas production – producer gas is manufactured by burning solid fuel in air and steam (in a furnace) to yield a mixture of burned and unburned gases. Thus treated, bagasse yields producer gas with a composition of around 11% carbon dioxide, 17% carbon monoxide, 6% methane, 6% hydrogen and 60% nitrogen, and a calorific value of about 5000 kJ/kg (3800 kJ/m^3). The use of producer gas (e.g. to fuel gas engines) is now a rather outdated technology. Similar processes can be used to obtain synthesis gas, a mixture of carbon monoxide and hydrogen (Alexander, 1985).

Use of bagasse in the manufacture of paper and related products

For many years, the fibrous component of bagasse has been exploited in a number of countries in the production of paper, cardboard, fibreboard and similar materials. Almazan (1994) estimated that bagasse-based paper contributed around 2 Mt to the world's annual paper output of 170 Mt. A key consideration is the detrimental effect of pith on the quality of pulp and requirement for chemical agents in processing (Paturau, 1989). Hence, depithing may be important if high quality pulp is to be obtained from bagasse, and the nature of the raw material also influences the suitability of different pulping techniques. Alkaline pulping processes are generally much favoured over acid processes for bagasse, and the most commonly used methods for pulping bagasse highlighted by Paturau (1989) are the soda process, the kraft (or sulphate) process, the Celdecor process, neutral sulphite or bisulphite processes, the Cusi process and the mechano-chemical process. Manufacture of paper from bagasse generally requires bleaching as well as pulping, and bagasse pulp may be blended with pulp from other sources, depending on the end product.

Like the cultivation and processing of cane for sugar, manufacture of pulp-based by-products from bagasse can involve significant water usage and generate substantial volumes of effluent. Paturau (1989) notes that good quality water is often needed for processing these by-products and that volumes consumed vary considerably according to the processes and final products involved. Water consumption estimates vary from around 100 m^3/t of pulp for production of unbleached pulp, to between around 200 and 450 m^3/t for bleached pulp. Further processing to the final product may require a further 50 m^3/t. Paturau (1989) notes that effluent disposal can represent a problem. As with cane processing, a feature of effluent produced by paper mills and similar operations is its relatively high biological oxygen demand (BOD), owing to its relatively high levels of organic material.

Bhaid (1990) considered the economy of water in small and medium size bagasse-based paper mills in India, noting that they need an ample water supply, which may not be available for half the year. Improved systems can reduce water demand, reduce pumping costs and allow the operation of smaller lagoons for effluent treatment. For example, between 1984 and 1987, Marathwada pulp mill decreased its water consumption from 160 to 100 m^3/t, while increasing its annual output from 7500 to 10,800 t. Penin *et al.* (1988) examined the physical and chemical properties and environmental impact of effluent from bagasse stored in bulk at the Panchito Gomez Toro paper factory. Here, anaerobic treatment of effluent, with or without solid wastes from the same factory, was recommended, and the possibility of using effluent as a fertilizer recognized. Paturau (1989) suggests that prospects for the production of fertilizer by-products would be enhanced by the use of ammonia-based pulping methods, but notes that such techniques have not proved suitable in most situations.

Bagasse can also be used in the manufacture of particle board (in sheets, or moulded for use in a wide range of situations), chipboard and cement board for use in the building industry (Paturau, 1989; Allen *et al.*, 1997). Unlike by-products based on pulps, manufacture of these materials generally involves dry processing, mixing particles or fibres with a bonding agent.

Use of bagasse in the manufacture of chemicals and activated carbon

Bagasse is a potentially valuable raw material for the production of a range of chemicals, including furfural, alpha-cellulose, xylitol and plastics (polymers). It can also be used as a source of activated carbon, a useful commodity across the chemical industry. Allen *et al.* (1997) note that a key, initial step in the use of bagasse as a source of many chemical feedstocks is a two-stage hydrolysis to yield glucose (from cellulose) and xylose (from xylan). Methods for fractionating bagasse into its cellulosic and lignin components, as a platform for the development of chemical feedstocks, are considered by Moens (2002). The following summaries of such uses of bagasse are based mainly on Paturau (1989).

Furfural ($C_5H_4O_2$), an aldehyde of furan, is a colourless, inflammable, volatile, aromatic liquid. It has a range of uses in the chemical industry, including as a selective solvent in the refining of high quality and vegetable oils, and as an intermediate in the manufacture of furfuryl alcohol, resins and tetrahydrofuran (which can be used in nylon production). Furfural can be derived from a range of pentosan-rich plant materials, including corn oats and the hulls of oats, rice, cotton seeds and groundnuts. Historically, the Quaker Oats Company dominated the world production of furfural. Around 22% of world furfural production is based on bagasse. The chemistry of furfural production is complex, but practical considerations include the need to use depithed bagasse and to deal with residual lignin, which can be burnt in the boiler furnace of the cane mill.

Alpha-cellulose is a purified, relatively insoluble form of cellulose, and is the main constituent of 'dissolving pulp', a material which can be used in the manufacture of rayon and acetate fibres, cellophane, plastics, explosives, photographic films, lacquers and fine papers. Historically, most dissolving pulp has been obtained from spruce pulp, but agricultural residues including bagasse can provide a source, although the economics of using them may be marginal.

A wide range of natural and synthetic plastics (polymers) have been turned to an enormous diversity of uses in human endeavours. In terms of bagasse, most attention in this area has focused on the lignin component of the material, which (for example) formed the basis of Valite, a bagasse-derived plastic which provided a substitute for shellac in the manufacture of gramophone records during the Second World War. However, the highly competitive market in plastics which has developed since that time has marginalized the viability of producing plastics from bagasse.

Xylitol is an easily purified sugar alcohol, which can be used as a sugar substitute in a range of specialized foodstuffs. It can be derived from xylan, via xylose, and can thus be obtained by the chemical processing of bagasse (see Gurgel *et al.*, 1995).

Other chemicals that can be derived from bagasse, but which are generally obtained by other means, include:

- Carboxymethyl cellulose (CMC), which has a range of uses, including as a dietary additive (thickener) and as a detergent promoter.
- Diacetyl (2 : 3 butanedione), which can be used as an aroma carrier in (for example) butter, vinegar, coffee and honey products.
- Methanol.
- Ammonia.
- Ethanol (see also Allen *et al.*, 1997).

Activated carbon is widely used in the chemical and allied industries, for the removal of impurities from an enormous range of reagents and products. Whilst other methods predominate, it is possible to obtain activated carbon from bagasse. Carbonization (as in the manufacture of charcoal, e.g. by destructive distillation – see above) is the initial step, followed by activation achieved by processing with steam or phosphoric acid.

Use of bagasse in mulches and soil amendments

Paturau (1989) notes that there is a small market for bagasse-based mulch, notably in the nursery industry in the growing of high value crops such as fruits and ornamental plants. Bagasse mulch may also find a use in other situations. For example, Manson (1981) describes its use to aid revegetation and stabilization of denuded land on road verges. Areas were sown with grass and supplied with NPK fertilizer, and a mulched plot produced vigorous grass cover, whilst an adjacent unmulched plot supported only sparse grass growth confined to moist depressions.

There has also been interest in the use of bagasse as a soil amendment and conditioner, either directly or as a component of composts integrating a range of organic materials including vinasse and livestock manures. Paturau (1989) considers that its low mineral nutrient content and slow decomposition rate results in most beneficial effects being obtained through modification of soil physical characteristics. In this context, for example, bagasse has been used in the stabilization of soils, as in the geotextile outlined by Finegan (1990) for control of soil erosion. However, there is also evidence that bagasse amendments can influence the biological characteristics of soils. Elnaghy *et al.* (1998) examined the effect of organic soil amendments on *Trichoderma harzianum* (a useful biological agent against several soil-borne fungal pathogens) in different Egyptian soil types. Amongst a range of amendments including bagasse, rice straw, *Eucalyptus* leaves and pigeon waste, bagasse in particular was found to greatly stimulate the *T. harzianum* population.

Use of bagasse in animal feed

Bagasse can be used in animal feeds, either as the primary component or by combination (particularly of pith) with molasses (Paturau, 1989). Bagasse has a digestibility to ruminants of around 25%, but this can be increased by appropriate chemical, thermomechanical or biological treatments to around 65%. Such treatments, followed by combination with molasses, protein and nutrients, can produce a livestock feed of equivalent nutritional value to a good quality lucerne hay (Allen *et al.*, 1997).

Other uses of bagasse

Dried bagasse, like other agricultural residues, has been used as litter (bedding material) for poultry. Primary considerations in the selection of poultry litter are high absorptive ability, low matting tendency, freedom from dust, cleanliness and manure value of the used litter. Paturau (1989) suggests that bagasse scores well against most of these criteria, with the possible exception of matting tendency, and that some 200,000 t of dried, screened and baled bagasse was marketed annually for this purpose, mostly to poultry producers in North America.

Sugarcane bagasse has also been investigated as a potential component of aquaculture sytems. Visscher and Duerr (1991) found potential for the use of bagasse-based feed in the culture of the marine shrimp *Penaeus vannamei*. Keshavanath *et al.* (2001) obtained disappointing results for bagasse (particularly when compared with bamboo) as an artificial substrate to enhance production of freshwater herbivorous fish in pond culture, although they also called for further research.

Paturau (1989) notes that almost every cane producing country has attempted to use bagasse as a component of building materials, for example, as a filler in various forms of cement or plaster. However, it appears to have no particular qualities to recommend it in this context, except (possibly) its low cost and weight. Allen *et al.* (1997) suggest that milled pith derived from bagasse could be produced in a suitable form for human consumption as dietary fibre in products such as bread, breakfast cereals and snack bars. Bagasse can be an excellent substrate for mushroom cultivation, with the cultivation residue potentially used in animal feed (UNEP, 1982).

Cane Tops/Trash

A very large tonnage of cane tops is available during the harvesting season, representing about 40% of the millable cane by weight (Paturau, 1989). Payne (1991) suggests that, for a 2-year cane crop at harvest, half of the fibre in the crop can be found in the tops and trash. Alexander (1985) estimates that cane fields can yield around 4.5 tons of cane tops per acre year, plus 5.0 tons of detached trash (leaves that have died, desiccated and fallen from the plant) and 1.5 tons of attached trash (dead leaves adhering to the stem).

Use of cane tops/trash as mulch

The potential benefits of retaining trash as a mulch in cane fields are outlined in Box 2.4, but cane trash mulching has also been explored in relation to other cropping systems. In India, Shaikh *et al.* (1994a,b) found that cane trash mulch at 5 t/ha failed to increase yield in rain-fed pearl millet (*Pennisetum glaucum*), although it did suppress weed populations, and Mishra *et al.* (2001) found that cane trash mulching reduced irrigation demands in aonla/guava orchards. In Réunion, Chabanne *et al.* (2001) found that sugarcane straw mulch increased above-ground biomass and essential oil yields in rose geranium (*Pelargonium asperum* (*Pelargonium graveolens*)) plots, where it also enhanced the biodiversity of soil macrofauna (seven taxa identified in bare soil plots; 13 taxa and substantially enhanced diplopod populations under mulch). Cane trash mulch has also been investigated in relation to the cultivation of tomatoes (Firake *et al.*, 1990), sunflowers (Das *et al.*, 1994) and soybeans (Jayapaul *et al.*, 1995, 1996).

Use of cane tops/trash in animal feeds

Cane tops have traditionally been widely used as livestock fodder, often by the families of the cane cutters (Paturau, 1989). They can be fed to livestock fresh, ensiled or in combination with other feeds, including other sugar processing wastes such as molasses and filter mud.

Other uses of cane tops/trash

SASA (2002) recommends the use of revets (bundles of cane tops) to line waterways until protective vegetation becomes established, as part of erosion control programmes.

Filter Mud

Filter mud (filter cake, scums, cachaza) is a major cane processing waste, typically recovered from press and vacuum filters when sludge from the clarification process is dewatered. Nasir and Quereshi (1999b) estimate that the Pakistan sugar industry, for example, produces around 1 Mt of filter cake each year. Paturau (1989) suggests that around 3.5% filter mud per weight of cane is typically produced, containing colloidal organic matter, some 15–30% fibre, 5–15% crude protein, 5–15% sugar, 5–15% crude wax and fats and 10–20% ash, including oxides of Si, Ca, P, Mg and K.

For disposal purposes, filter mud can be handled dry (e.g. transferred to holding bins or trucks for removal from the mill), or it may be mixed with water (slurried) and then discharged directly, settled (and the supernatant liquid discharged) or impounded (e.g. UNEP, 1982). In addition to its possible value as a fertilizer/soil amendment, various other uses have been recognized for filter mud, notably in animal feeds and as a source of wax (Paturau, 1989).

Use of filter mud as a fertilizer/ soil amendment

Filter mud is generally disposed of on land and incorporated into soils as a fertilizer, often on cane fields some 6 weeks ahead of planting. In this context, its value has primarily been ascribed to its phosphate content, at around 1% by weight (Paturau, 1989). However, this use of filter mud can result in odour problems, and inputs should be based on due consideration of the character of the recipient soil if the risk of wider environmental pollution is to be minimized (UNEP, 1982).

Zia *et al.* (1999) investigated various soil amendments for use in conjunction with brackish groundwater for irrigation of wheat in Faisalabad, Pakistan. They found that, in combination with effective microorganisms (EM), sugarcane filter cake was superior to farmyard or poultry manure in sustaining crop yield and soil properties. Van Antwerpen *et al.* (2003) also found that filter cake was superior to a range of other soil amendments, in its ability to improve soil biological properties. It has also been suggested that applications of filter cake (or composted, dewaxed filter mud) can promote the conservation of soil moisture (Paturau, 1989). Jayapaul *et al.* (1995, 1996) examined its use in this context in the cultivation of irrigated soybean (*Glycine max* L. Merr) at Madurai, Tamil Nadu. Such effects may be attributable to the ability of filter cake to increase soil organic matter content (also noted in Puerto Rico by Lugo-Lopez *et al.*, 1981). Yield benefits were recorded by Scandaliaris *et al.* (1995) in experiments with filter cake as a fertilizer on cane in Argentina. These authors concluded that, for ratoons, the filter cake could be applied in solid form at 25 t/ha (equivalent to around 90 kg N), or distributed in irrigation water. For plant cane, they recommended 100 t/ha worked into the ground or 25 t/ha placed in furrows before planting. However, when comparing filter cake with municipal solid waste plus biosolids, Alva *et al.* (1999) found that leaching of most elements (including N) was greater from soil amended with filter cake than with the other materials, although leachate nitrate concentrations did not exceed drinking-water quality standards.

Filter cake can also be used in combination with other wastes in the preparation of soil amendments. For example, combination with molasses in a ratio of around 7 : 4 gives a free-flowing material, easily applied to fields from tankers (Paturau, 1989). Marie-Jeanne (1993) suggests that fly ash, recovered from chimney gases and used as a filtration aid in the sugar mill, can subsequently be combined with filter cake and used as a soil amendment. Anon. (1991) reports that filter cake can be used as a matrix upon which vinasse can be degraded by solid-state fermentation to produce an organic fertilizer. Use of filter cake in combination with distillery waste is considered further in the section dealing with the use of vinasse as a fertilizer/soil amendment.

Use of filter mud as a source of wax

Cane wax, the lipid fraction of sugarcane, comprises waxy lipids (true wax, derived mostly from the whitish waxy coating on the stalk) at around 0.12% of cane by weight, and fatty lipids at around 0.06% (Paturau, 1989). Quantities of these materials become separated from the bagasse during cane processing, and are occluded from the juice during clarification, ultimately being deposited in the filter mud. Using appropriate solvents, cane wax can be recovered from filter cake, and the waxy lipids separated from the fatty lipids. Various methods can then be used for refinement and bleaching of the wax, which can be used in materials like polishes. Although wax is a potentially valuable by-product, extraction and processing costs are relatively high, leaving a rather small profit margin (Paturau, 1989). However, Allen *et al.* (1997) are confident that new technologies and an increasing market for natural waxes will increase the viability of cane wax as a by-product from filter cake (or directly from the cane stalk epidermis).

Use of filter mud in animal feeds

Filter mud can be integrated into animal feeds, for example, in combination with other waste products of cane sugar production, such as molasses and cane tops (Paturau, 1989).

Boiler and Fly Ash

Boiler ash is the residue left in the boiler grate after bagasse burning, whilst fly ash (collected by particulate emissions control systems in the chimney stack) may also need to be dealt with. Paturau (1989) estimates that bagasse ash is generated at a rate of 0.3% of the weight of cane processed, and provides a summary of the chemical constitution of bagasse ash, based on data from a number of countries. Both boiler and fly ash may be handled dry and retained on site by impounding (UNEP, 1982), but stockpiles may present an environmental hazard. Bloesch *et al.* (2003) studied the changes in nutrient value and potential environmental impacts of stockpiled mill mud/ash at a sugar mill in New South Wales, Australia. Stockpiles were very permeable, with 31–48% of rainfall passing through as leachate, but with negligible runoff. Leachate concentrations of mineral N and P were high, and very high BOD levels in the un-attenuated leachate were of concern. The excess liquid initially draining from the stockpiles had the greatest oxygen depletion potential. Such findings have implications for the siting of stockpiles. Even though there were losses of 22–33% N and 6–16% P, loss of mass due to carbon mineralization resulted in overall N and P enrichment.

Uses of boiler and fly ash

Like other waste materials from cane processing, bagasse ash has been used as a fertilizer, but other uses have also been investigated. These include the use of the material in glass manufacture, as a partial substitute for cement and as a filtration aid (Paturau, 1989). In relation to the latter, Marie-Jeanne (1993) noted that fly ash extracted from boiler chimney gas could be used as a filtration aid in the sugar mill itself. Gupta *et al.* (2002) report on the use of bagasse fly ash beyond the sugar factory for the removal of the pesticides lindane and malathion from waste water. They found that 97–98% removal was possible under optimum conditions, demonstrating that fly ash was inexpensive and effective for the task.

Molasses

Final or blackstrap molasses is the residue produced after repeated crystallization of sugar – the waste syrup from which no further sucrose can be extracted by these means. This is the form of molasses considered below. High test molasses, a concentrated form of partly inverted cane juice (a syrup used in alcohol production), is a specific product of certain cane mills.

Molasses is an important raw material, particularly for the fermentation industry, although it also has other applications. In South Africa, for example, cane mills produce around 850,000 t molasses each year, around 75% of which is traded locally and the remainder exported (Wynne and Meyer, 2002). The yield of final molasses is typically around 2.7% by weight of cane, and its precise composition is somewhat variable. Broadly speaking, molasses is a complex mixture of water (say, at 20%), sucrose (30–35%), glucose (4–7%), fructose (6–9%), other reducing substances (3%), ash (oxides of e.g. K, Ca, Mg, Si, Na, Fe and Al, with SO_3, Cl and P_2O_5 acids, totalling around 12–14%), nitrogenous compounds (crude protein and amino acids, 4.5%), lipids (0.4%) and traces of pigments and vitamins (Paturau, 1989; Wynne and Meyer, 2002).

Use of molasses in the production of yeast/single-cell protein

Paturau (1989) notes that molasses may be used in the production of baker's yeast (*Saccharomyces cerevisiae*) or feed yeast (various species, including *S. cerevisiae*, but mostly *Torulopsis utilis*) for consumption by humans or livestock. It may also be used in the production of high protein dietary supplements for humans or livestock, derived from a range of microorganisms (so-called single-cell protein, SCP). Feed yeast (generally as *S. cerevisiae*) can be recovered as a by-product of ethanol manufacture by the fermentation of molasses, which is considered in more detail below. However, yeast may also be the primary product from the processing of molasses.

Use of molasses in animal feeds

Paturau (1989) notes that the most important characteristic of molasses as animal feed is its carbohydrate content (mostly sugars), although it can also be a useful source of trace elements and B-complex vitamins; in contrast, the digestible protein content is negligible. However, molasses also has the advantage of high palatability, and can be used as a binding agent, so it can be used to induce livestock to eat poor quality roughage and to make dry feed more easy to handle. Bagasse can be combined with molasses, and this has provided a useful role for the pith (which may have to be removed prior to the use of the fibrous component in other bagasse by-products). Animal feeds based on combinations of molasses and cane tops, or cane tops and filter mud, have also been investigated. Whilst some molasses is used directly by livestock farmers, a greater quantity is used in the production of formula feeds.

Molasses has also been investigated as a component in the ensiling process of other (non-cane) plant wastes. For example, Megias *et al.* (1999) studied canned artichoke (*Cynara scolymus* L.) by-product ensiling, using formic acid, cane sugar molasses and sodium chloride treatments. The nutritive value of the final silage product was greatest with sodium chloride. Total effluent production was least with cane sugar molasses, but effluent chemical oxygen demand (COD) and conductivity were both relatively high with cane sugar molasses and sodium chloride treatments.

Use of molasses as a fertilizer/ soil amendment

The use of molasses as a fertilizer in cane cultivation has a long history (e.g. see Doty, 1933). Paturau (1989) suggests that it is generally applied in the furrows, 2 weeks ahead of planting, at the rate of 10–20 t/ha (equivalent to around 500–1000 kg K_2O/ha). However, the extent of its use has historically varied with the cost and availability of other sources of organic and inorganic fertilizers. Wynne and Meyer (2002) suggest that, in order to provide a balanced nutrient input, application of molasses should be accompanied by the application of nitrogenous compounds such as urea. They summarize the advantages and disadvantages of molasses use as a fertilizer as follows:

Advantages

- A good source of K.
- Also provides numerous trace elements in appreciable amounts.
- Increases soil organic matter content and microbial activity, enhancing nitrification.
- Improves soil aggregation and reduces surface crusting.

- Potential 5–10% improvement in cane yields on low fertility soils.

Disadvantages

- Variable nutrient content, creating difficulties for consistent and even application.
- Difficult to handle and apply, particularly on sloping land.
- Need for storage (must be collected rateably from the mill, but cannot necessarily be so applied to the field).
- Risk of soil and water pollution if not properly applied.

The particular value of molasses in improving soil structure is also noted by authors including Lugo-Lopez *et al.* (1981). It has also been suggested that the increased soil microbial activity arising from molasses application can assist in the control of certain soil pests, by enhancing levels of nematophagous fungi (Tianco, 1983).

Wynne and Meyer (2002) have recently assessed the economics of using molasses (in combination with urea) as a fertilizer in the South African sugar industry. Key factors in determining whether this practice is economically viable include the exchange rate (and its effect on prices for imported fertilizer and exported molasses) and competition for molasses with distillers. However, it is noted that molasses used in ethanol production ultimately gives rise to vinasse, which also has potential value as a fertilizer (see below).

Other uses of molasses

Cane molasses has also been exploited, on a limited scale, as a raw material in the production of dextran, monosodium glutamate, L-lysine, xanthan gum, aconitic acid and itaconic acid (see Paturau, 1989).

Use of molasses in fermentation/ ethanol production

Fermentation is most widely recognized as the conversion of sugar to ethanol and carbon dioxide, by the action of enzymes produced by yeasts. However, a wide range of microorganisms and enzymes may be involved in fermentation in a wider sense, and the process can result in a broad spectrum of organic compounds. For example, molasses is used as a raw material in fermentation operations producing acetic acid and vinegar, butanol and acetone, lactic acid, citric acid and glycerol (see Paturau, 1989). None the less, the production of ethanol by fermentation (and subsequent distillation) remains a predominant activity associated with the cane sugar industry. Traditionally, such activities have centred around the production of rum from molasses. However, ethanol has a wide range of other uses (e.g. as a fuel, organic solvent or chemical intermediate in the production of other compounds) and various means by which the sugar industry might exploit these have also been investigated. Particularly where promising economic opportunities have been identified (e.g. in relation to fuel alcohol production in Brazil), direct fermentation of cane juice has also been explored. The particular issue of fuel alcohol production is considered separately below.

Paturau (1989) notes that rum is generally defined as an alcoholic distillate from the fermentation of sugarcane juice or cane sugar by-products, possessing an identifiable flavour and aroma, based on a characteristic combination of alcohols, esters, aldehydes and organic acids. Cane molasses is generally accepted as the most suitable and economical raw material for rum production, and a range of criteria can be applied to identify the grades of molasses which result in the best rum. Industrial ethanol is produced in various grades and has various uses, but the processes involved in its production are largely similar to those employed in the manufacture of rum. A 'mash' is fermented, based on molasses diluted to around 14–18% sugar, a slightly higher concentration than for rum production. Nutrients are added (and the pH adjusted) as necessary to enhance the fermentation by yeast, generally a strain of *S. cerevisiae*, also widely used in the rum industry. Fermentation produces a 'beer' of up to around 13% alcohol. The ethanol thus obtained is purified by distillation, with stills in industrial use typically operating at an efficiency of around 97.5%. Although figures vary (e.g. according to the composition of molasses and strains of yeast used), the production of 10 l of industrial alcohol might be expected to consume around 38 kg of molasses. Further refinement (dehydration) is required to produce absolute alcohol, a purified form of ethanol used in the pharmaceutical industry, or for use in internal combustion engines.

The processes described above represent a highly summarized and simplified description of the production of ethanol from molasses. Two considerations, in particular, should be borne in mind in this context: (i) fermentation and distillation are evolving technologies, with new developments potentially affecting the economics and environmental impacts of the industry; and (ii) the economics and hence patterns of output of the production of industrial ethanol from sugarcane and its by-products are substantially influenced by external factors, notably world trade in oil. The latter is particularly important, as it affects the viability of using cane- or molasses-derived ethanol (rather than ethylene from the petrochemical industry) as a chemical intermediate, and influences the viability of using this ethanol as a fuel source (in competition with petrol). With such issues in mind, Paturau (1989) concluded that ethanol produced from molasses was generally insufficiently competitive to replace either petrol or ethylene, except under unusually favourable commercial conditions. However, Allen *et al.* (1997) predict growth in the use of ethanol from molasses as a feedstock in the (petro)chemical industry.

The scale of distillery operations associated with the sugar industry justifies a brief consideration here of their environmental impacts.

Environmental impacts of distillery operations

The production of ethanol from cane juice or molasses has the potential to consume significant quantities of water and generate environmental pollutants. Distillery operations generate a number of waste products, notably vinasse (stillage, slops, spent wash, dunder), the effluent arising from alcohol production. These wastes can contribute to environmental impacts, but can also find use as by-products of the industry. For example, fusel oil is a by-product of distillation, a mixture mostly of higher alcohols, comprising around 62.5% amyl and iso-amyl alcohol, 15% isobutyl alcohol, 12.5% *n*-propyl alcohol, 5% ethanol and 5% other residues. Around 1.1 l fusel oil is produced per 1000 kg molasses processed. It is not generally refined, and may present disposal problems, although it does have some commercial value as a lacquer solvent (Paturau, 1989).

Environmental impacts on air quality

Carbon dioxide is produced during fermentation. Paturau (1989) estimates that distillery production rates are around 160 kg carbon dioxide per 1000 kg of molasses processed (theoretically, 46.6 kg carbon dioxide per 100 kg of fermented sugars). Of this, Paturau (1989) estimates that 80–85% is recovered and processed to yield liquid CO_2 (used in the carbonation of beverages, fire extinguishers and food preservatives) or solid CO_2 (dry ice). Distillery operations also result in odour problems, for example, as a consequence of the decomposition of nitrogenous material in the fermented mash. Such material can be removed by decantation or filtration.

Water consumption

The production of 10,000 l of industrial ethanol might be expected to consume around 1 Ml of water in the fermentation stage and another 1 Ml in the distillation stage (Paturau, 1989). This does not include steam used in distillery operations, which (along with electricity) may be supplied by a cane factory where this operates on the same site as the distillery. However, quantities of water consumed vary according to the particular methods used by the distillery. The Biostil process of fermentation, for example, uses only 60% of the process water required by traditional methods (Paturau, 1989).

Environmental impacts on water and soil quality

Cortez *et al.* (1998) consider that around 10–15 l of vinasse are produced for every 1 l of ethanol produced, and that its polluting characteristics are a major concern in most cane growing countries with an associated distilling industry. Such concern is based largely on the substantial volumes produced and the high BOD of the effluent. Some authors also note the relatively high acidity of vinasse (giving values of pH 4.0–4.5), and the problems that this presents (e.g. Ng Kee Kwong and Paul, 1997; Nasir and Quereshi, 1999a). Alexander (1985) cites research which estimates that the 'pollutant value' of 1 day's vinasse output from Puerto Rican distilleries in 1980 (710,000 gallons) was equivalent to the daily sewage output of a city with over a million inhabitants. Paturau (1989) estimates vinasse BOD at around 25,000–50,000 ppm, with a high organic matter content (around 7.5 g/ml), and an approximate composition (excluding water) of: approximately 29% mineral matter; 21% gums; 17% wax, phenolic bodies, lignin, etc.; 11% reducing sugars; 9% proteins; 5.5% glycerol; 4.5% lactic acid; 1.5% volatile acids; 1.5% other organic acids.

Alexander (1985) notes that rum distillery wastes were historically discharged into the Atlantic Ocean and Caribbean Sea in massive quantities, challenging the potential of even these substantial water bodies to dilute and disperse the material. Particularly where relatively small distilleries are surrounded by cane fields, it has also historically been common practice to pump vinasse on to the land as a fertilizer (Paturau, 1989; Wang *et al.*, 1996). The potential value of vinasse in this respect is considered further below; however, it should be noted that uncontrolled dumping on land can result in environmental pollution. Chapman and Usher (1996), for example, report that spraying of large quantities of excess vinasse on to an extensive land disposal site in the Mackay district of Australia sometimes resulted in unmanageable environmental problems, including runoff into marine wetlands, odours and fly breeding.

As with cane factory effluents, a range of methods are available for treating distillery effluents prior to their disposal. These include aerobic treatment using activated sludge, anaerobic bacterial digestion and concentration by evaporation (Paturau, 1989). The latter two approaches have the potential to yield useful by-products. Anaerobic digestion can be harnessed to produce biogas (a mixture of methane and carbon dioxide – see under bagasse) and fertilizer sludge (Alexander, 1985; Paturau, 1989; Lettinga and van Haandel, 1993; Cortez *et al.*, 1998). Concentration of vinasse, on the other hand, allows recovery of water which can be fed back into the distillery (e.g. for molasses dilution), and yields a concentrated syrup (concentrated molasses solubles/solids, CMS) of around 60% solids. The CMS can be neutralized by treatment with lime if necessary, and then used as a fertilizer, in animal feed (possibly combined with cane tops and/or bagasse) or as fuel in an appropriately designed furnace (the resultant ash having potential uses as a potash-rich fertilizer, or as cullet in glass production). The process of producing CMS incurs additional costs for the distillery operations. However, the Biostil process of fermentation produces relatively concentrated vinasse that can be used directly without further concentration by evaporation (Paturau, 1989). Alternatively, a fuel material can be derived by mixing vinasse with bagasse pith, and drying the resultant material with boiler flue gases.

In comparing vinasse treatment methods, Paturau (1989) suggests that aerobic treatment may have prohibitive energy costs for distilleries, and notes that anaerobic treatment (although a more viable option) is not sufficient in itself to entirely eliminate potential pollution hazards. It is estimated that, even where treatment achieves 95% breakdown of organic pollutants (say, reducing the BOD from 25,000 to 1250 ppm), the resultant effluent remains some three times more polluting than domestic sewage. A follow-up treatment, using an aerobic process or by simple lagooning, is recommended. The efficacy of a range of microorganisms for the treatment of vinasse continues to be investigated. For example, Angayarkanni *et al.* (2003) report on the biotreatment of distillery effluent in India, using the fungus *Aspergillus niveus* (*Fennelia nivea*), with various carbon sources, including bagasse, molasses and sucrose.

In Kenya, the potential for the treatment of distillery effluent by artificial wetlands has been investigated, using *Cyperus papyrus*, *Cyperus* sp., *Phragmites mauritianus* and *Echinochloa pyramidalis*, but with only limited success (Thuresson, M., 2001; Thuresson, S., 2001).

Other opportunities for the utilization or treatment of vinasse have also been investigated. For example, Herrera Coello and Diaz Rodriguez (1998) found that vinasse could partially replace molasses in the production of single-cell protein for animal feed (derived from the yeast *Kluyveromyces fragilis*).

Use of vinasse as a fertilizer/soil amendment

There is (rightly) concern over the potential polluting effects of uncontrolled dumping of vinasse on the land, or its inappropriate use as a fertilizer. For example, Szmrecsanyi (1994) concludes that the indiscriminate use of large amounts of vinasse as a fertilizer can lead, for example, to salinization and river pollution. However, there is also a body of evidence which suggests beneficial effects on soil quality from appropriately managed application of vinasse. These effects are ascribed, in particular, to the high levels of organic matter (e.g. Nasir and Quereshi, 1999a) and K (e.g. Chang and Li, 1989; Ng Kee Kwong and Paul, 1997) in vinasse. In addition to its potential role in enhancing soil quality, it has been suggested that vinasse may have insecticidal qualities that might be harnessed in agriculture (Sundaramurthy, 1998).

Examples of the use of vinasse as a fertilizer include the work of Nasir and Quereshi (1999a), who compared the application of vinasse (at 740 t/ha + 50 kg N/ha) 4 months before cane planting with the application of 170–100–50 kg NPK/ha at planting. The vinasse treatment improved the original soil quality through a reduction in pH (from 8.10 to 7.5). However, other studies have suggested that changes in soil pH effected by vinasse tend to be only short-term (Rao, 1983; Cruz *et al.*, 1991; Ng Ke Kwong and Paul, 1997). Nasir and Quereshi (1999a) also found increased organic matter content (from 0.3 to 0.9%) and enhanced NPK levels in soils following application of vinasse + N. Cane yields were around 35% greater under the vinasse treatment than with the fertilizer treatment, and there was evidence of small increases in juice quality parameters. Similar experiments in Venezuela suggested that vinasse could substitute for around 50% of N, 75% of P and 100% of K otherwise obtained from inorganic fertilizers in cane cultivation (Gomez Toro, 1996). Chang and Li (1989) found that a single vinasse application (at 50 t/ha) to fields in Taiwan could provide sufficient K for at least two cane crops.

Vinasse can also be applied in dilute form in irrigation water, as part of a fertigation strategy (Wang *et al.*, 1996). For example, Booth and Lightfoot (1990) reported the value of vinasse in irrigation waters for maintaining nutrient levels in cane growing soils in Zimbabwe. Nasir and Quereshi (1999a) produced evidence of a beneficial effect on yields of the application of dilute vinasse in irrigation water to standing crops of cane, wheat, rice and fodder maize.

In contrast to this strategy of diluting vinasse, Chapman and Usher (1996) suggest that one of the problems associated with vinasse application as a fertilizer in the Mackay district of Australia had historically been that the material was already too dilute at the point of production. This resulted in uneconomical transport costs, and it was only through the production of a more concentrated form of vinasse (biodunder), when the Biostil process (see above) was adopted at the local distillery, that use as a fertilizer became a viable option. Enriched with standard nutrients, this vinasse has been successfully marketed for use on cane fields, at 55–95% of the cost of equivalent quantities of KCl fertilizer. Chapman *et al.* (1995) note an additional potential benefit from the use of this form of vinasse in fertilizer application, in that dissolution of urea in biodunder, rather than application of granular urea, decreased N losses through volatilization of ammonia.

Vinasse can also be used as a fertilizer/soil amendment in combination with filter cake. It has been suggested that this combination is attractive because the vinasse supplies the K and S, while the filter cake supplies the N and P that would otherwise be derived from inorganic fertilizers (Korndorfer and Anderson, 1997). Complementarity is also noted by Muraoka *et al.* (1995), who comment on the relatively high organic matter content of both materials, and the balance between high P levels in filter cake and high K and Ca levels in vinasse. This nutrient matching may also contribute to the perceived value of vinasse/filter cake combinations in animal feed (e.g. Zamora *et al.*, 1996). Ng Kee Kwong and Paul (1997) investigated combinations of filter cake applied at 12 t/ha at the time of planting, with different levels of either inorganic KCl fertilizer or vinasse, in cane cultivation in Mauritius. They found that vinasse was a perfectly adequate substitute for KCl, and that a single large dose (equivalent to 360 kg K_2O/ha) of vinasse, banded in furrows at the time of planting, could provide all the K requirements of the plant crop and five ratoons. Nasir and Quereshi (1999b) concluded that application of a vinasse/filter cake biocompost at 25 t/ha could produce increased cane yields whilst replacing all the K and half the N and P that would otherwise be applied as inorganic fertilizers.

The use of vinasse in combination with other amendments may also have beneficial effects on soil quality. For example, Nasir and Quereshi (1999a) suggest that vinasse applied with gypsum can assist in the treatment of sodic soils.

Other uses of vinasse

Vinasse has also been used in Brazil, in combination with soil, in the manufacture of bricks (Freire and Trevisan, 2002).

Fuel Alcohol from Sugarcane

As noted above, rum and industrial alcohol can be produced by fermentation and distillation of molasses. Ethanol can also be derived by the same methods directly from cane juice, and this approach has been investigated particularly in relation to fuel alcohol production. Ethanol gives a smokeless and colourless burn, and yields around 29,700 kJ/kg, slightly more than coal (Paturau, 1989).

Use of plant materials and derivatives in the production of fuel alcohol (bioethanol) has attracted considerable interest in recent years. Sugarcane and its by-products have been a major focus, although maize has also been used, particularly in the USA (Jolly, 2001), and other crops, including cassava (Verschuur and van Wijk, 1990), may also have potential in this role. Production of bioethanol from woody plant material is also possible, and would be more economically attractive given appropriate biotechnological advances for efficient degradation of lignocellulose and fermentation of relatively complex sugars (Verschuur and van Wijk, 1990; Zaldivar *et al.*, 2001).

Interest in bioethanol production, especially from sugarcane and its by-products, has been particularly acute in Brazil (UNEP, 1982), which launched a National Alcohol Programme in 1975. However, the technology has also been explored elsewhere, including other parts of Latin America (Cardenas, 1993; Reeser *et al.*, 1994), India (Verschuur and van Wijk, 1990; Sharma and Goel, 1997), parts of Africa (Verschuur and van Wijk, 1990), and Australia, Mexico, Thailand and the USA (Jolly, 2001).

The primary arguments in favour of bioethanol as a fuel (particularly for vehicles, either as pure alcohol, or mixed with petrol as 'gasohol' – Goldemberg *et al.*, 1993) is that it results in less air pollution than fossil fuels (Verschuur and van Wijk, 1990; Goldemberg *et al.*, 1993; Sharma and Goel, 1997; Kammerbauer *et al.*, 1999), and represents a renewable resource (Reeser *et al.*, 1994). However, fuel alcohol has also been recognized for its potential to reduce dependence on foreign oil, enhance local employment opportunities and create 'added value' in a sugar industry afflicted by unfavourable world commodity prices (Goldemberg *et al.*, 1993; Sharma and Goel, 1997).

Nastari (1992) describes bioethanol as environmentally friendly at all levels of production and use. However, concerns have been raised over the potential negative impacts of the bioethanol industry. In Brazil, Madeley (1981) considered that production of ethanol was generating enormous amounts of waste and causing serious pollution, particularly of water, and that peasant farmers were being displaced by large landowners who had been encouraged by government incentives to expand sugar hectarages. Cortez *et al.* (1998) note that river pollution near distilleries was a problem in Brazil, even before the National Alcohol Programme was launched, when much less effluent was being produced. In addition to the generation of potentially polluting residues, UNEP (1982) notes that the economics of fuel production through fermentation of sugar crops is complex and needs careful consideration. Fry (1997) noted the predominance of ethanol production from sugarcane in Brazil, but suggested that removal of government incentives for fuel alcohol production would result in a shift back towards production of sugar, as ethanol becomes an increasingly uncompetitive product.

SUGAR BEET

The value of sugar beet by-products (principally tops, pulp and molasses) was recognized very early in the cultivation of the plant (Winner, 1993). Cooke and Scott (1993) note that these materials are extensively used as animal feeds (either separately or in combinations), and that this compares favourably with the level of by-product utilization in the cane sugar industry. However, beet by-products have also been used in a variety of other ways. Beet molasses (like that produced from cane) can be used in fermentation to yield alcohol and waste products with potential by-product uses. Draycott and Christenson (2003) note that a range of beet processing wastes are returned to the land as soil amendments in different countries, such as pulp-based wastes (including raffinate), lime and post-distillation waste (vinasse). Lime is used principally to adjust the pH of acid soils, while other wastes provide a range of nutrients and small amounts of organic matter.

Sugar Beet Tops

Use of sugar beet tops in animal feed

Harland (1993) notes that, depending on harvesting method, beet tops may comprise leaves, or leaves and root crowns. Fresh tops can be grazed in the field (common practice when they are fed to sheep) or removed from the field to be fed directly to livestock or to be ensiled. Beet tops are very palatable and may be fed to all ruminants (and sparingly to pigs). On an energy basis, 10 kg beet top silage is equivalent to 1.5 kg barley. The yield of tops is generally similar to that of roots, but varies according to harvesting time, variety, etc. In the UK, maximum top yield (generally obtained by harvesting in September) is around 50 t/ha fresh weight (5–6 t/ha dry matter).

The ensilage of beet tops results in the production of effluents, which can cause pollution problems if not properly managed. Nuttall and Stevens (1983) report effluent discharges of 194–333 l/t, with most effluent produced in the first 7 days.

Harland (1993) notes four principal requirements for the production of good silage from beet tops, whilst minimizing potential environmental impacts from silage effluents:

- Tops should be removed from the field during harvest operations, to minimize contamination by soil.
- They should be wilted, to reduce effluent production and to aid compaction in silos.
- Silos should be appropriately sited for proper effluent disposal.
- Silos should be filled correctly, to ensure that air is excluded.

Other uses of sugar beet tops

Sugar beet tops can be left on the soil surface of fields, as a source of nitrogen, a practice which has implications for soil quality and which may increase the risk of nitrate leaching (e.g. see Destain *et al.*, 1991; Christensen, 2004). The production of methane from sugar beet leaves has also been investigated, on a laboratory scale (e.g. see Zauner, 1988).

Sugar Beet Pulp

Use of sugar beet pulp in animal feed

Harland (1993) notes that, following extraction of sucrose in the initial stages of processing, wet pulp is produced (in the form of spent cossettes), with a dry matter content of just 6–12%. This can be used directly as animal feed, or pressed first to increase the dry matter content to 18–30%. Pressed pulp can also be ensiled, further dried (to yield dried plain sugar beet pulp, containing 87–92% dry matter), mixed with molasses, or mixed with molasses and then dried. Beet pulp provides livestock with energy (mostly derived from structural carbohydrates) and digestible fibre. It is suitable for feeding to ruminants, and more sparingly to pigs. Used in combination with other feeds (such as those based on straw and hay), it can increase their digestibility, making it a valuable addition to livestock feed rations.

Harland (1993) notes that wet pulp can be difficult to handle or store, but is none the less used as animal feed in some beet growing countries (including Denmark). Pressed pulp is more widely used throughout Europe as animal feed. It should be used within 5–7 days, if mould growth and spoilage are to be avoided, or ensiled (e.g. Kunteova, 1997) with appropriate care, to maximize the quality and quantity of silage recovered. Because of its relatively low water content, dried plain sugar beet pulp is relatively stable, and can (generally in pellet form) be used for feed directly or stored for up to a year.

De Brabander and Boucque (1992) suggest that use of feeds like sugar beet pulp, which reduce nitrogen excretion and total faecal dry matter production by cattle when compared with grass-based feeds, can help to reduce the environmental impacts associated with livestock farming.

Use of sugar beet pulp as fuel

The use of beet pulp as a source of fuel has been considered by a range of authors (e.g. Otorowski, 1990; Mantovani and Vaccari, 2001). Like bagasse, beet pulp can be used to produce fuels such as biogas by fermentation, or can be burned directly as fuel. Otorowski (1990) notes that beet pulp for burning must contain < 50% moisture, and suggests that direct burning of predried beet pulp can result in better heat efficiency than conversion to biogas by anaerobic fermentation, estimating that dried pulp (90% dry matter) generates half as much heat as an equal weight of fuel oil in an efficient boiler. UNEP (1982) notes that it may be possible to obtain alcohol from beet pulp by microbial fermentation; this would be an attractive proposition if it could profitably be conducted using wet pulp.

Drying sugar beet pulp

As noted above, beet pulp is often dried prior to being used as animal feed, and must be dried if it is to be burned as fuel. Pulp pressing is a very energy-hungry activity (UNEP, 1982), and there is environmental concern over emissions from pulp driers. Thielecke *et al.* (1991) identified waste gas from pulp drying as the most important source of emissions of organic compounds at a beet sugar factory. The total organic carbon absorbable in NaOH solution typically amounted to 200–350 mg/m^3 (up to a maximum of 665 mg/m^3), and increased with increasing molasses dose and higher dryer inlet temperature. The main components were 60–80% acids (50–60% acetic) and 7–10% carbonyl compounds (acetaldehyde, benzaldehyde, furfural and ten others). Such emissions can be controlled by the use of equipment such as scrubbers (e.g. Thielecke *et al.*, 1991; Buchholz *et al.*, 1992). Jensen (1995, 2002, 2003) discusses the use of alternative steam drying methods, considered more energy efficient and capable of reducing the atmospheric pollution potential of beet drying (relative, for example, to drying in rotating drums). Methods for drying beet pulp, including aspects of their environmental impacts, are also discussed by Accorsi and Zama (1996).

Other uses of sugar beet pulp

Harland (1993) notes that sugar beet pulp has been investigated as a possible component in human diets, where it may yield health benefits associated with its high fibre content, and Stekar (1999) suggests that beet pulp is a useful additive to silage, reducing the production of potentially highly polluting effluents.

Lime Mud

Storage of lime-rich filter mud produced as a waste material from clarification in beet sugar factories may present problems. Draycott and Christenson (2003) note that lime from beet processing had been stockpiled on site in Michigan (USA) for over 100 years, but that new environmental regulations and limited storage capacity compelled processors to change this practice.

Use of lime mud as a soil amendment

Lime recovered from beet processing has an important potential by-product role as a soil amendment. As the crop is very sensitive to soil acidity, lime is commonly applied to beet fields to ensure a close to neutral pH. Soils should be tested well in advance of sowing, and more than one lime application (followed by ploughing to ensure thorough incorporation with the soil) may be required (Draycott and Christenson, 2003). Christenson *et al.* (2000) report on the responses of various crops to soil additions of waste lime from beet processing.

Sugar Beet Molasses

The residual syrup (molasses) from sugar beet processing is used principally in animal feed or for fermentation. Roughly half of its dry matter content is sugar, which conventional processing is unable to extract by crystallization (e.g. Harland, 1993). Molasses can be desugared by other means, to produce raffinate, which can be used in animal feeds or as a soil amendment (e.g. Goos *et al.*, 2001).

Use of sugar beet molasses in animal feed

Harland (1993) notes that the high viscosity of molasses means that it is difficult to handle and is generally diluted before use by farms and feedmills. The precise chemical composition of molasses is variable, but its relatively high energy content and palatability make it suitable for inclusion in the diets of a range of livestock. It is often provided as one element of a compound feed. It has been argued (e.g. Araba and Byers, 2002) that the use of sugar beet molasses in cattle feed results in beef production with reduced environmental impact, on the grounds that traditional grain-based feeds consume more fossil energy in their production than molasses. However, the complex combinations of food material which are typically included in livestock diets make it difficult to assess environmental impacts of this kind in a meaningful way.

Molasses may be used in animal feed in combination with pressed or dried beet pulp. Such products are widely used in Europe, in various forms, and provide a major outlet for beet factory wastes (Harland, 1993). A commonly used form is dried molassed sugar beet feed (DMSBF), which stores well for up to a year, is palatable to (and suitable for) a wide range of livestock and is generally assumed to have a similar energy value to that of barley. DMSBF combines the rapidly available energy from the sugar in molasses with the slower release energy of digestible fibre in the pulp.

Beet molasses is also often used as an additive to silage, particularly that based on low quality grasses, where it provides a valuable addition of sugar, aiding fermentation (Harland, 1993). Interestingly from an environmental perspective, DMSBF has also been used as an additive to herbage silage, partly to enhance the ensilage process, but also to reduce potential pollution from silage effluents. Harland (1993) notes that DMSBF can absorb up to three to four times its own weight in effluent, thus retaining substances which are of value in the final animal feed product and reducing the discharge of pollutants. Interest in this use of DMSBF was renewed in the late 1980s (particularly in the UK and Ireland), when concern over silage effluent pollution and numbers of related prosecutions increased.

Fermentation of sugar beet molasses

Like the molasses produced in cane sugar processing, beet molasses can be used in the production of alcohol by fermentation (although it is not suitable for the manufacture of rum), with vinasse produced as a waste product. This poses a pollution risk if discharged as effluent, but it can be treated to reduce its environmental impact (e.g. in an anaerobic reactor, with the potential to produce fuel gas – Jenicek *et al.*, 1994) or can be utilized directly as a by-product. Harland (1993) notes that vinasse from the fermentation of beet molasses is very low in sugar but rich in crude protein and potassium. Madejon *et al.* (2001b) noted that the annual production of vinasse from the distillation of fermented beet molasses in south-west Spain was around 50,000 t. The vinasse was characterized by a high organic matter content (BOD = 60–70 kg O_2/l) and high salinity (electrical conductivity (EC) = 25–30 dS/m). Historically, vinasse had been held in evaporation ponds, resulting in groundwater contamination, odours and nuisance from insects. More recently, the trend had been to use it, often in concentrated form, in animal feeds (e.g. see Weigand and Kirchgessner, 1980; Haaskma and Vecchiettini, 1988), in the production of potassium salts or as a component in composts (see below).

Use of sugar beet vinasse as a fertilizer/soil amendment

There is interest in the use of beet vinasse as a fertilizer/soil amendment, including its potential value in this respect to organic agriculture (e.g. Schmitz and Fischer, 2003). Debruck and Lewicki (1997) estimate that beet vinasse concentrated to 68° Brix contains 8–9% K_2O and 2–3% total N, so application of 2.5 t/ha is roughly equivalent to 80 kg N/ha and 200–225 kg K_2O/ha. These authors found that application of concentrated beet vinasse at 2.7–3.5 t/ha improved yields in a range of crops by 10–25% compared with untreated controls. Other relevant publications in this area include those by Kunteova (1997), and a series by Madejon and colleagues based on work in Spain (Madejon *et al.*, 1995, 1996, 2001a,b; Diaz *et al.*, 2002a,b). These latter studies focus on co-composting of concentrated depotassified beet vinasse with other agro-processing wastes (such as grape marc, olive pressed cake and cotton gin trash) where direct application of vinasse is constrained by its relatively high density and salinity. Generally, the resulting compost increased yields of study crops, as did inorganic fertilizer. Unlike inorganic fertilizer, however, the composts also tended to increase soil organic matter content. However, there was also evidence that, under certain conditions, vinasse-based compost could result in increased leaching when compared with inorganic fertilizer (2–180 mg/l versus 3–25 mg/l nitrate in drainage water).

Sugar Beet as Biofuel

Sugarcane has long been explored as a source of fuel alcohol in Brazil and elsewhere (see above), and there was also early interest in using beet in a similar way, notably in France (Winner, 1993). Winner (1993) notes that a beet sugar yield of 8–10 t/ha corresponds to an output of 5000–6000 l/ha of ethanol. More recent studies have continued to consider the potential of sugar beet as a source of ethanol for fuel (e.g. Serase *et al.*, 1993; Jossart *et al.*, 1995; Kunteova, 1996; Mach and Svatos, 1998). However, others suggest that using beet as a biomass fuel for the production of electricity, rather than as a source of bioethanol for vehicles, is ecologically and economically preferable (Hanegraaf *et al.*, 1998).

Other Sugar Beet By-products

Gryllus and Anyos (1993) describe how ammonia-rich condensates from one sugar beet factory were treated, reducing emissions of this pollutant gas and yielding 5.6 t/day $(NH_4)_2SO_4$ as a fertilizer by-product.

References

Abbas Keshavarz (1998) Drainage and irrigation network for sugarcane production in Iran. In *Drainage in the 21st Century: Food Production and the Environment. Proceedings of the 7th International Drainage Symposium* (ed. L.C. Brown), pp. 314–320. American Society of Agricultural Engineers (ASAE), St Joseph.

Abedi-Koupai, J., Afyuni, M., Mostafazadeh, B. and Bagheri, M.R. (2001) Influence of treated wastewater and irrigation systems on soil physical properties in Isfahan province. In *International Workshop on Wastewater Reuse Management, 52nd IEC Meeting of the International Commission on Irrigation and Drainage* (eds R. Ragab, G. Pearce, J.C. Kim, S. Nairizi and A. Hamdy), pp. 165–173. Korean National Committee on Irrigation and Drainage (KCID), Gyonggi-do.

Abotal, R.D. and Cabigon, L.G. (2001) Analysis of metals in the wastewater of Crystal Sugar Company, Inc., North Poblacion, Maramag, Bukidnon. *CMU Journal of Science* Vol. 9, pp. 38–54.

Abruna-Rodriquez, F. and Vicente-Chandler, J. (1967) Sugarcane yields as related to acidity of a humid tropical Ultisol. *Agronomy Journal*, Vol. 59, pp. 330–332.

Accorsi, C.A. and Zama, F. (1996) Low pressure steam semi-industrial plant for pulp drying. *Zuckerindustrie*, Vol. 121, pp. 251–255.

Agbossou, K.E., Baldy, C. and Tiercelin, J.R. (1995) The importance of extra irrigation for improving sugar-cane production in Benin. *Secheresse*, Vol. 6, pp. 325–330.

Agrawal, M.L. (2001) Research strategies for improving the productivity and quality of sugarcane under rainfed and deficient irrigation conditions in Uttar Pradesh, India. *Sugar Technology*, Vol. 3, pp. 1–3.

Ah-Koon, D., Ng Cheong, L.R., Teeluck, M., Ratna, D. and Muthy, K. (2000a) Strategies for deficit irrigation of sugar cane in Mauritius. *International Sugar Journal* Vol. 102, pp. 648–650.

Ah-Koon, D., Ng Cheong, L.R., Teeluck, M., Ratna, D. and Muthy, K. (2000b) Strategies for deficit irrigation of sugar cane in Mauritius. *Sugar Cane International*, February, pp. 16–18.

Ahluwalia, M.S., Singh, K.J., Baldev Singh and Sharma, K.P. (1998) Influence of drip irrigation on water use and yield of sugarcane. *International Water and Irrigation Review* Vol. 18, pp. 12–17.

Ahmed, A. (1991) Irrigation hazards due to poor management: a case study of Kano River Project, Nigeria. In *Techniques for Environmentally Sound Water Resources Development. African Regional Symposium Proceedings* (ed. R. Wooldridge), pp. 21–33. Pentech Press, London.

Ahmad, A. and Singh, P.P. (1991) Environmental impact assessment for sustainable development: Chittaurgarh irrigation project in outer Himalayas. *Ambio* Vol. 20, pp. 228–302.

Ahmad, N. (1996) Management of Vertisols in rainfed conditions. In *Vertisols and Technologies for their Management* (eds N. Ahmad and A. Mermut), pp. 363–428. Developments in Soil Science 24, Elsevier, Amsterdam.

Akhtar, M. and Silva, J.A. (1999) Assessment of crop residue cover with photographic slide and point intercept methods. *Tropical Agriculture* Vol. 76, pp. 205–207.

Akinerdem, F. (1992) Effect of drought and number of irrigation treatments on yield and quality of sugarbeet (*Beta vulgaris saccharifera* L.). *Doga, Turk Tarim ve Ormancilik Dergisi* Vol. 16, pp. 757–762.

Alaban, R.A., Barredo, F.C. and Aguirre, A.L. (1990) An assessment of some indicators and determinants of farm productivity and soil fertility in the VMC district: trends, associations, interactions 1969–1990. *PHILSUTECH Proceedings*, Vol. 37, pp. 64–83.

Alexander, A.G. (1985) *The Energy Cane Alternative*. Elsevier, Amsterdam.

Ali, A. and Ahmad, M.S. (1993) Environmental chemistry of groundwater in parts of the Gandak basin, India. *Journal of Environmental Hydrology*, Vol. 1, No. 4, pp. 31–36.

Ali, A.D., Reagan, T.E. and Flynn, J.L. (1984) Influence of selected weedy and weed-free sugarcane habitats on diet composition and foraging activity of the imported fire ant (Hymenoptera: Formicidae). *Environmental Entomology*, Vol. 13, No. 4, pp. 1037–1041.

Ali, M.M. and Soltan, M.E. (1996) The impact of three industrial effluents on submerged aquatic plants in the River Nile, Egypt. *Hydrobiologia*, Vol. 340, No. 1/3, pp. 77–83.

Allen, C.J., Mackay, M.J., Aylward, J.H. and Campbell, J.A. (1997) New technologies for sugar milling and by-product modification. In *Intensive Sugarcane Production: Meeting the Challenges Beyond 2000* (eds B.A. Keating and J.R. Wilson), pp. 267–285. CAB International, Wallingford, UK.

Allison, M.F. (1989) Sugar beet, straw incorporation and nitrogen. *British Sugar Beet Review*, Vol. 57, pp. 37–39.

Allison, M.F. and Armstrong, M.J. (1991) The nitrate leaching problem – are catch crops the solution? *British Sugar Beet Review* Vol. 59, pp. 8–11.

Allison, M.F., Jaggard, K.W. and Last, P.J. (1992) Effects of straw incorporation on the yield, nitrogen fertilizer and insecticide requirements of sugarbeet (*Beta vulgaris*). *Journal of Agricultural Science, Cambridge*, Vol. 118, pp. 199–206.

Allison, M.F., Brown, S. and Pilbrow, J. (1993) Nitrogen catch crops and sugar beet. *British Sugar Beet Review*, Vol. 61, pp. 16–19.

Allison, M.F., Armstrong, M.J., Jaggard, K.W., Todd, A.D. and Milford, G.F.J. (1996) An analysis of the agronomic, economic and environmental effects of applying N fertilizer to sugarbeet (*Beta vulgaris*). *Journal of Agricultural Science*, Vol. 127, No. 4, pp. 475–486.

Allison, M.F., Armstrong, M.J., Jaggard, K.W. and Todd, A.D. (1998a) Integration of nitrate cover crops into sugar beet rotations. I. Management and effectiveness of nitrate cover crops. *Journal of Agricultural Science, Cambridge*, Vol. 130, pp. 53–60.

Allison, M.F., Armstrong, M.J., Jaggard, K.W. and Todd, A.D. (1998b) Integration of nitrate cover crops into sugar beet rotations. II. Effect of cover crops on growth, yield and N requirements of sugar beet. *Journal of Agricultural Science, Cambridge*, Vol. 130, pp. 61–67.

Allsopp, P.G. and Manners, J.M. (1997) Novel approaches for managing pests and diseases in sugarcane. In *Intensive Sugarcane Production: Meeting the Challenges Beyond 2000* (eds B.A. Keating and J.R. Wilson), pp. 173–188. CAB International, Wallingford, UK.

Almazan, O. (1994) Past, present and future of sugarcane by-products in Cuba. *Cooperative Sugar*, Vol. 26, pp. 253–263.

Altaf-ur-Rehman and Said Rehman (1998) Effect of nitrogen fertilization and irrigation on the juice quality of sugarcane. *Pakistan Sugar Journal*, Vol. 12, No. 2, pp. 24–28.

Alva, A.K., Prakash, O. and Paramasivam, S. (1999) Leaching of nitrogen forms, cations, and metals as influenced by compost amendment to a candler fine sand. *Journal of Environmental Science and Health. Part A, Toxic/Hazardous Substances and Environmental Engineering*, Vol. A34, No. 7, pp. 1473–1483.

Amador, D.B. and Viana, V.M. (2000) 'Low forests' dynamics in a forest remnant restoration. *Scientia Forestalis*, No. 57, pp. 69–85.

Amitabh, P.C.S., Sharma, N.K. and Poras, A.S. (1999) Environmental pollution control in sugar mill. In *Proceedings of the 61st Annual Convention of the Sugar Technologists' Association of India, New Delhi, India, 7–9 September, 1999*, pp. c27–c34. Sugar Technologists Association of India, New Delhi.

Anderson, D.L. and Dean, J.L. (1986) Relationship of rust severity and plant nutrients in sugarcane. *Phytopathology* Vol. 76, pp. 581–585.

Anderson, D.L. and Rosendahl, P. (1997) The development and application of environmental policy: the Everglades and Florida sugar industry. In *Intensive Sugarcane Production – Meeting the Challenges Beyond 2000* (eds B.A. Keating and J.R. Wilson), pp. 381–402. CAB International, Wallingford, UK.

Anderson, D.L. and Rosendahl, P.C. (1998) Development and management of land/water resources: the Everglades, agriculture, and south Florida. *Journal of the American Water Resources Association*, Vol. 34, No. 2, pp. 235–249.

Anderson, J.M. (1994) Functional attributes of biodiversity in land use systems. In *Soil Resilience and Sustainable Land Use* (eds D.J. Greenland and I. Szabolcs), pp. 267–290. CAB International, Wallingford, UK.

Anecksamphant, C. and Wichaidit, P. (2002) Utilization of water for rainfed agricultural areas in Northeast Thailand. In *JIRCAS International Symposium Series, No. 10* (eds M. Yajima, K. Okada and N. Matsumoto) pp. 87–91. Japan International Research Centre for Agricultural Sciences, Tsukuba-shi.

Angayarkanni, J., Palaniswamy, M. and Swaminathan, K. (2003) Biotreatment of distillery effluent using *Aspergillus niveus*. *Bulletin of Environmental Contamination and Toxicology*, Vol. 70, No. 2, pp. 268–277.

Anon. (1984) *Integrated Development Plan, Dodoma Region [Tanzania]: Development of Small Scale Irrigation* (2 vols.), 375 pp. Agrar-und Hydrotechnik GMBH, Essen, German Federal Republic.

Anon. (1991) Alfa Laval launches bioearth process. *Bharatiya Sugar*, Vol. 16, No. 6, p. 20.

Anon. (1992) Has sugar harmed the Everglades? *Sugar y Azucar*, Vol. 87, No. 3, pp. 32–33.

Antonelli, M., Petrini, A. and Santilocchi, R. (2001) Conservation tillage of durum wheat in Central Italy. In *Conservation Agriculture, a Worldwide Challenge. First World Congress on Conservation Agriculture, Madrid, Spain, 1–5 October 2001*, Vol. 2 (eds L. Garcia-Torres, J. Benites and A. Martinez-Vilela), pp. 51–54. XUL, Cordoba.

Anwar, M., Mubarik Ahmed and Rizvi, M.H. (2002) Effect of lodging and preharvest irrigation on juice quality of sugarcane varieties in Nawabshah, Sindh. *Pakistan Sugar Journal*, Vol. 17, No. 3, pp. 22–27.

Aoki, T. and Salleh, B.B. (2001) Analyses on the diversity of fungal communities found on sugarcane and rice plants in Malaysia. *Soil Microorganisms*, Vol. 55, No. 2, pp. 81–91.

Araba, A. and Byers, F.M. (2002) Environmentally friendly beef production from two genotypes fed cereal/molasses diets. *Livestock Production Science*, Vol. 77, No. 2/3, pp. 301–309.

Arain, M.A., Mubarik Ahmed and Khan, M.A. (2000) Some physico-chemical characteristics of soil in sugarcane cultivated areas of Nawabshah, Sindh, Pakistan. *Pakistan Journal of Botany*, Vol. 32, pp. 93–100.

Araji, A.A. and Hafez, S.L. (2001) The economic and environmental impacts of nematode biocontrol methods: an ex-ante approach. *Nematropica*, Vol. 31, No. 2, pp. 181–192.

Arakel, A.V., Hill, C.M., Piorewicz, J. and Connor, T.B. (1989) Hydro-sedimentology of the Johnstone River estuary. In *Sediment Water Interaction* (eds P.G. Sly and B.T. Hart). *Hydrobiologia*, Vol. 176/177, pp. 51–60.

Arakel, A.V., Loder, T., McConchie, D. and Pailles, C. (1993) Environmental consequences of land degradation in coastal drainage basins of north Queensland, Australia: influence of farming practices. *Land Degradation and Rehabilitation*, Vol. 4, No. 2, pp. 99–112.

Araujo, A.R.B., Teixeira, M.I.J.G. and Rodrigues, R.R. (1999) Floristics and phytosociology of a section of cerrado in the municipality of Franca. *Naturalia (São Paulo)*, Vol. 24, pp. 153–170.

Armas, F.M., Garcia Batista, R., Alvarez, B.Q. and Alvarez Velasco, R. (1991) Effect of intensive sugarcane cultivation on some physical properties of a fersiallitic soil. *ATAC, Revista de la Asociacion de Tecnicos Azucareros de Cuba*, No. 3–4, pp. 27–36.

Arnold, J. (2003) The 2002 Campaign – ZAF (Zuckerfabriken Aarberg und Frauenfeld AG). *Zuckerindustrie*, Vol. 128, pp. 355–357.

Arroyo, J.M., Urbano, P., Rojo, C. and Gonzalez, F. (1999) Irrigation of sugarbeet: comparative study between drip irrigation and sprinklers. *Agricultura, Revista Agropecuaria*, Vol. 68, No. 803, pp. 490–493.

Arthington, A.H., Marshall, J.C., Rayment, G.E., Hunter, H.M. and Bunn, S.E. (1997) Potential impact of sugarcane production on riparian and freshwater environments. In *Intensive Sugarcane Production: Meeting the Challenges Beyond 2000* (eds B.A. Keating and J.R. Wilson), pp. 403–421. CAB International, Wallingford, UK.

Artigao, A., Ortega, J.F., Tarjuelo, J.M. and de Juan, J.A. (2002) The impact of irrigation application upon soil physical degradation in Castilla-La Mancha (Spain). *Advances in Geoecology*, No. 35, pp. 83–90.

Arzola Pina, N. and Yera Martin, B. (1995a) Perspectives on the use of the remnants from the paper mill 'Sergio Gonzalez' in the irrigation of sugarcane. *Centro Azucar*, Vol. 22, No. 2, pp. 23–30.

Arzola Pina, N. and Yera Martin, B. (1995b) Making profitable the remnants of torula ferments in the ferti-irrigation of sugar cane. *Centro Azucar*, Vol. 22, No. 2, pp. 73–77.

Arzola Pina, N. and Yera Martin, B. (1995c) Possibilities of irrigation of sugarcane with the effluent of the bagasse and the liquid residual of the sugar production. *Centro Azucar*, Vol. 22, No. 2, pp. 87–91.

Asad, M.T., Kheradnam, M., Kamkar-Haghighi, A.A., Karimian, N.A. and Farsinejad, K. (2000) Sugarbeet response to N and irrigation levels and time of N application. *Iranian Journal of Agricultural Sciences*, Vol. 31, No. 3, pp. 427–443.

Ascough, G.W. and Kiker, G.A. (2002) The effect of irrigation uniformity on irrigation water requirements. *Water SA*, Vol. 28, No. 2, pp. 235–241.

Asher, M.J.C., Luterbacher, M.C. and Frese, L. (2001) Wild *Beta* species as a source of resistance to sugar-beet pests and diseases. *Proceedings of the Institut International de Recherches Betteravières*, Vol. 64, pp. 141–152.

Ashok Varma (2002) Biological control of sugarcane insect pests in India. In *Sugarcane Crop Management* (eds S.B. Singh, G.P. Rao and S. Easwaramoorthy), pp. 677–695. Sci Tech Publishing, Houston.

Askew, M.F. and Holmes, C.A. (2002) The potential for biomass and energy crops in agriculture in Europe, in land use, policy and rural economy terms. *International Sugar Journal*, Vol. 104, pp. 482, 486, 488, 490–492.

Aso, G., Ruiz, R.M., Paz, D. and Cardenas, G.J. (2001) Analysis of possible water reuse in sugar industry. *Revista Industrial y Agricola de Tucuman*, Vol. 78, No. 1/2, pp. 71–79.

Atkinson, B. (2004) The CDM, Kyoto Protocol and the sugar, ethanol and bio-fuels industry. *International Sugar Journal*, Vol. 106, pp. 175–177.

Atkinson, P.R. and Nuss, K.J. (1989) Associations between host-plant nitrogen and infestations of the sugarcane borer, *Eldana saccharina* Walker (Lepidoptera: Pyralidae). *Bulletin of Entomological Research*, Vol. 79, pp. 489–506.

Attard, S.J., Inman-Bamber, N.G. and Engelke, J. (2003) Irrigation scheduling in sugarcane based on atmospheric evaporative demand. *Proceedings of the Australian Society of Sugar Cane Technologists*, Vol. 25, p. 35.

Autrey, L.J.C. (1999) ISSCT environmental symposium sustainable agricultural practices. In *Proceedings of the International Society of Sugar Cane Technologists*, Vol. 24, pp. 48–51.

Avillez, F. and Ramos Rocha, J.A. (1988) Agricultural production on irrigated land in Portugal: present situation and future development; marketing and processing of irrigated agricultural products in Portugal. *Revista de Estudios Agro-Sociales*, Vol. 37, No. 143, pp. 105–128, 129–149.

Avram-Waganoff, P. (1990) Technology transfer between the beet and cane sugar industries. *Zuckerindustrie*, Vol. 115, No. 7, pp. 544–550.

Azhar Javed, M. and Tariq, M. (2003) *Water Crises and Production Gaps in Punjab, Risk Analysis and Mitigation Measures*. Science International, Lahore.

Azzazy, N.B. (1998) Effect of sowing date, irrigation interval and nitrogen fertilization on yield and quality of sugar beet under Upper Egypt conditions. *Egyptian Journal of Agricultural Research*, Vol. 76, No. 3, pp. 1099–1112.

Azzazy, N.B., Mohamed, H.A. and El-Geddawy, I.H. (1999) Effect of delivery delay on quality parameters of some sugarcane varieties under drip and surface irrigation. *Egyptian Journal of Agricultural Research*, Vol. 77, No. 2, pp. 745–768.

Azzopardi, M., Salvetti, M., Kaddatz, P., Schmidtke, K. and Quirk, R. (2002) Compass: helping Queensland and New South Wales' cane growers move in the right direction. *Proceedings of the Australian Society of Sugar Cane Technologists*, Vol. 24, pp. 304–310.

Baguette, M. and Hance, T. (1997) Carabid beetles and agricultural practices: influence of soil ploughing. *Biological Agriculture and Horticulture*, Vol. 15, No. 1–4, pp. 185–190.

Bailey, R. (1990) *Irrigated Crops and their Management*, 192 pp. Farming Press Books, Ipswich.

Bailey, R.J. and Minhinick, J. (1993) The agricultural requirement for water, with particular reference to irrigation. In *Agriculture and the Environment* (ed. G.J. Jones), pp. 107–116. Ellis Horwood Ltd, Chichester.

Baker, P., Haylock, L.A., Garner, B.H., Sands, R.J.N. and Dewar, A.M. (2002) The effects of insecticide seed treatments on beneficial invertebrates in sugar beet. In *BCPC Conference: Pests and Diseases*, Vols 1 and 2. *Proceedings of an International Conference Held at the Brighton Hilton Metropole Hotel, Brighton, UK, 18–21 November 2002*, pp. 653–658. British Crop Protection Council, Farnham.

Bakewell, D. (1980) Wind erosion control. An assessment of methods available. *British Sugar Beet Review*, Vol. 48, No. 3, pp. 7–9.

Bakker, D.M., Raine, S.R. and Robertson, M.J. (1997) A preliminary investigation of alternate furrow irrigation for sugar cane production. *Proceedings of the Australian Society of Sugar Cane Technologists*, Vol. 19, pp. 302–309.

Bakker, H. (1999) *Sugar Cane Cultivation and Management*. Kluwer Academic/Plenum Publishers, New York.

Balakrishnan, R. and Nair, N.V. (2003) Strategies for developing core collections of sugarcane (*Saccharum officinarum* L.) germplasm – comparison of sampling from diversity groups constituted by three different methods. *Plant Genetic Resources Newsletter* No. 134, pp. 33–41.

Balakrishnan, R., Nair, N.V. and Sreenivasan, T.V. (2000) A method for establishing a core collection of *Saccharum officinarum* L. germplasm based on quantitative-morphological data. *Genetic Resources and Crop Evolution*, Vol. 47, pp. 1–9.

Balasubramanian, V. and Kanehiro, Y. (1976) Denitrification potential and pattern of gaseous N loss in tropical Hawaiian soils. *Tropical Agriculture (Trinidad)*, Vol. 53, pp. 293–303.

Baliddawa, C.W. (1985) Plant species diversity and crop pest control. An analytical review. *Insect Science and its Application*, Vol. 6, No. 4, pp. 479–487.

Ballantyne, I. (1998) Environment and efficiency – a contradiction? In *7th ISO Seminar 'Conflictual issues in world sugar', London, UK, 24–25 November, 1998*, 10 pp. International Sugar Organisation, London.

Bangar, K.S., Sharma, S.R. and Rathore, O.P. (1993) Economy in fertilizer budget in sugarcane through biofertilizers and organic amendments under limited water supply condition. *Indian Sugar*, Vol. 43, No. 4, pp. 237–242.

Baran, R., Tuo, K. and Parriaud, J. (1999) Geographic information systems, a sugarcane irrigation management tool. *Agriculture et Developpement*, No. 24, pp. 141–146.

Barberi, P. (1997) Control of weed infestations by non-chemical means. *Informatore Agrario*, Vol. 53, No. 11, pp. 25–32.

Barberi, P., Silvestri, N. and Bonari, E. (1995) Weed flora development in different cropping systems. II. Effects on weed community structure. *Rivista di Agronomia*, Vol. 29, No. 4, pp. 523–532.

Bartocci, F., Siciliano, G., Guzzini, A. and Trucchi, A. (2001) Monitoring of nitrogen content in soils and sugarbeets leafs in Central Italy. *Proceedings of the Institut International de Recherches Betteravières*, Vol. 64, pp. 413–418.

Bartsch, D., Lehnen, M., Clegg, J., Pohl-Orf, M., Schuphan, I. and Ellstrand, N.C. (1999) Impact of gene flow from cultivated beet on genetic diversity of wild sea beet populations. *Molecular Ecology*, Vol. 8, pp. 1733–1741.

Bartsch, D., Slyvchenko, O., Lehnen, M., Mucher, T., Saeglitz, C., Driessen, S., Wehres, U., Hoffmann, A., Gathmann, A. and Schuphan, I. (2002) Gene flow from transgenic crops – implication for plant genetic resources. *Byulleten' Gosudarstvennogo Nikitskogo Botanicheskogo Sada*, No. 85, pp. 48–55, 67–68.

Bartsch, D., Cuguen, J., Biancardi, E. and Sweet, J. (2003) Environmental implications of gene flow from sugar beet to wild beet – current status and future research needs. *Environmental Biosafety Research*, Vol. 2, pp. 105–115.

Baruah, A.K., Sharma, R.N. and Borah, G.C. (1993) Impact of sugar mill and distillery effluents on water quality of River Gelabil, Assam. *Indian Journal of Environmental Health*, Vol. 35, No. 4, pp. 288–293.

Bastiaanssen, W., Ahmad, M.-u.-D. and Tahir, Z. (2003) Upscaling water productivity in irrigated agriculture using remote-sensing and GIS technologies. In *Water Productivity in Agriculture: Limits and Opportunities for Improvement* (eds J.W. Kijne, R. Barker and D. Molden), pp. 289–300. CAB International, Wallingford, UK.

Bastow, J., Peck, W., Astill, C., Bakewell, D. and Roebuck, D. (1978) Wind erosion. Three methods of lessening the damage caused by the big blow. *British Sugar Beet Review*, Vol. 46, No. 3, pp. 14–17.

Basu, P. and Talukdar, J. (1997) An environment friendly boiler for generation of power of process steam in sugar industries. In *Proceedings of the 59th Annual Convention of the Sugar Technologists' Association of India, Goa, India, 26–28 September 1997*, pp. E111–E119. Sugar Technologists Association of India, New Delhi.

Bateman, R.P. (2003) Rational pesticide use: spatially and temporally targeted application of specific products. In *Optimizing Pesticide Use* (ed. M. Wilson). John Wiley & Sons, Chichester, pp. 131–159.

Bazza, M. and Tayaa, M. (1999) Contribution to improve sugar beet deficit-irrigation. In *Crop Yield Response to Deficit Irrigation* (eds C. Kirda, P. Moutonnet, C. Hera and D.R. Nielsen), pp. 148–160. Kluwer Academic Publishers, Dordrecht.

Becker, K.W. and Bruss, A. (1996) Effects of long-term fertilizer N reduction on winter grain and sugar beet yields. *Journal of Agronomy and Crop Science*, Vol. 177, No. 5, pp. 339–354.

Becker, T., Schmidt, H. and Steffen, G. (1989) *Effects on the Farm Business of a Ban on Certain Plant Protection Materials in Agriculture. Effects of a Ban on the Use of Particular Active Substances in Plant Protection Materials on Farm Businesses with Various Types of Farming System.* Reihe A, Angewandte Wissenschaft No. 373, Schriftenreihe des Bundesministers fur Ernahrung, Landwirtschaft und Forsten. Landwirtschaftsverlag GmbH, Munster-Hiltrup.

Bedford, S.E. and Usher, M.B. (1994) Distribution of arthropod species across the margins of farm woodlands. *Agriculture, Ecosystems and Environment*, Vol. 48, No. 3, pp. 295–305.

Beeharry, R.P. (2001) Carbon balance of sugarcane bioenergy systems. *Biomass and Bioenergy*, Vol. 20, pp. 361–370.

Bekker, W. (1992) Go ahead for start on TSB's new Komati Mill. *South African Sugar Journal*, Vol. 76, No. 11, pp. 350–351.

Bell, J.N.B. (1984) Air pollution problems in western Europe. In *Gaseous Air Pollutants and Plant*

Metabolism (eds M.J. Koziol and F.R. Whatley), pp. 3–24. Butterworths, London.

Bengtson, R.L., Selim, H.M. and Ricaud, R. (1998) Water quality from sugarcane production on alluvial soils. *Transactions of the ASAE*, Vol. 41, No. 5, pp. 1331–1336.

Benhnini, A. (1991) Management and economics of water in sugar manufacture. *Sucrerie Maghrébine*, No. 46/47, pp. 32–34.

Bennewicz, J., Kaczorowski, G. and Barczak, T. (2001) Aphids in midfield thickets in Lower Vistula Valley Landscape Park. *Journal of Plant Protection Research*, Vol. 41, No. 4, pp. 348–362.

Berbel, J. and Gomez-Limon, J.A. (2000) The impact of water-pricing policy in Spain: an analysis of three irrigated areas. *Agricultural Water Management*, Vol. 43, No. 2, pp. 219–238.

Berding, N. and Skinner, J.C. (1987) Traditional breeding methods. In *Proceedings Copersucar International Sugarcane Breeding Workshop*, pp. 269–320. Copersucar Technology Centre, Piracicaba, Brazil.

Berg, M., Haas, G. and Kopke, U. (1997) Water protection areas: comparison of nitrate leaching in organic, integrated and conventional farming systems. *Schriftenreihe – Institut für Organischen Landbau*, Vol. 4, pp. 28–34.

Bernstein, L., Francois, L.E. and Clark, R.A. (1966) Salt tolerance of NCo varieties of sugarcane. 1 – Sprouting growth and yield. *Agronomy Journal* Vol. 58, pp. 489–493.

Berthelot, P.B. and Robertson, C.A. (1990) A comparative study of the financial and economic viability of drip and overhead irrigation of sugarcane in Mauritius. *Agricultural Water Management*, Vol. 17, No. 1–3, pp. 307–315.

Bhaid, S.B. (1990) Economy of water in small and medium size bagasse based paper mills: a practical approach. In *Thirty-eighth Annual Convention 1988: Deccan Sugar Technologists' Association (India)*, Part 1, pp. B23–B28. Deccan Sugar Technologists' Association, Pune.

Bhale, V.M., Dhoble, M.V., Gunjkar, M.U. and Shinde, B.M. (2002) Scheduling irrigation in sugarcane based intercropping system. In *Proceedings of the 64th Annual Convention of the Sugar Technologists' Association of India, Cochin (Kerala), India, 17–19 August 2002*, pp. A48–A57. Sugar Technologists Association of India, New Delhi.

Bhoi, P.G., Bankar, M.C., Raskar, B.S. and Shinde, S.H. (1999) Effect of fertigation and planting techniques on yield and quality of suru sugarcane under drip irrigation. *Indian Sugar*, Vol. 49, No. 7, pp. 487–492.

Biagi, B. (1986) Development of a data base of land characteristics and computerized analysis of actual and potential land degradation risks. In *Soil Erosion in the European Community – Impact of Changing Agriculture* (eds G. Chisci and R.P.C. Morgan), pp. 125–142. A.A. Balkema, Rotterdam.

Bielas, B. (2001) The first stage in waste water purification – open or closed systems. *Gazeta Cukrownicza*, Vol. 109, No. 9, pp. 155–156.

Biggs, J.S., Keating, B.A. and Thorburn, P.J. (2000) Time trends of nitrate in groundwaters under intensive agriculture in the Bundaberg region. *Proceedings of the Australian Society of Sugar Cane Technologists*, Vol. 22, pp. 296–301.

Birch, R.G. (1997) Transgenic sugarcane: opportunities and limitations. In *Intensive Sugarcane Production: Meeting the Challenges Beyond 2000* (eds B.A. Keating and J.R. Wilson), pp. 125–140. CAB International, Wallingford, UK.

Birkas, M. (2001) Environmentally protective solution for the system of cultivation of sugarbeet. *Cukoripar*, Vol. 54, No. 2, pp. 60–64.

Bishop, R.T. and Kruger, E.M. (1979) Soil modification with copolymer emulsions. *Proceedings of the South African Sugar Technologists' Association*, Vol. 53, pp. 177–181.

Bizik, J. (1989) Long-term observation of the movement of soil solution and nitrate in a chernozem profile and the influence of irrigation and fertilization. Protection of water quality from harmful emissions with special regard to nitrate and heavy metals. In *Proceedings of the 5th International Symposium of CIEC* (eds E. Welte and I. Szabolcs) pp. 119–122. CIEC, Budapest.

Blackburn, F. (1984) *Sugarcane*. Longman Group, Harlow.

Blagodatsky, S.A., Blagodatskaya, E.V. and Anderson, T.H. (2002) Kinetic respiration analysis as a tool for monitoring of soil microbial communities under environmental impact. *Mitteilungen der Deutschen Bodenkundlichen Gesellschaft*, Vol. 99, pp. 145–146.

Blair, N. (2000) Impact of cultivation and sugar-cane green trash management on carbon fractions and aggregate stability for a Chromic Luvisol in Queensland, Australia. *Soil and Tillage Research*, Vol. 55, No. 3/4, pp. 183–191.

Bloesch, P.M., Barry, G.A., Rayment, G.E. and Beattie, R.N. (2003) Stockpiled mill mud/ash: environmental implications and changes in nutrient value with age. *Proceedings of the Australian Society of Sugar Cane Technologists*, Vol. 25, p. 8.

Boddey, R.M., Urquiaga, S., Reis, V. and Dobereiner, J. (1991) Biological nitrogen fixation associated with sugar cane. *Plant and Soil*, Vol. 137, pp. 111–117.

Bohl, H.P., Bonnett, G.D., Fanning, D.J., Rayment, G.E. and Davidson, A.B. (2002) Biological oxygen demand and sugars in irrigation water runoff from sugarcane fields. *Proceedings of the Australian Society of Sugar Cane Technologists*, Vol. 24, pp. 297–303.

Boinchan, B. and Lykov, A. (1999) Results of long-term trials with fertilizers. *Mezhdunarodnyi Sel'skokhozyaistvennyi Zhurnal*, No. 6, pp. 42–45.

Bonari, E., Mazzoncini, M., Masoni, A., Bertolacci, M. and Perioli, R. (1992) Yield response of maize (*Zea mays L.*), soyabeans (*Glycine max L.*), sunflowers (*Helianthus annuus L.*) and sugarbeet (*Beta vulgaris L.*) to different irrigation regimes in north-central Italy. *Rivista di Agronomia*, Vol. 26, No. 4, suppl., pp. 676–684.

Booij, C.J.H. and Noorlander, J. (1992) Farming systems and insect predators. *Agriculture, Ecosystems and Environment*, Vol. 40, No. 1–4, pp. 125–135.

Boonthum, A., Aree, J., Lairungreung, C. and Sophanodora, D. (1997) Effects of leaving burnt cane in the field on yield and quality of sugarcane. *Kasetsart Journal, Natural Sciences*, Vol. 31, pp. 224–231.

Booth, R.J. and Lightfoot, C.J. (1990) The reticulation of ethanol stillage through irrigation systems and its use for fertilization of sugarcane in Zimbabwe. *Agricultural Water Management*, Vol. 17, pp. 49–58.

Borhidi, A., Salabarria Fernandez, D. and Voros, L. (1986) Hydrobotanical studies in the framework of Cuban Zapata Project. *Acta Botanica Hungarica*, Vol. 32, No. 1–4, pp. 119–133.

Borin, M., Bonaiti, G. and Giardini, L. (2001) Controlled drainage and wetlands to reduce agricultural pollution: a lysimetric study. *Journal of Environmental Quality*, Vol. 30, pp. 1330–1340.

Borowczak, F. (1996) Influence of weather conditions, sprinkler irrigation and nitrogen fertilizer application on the yield of sugarbeet seeds. *Zeszyty Problemowe Postepow Nauk Rolniczych*, Vol. 438, pp. 69–75.

Borowczak, F. and Grzes, S. (2002a) Influence of irrigation, foliar fertilization and nitrogen application on the yields and economic effects of sugar beet cultivation for seeds. *Biuletyn Instytutu Hodowli i Aklimatyzacji Roslin*, No. 222, pp. 155–163.

Borowczak, F. and Grzes, S. (2002b) Influence of irrigation, foliar fertilization and nitrogen application on root yields and on economic effects of sugar beet cultivation. *Biuletyn Instytutu Hodowli i Aklimatyzacji Roslin*, No. 222, pp. 203–213.

Borowiec, S. and Zablocki, Z. (1989) The effect of cultivated plants on nitrate-nitrogen concentrations in drainage water. Protection of water quality from harmful emissions with special regard to nitrate and heavy metals. In *Proceedings of the 5th International Symposium of CIEC* (eds E. Welte and I. Szabolcs), pp. 259–262. CIEC, Budapest.

Bos, J.F.F.P. and van de Ven, G.W.J. (2000) Mixing specialized farming systems in Flevoland (The Netherlands): agronomic, environmental and socio-economic effects. *Netherlands Journal of Agricultural Science*, Vol. 48, No. 3/4, pp. 185–200.

Boschi, V. and Chisci, G. (1978) Influence of the crop and of the watershed on the runoff and erosion in hilly clay areas. *Genio Rurale*, Vol. 41, No. 4, pp. 7–16.

Boschi, V., Chisci, G. and Ghelfi, R. (1985) The introduction of a lucerne ley in crop rotations on hilly lands: effects on runoff and soil erosion. *Rivista di Agronomia*, Vol. 18, No. 3/4, pp. 199–215.

Bosworth, M. (2000) Billington's – the unrefined sugar specialists. *International Sugar Journal*, Vol. 102, pp. 654–655.

Bowman, G., Hicks, W., Fitzpatrick, R. and Davies, P. (2000) Remediation options for the acid sulfate soil 'hotspot' at East Trinity Inlet, Cairns, North Queensland. In *Proceedings of a Workshop on Remediation and Assessment of Broadacre Acid Sulphate Soils* (ed. P. Slavich), pp. 130–145. NSW Agriculture, Wollongbar.

Braddock, D. and Downs, P. (2001) Wastewater irrigation – a strategy for increasing sugarcane production. *Proceedings of the International Society of Sugar Cane Technologists*, Vol. 24, pp. 171–173.

Bramley, R.G.V., Ellis, N., Nable, R.O. and Garside, A.L. (1996) Changes in soil chemical properties under long-term sugar cane monoculture and their possible role in sugar yield decline. *Australian Journal of Soil Research*, Vol. 34, No. 6, pp. 967–984.

Bramm, A. (1988) *Rotations, Analysis and Outlook. Use of a Greater Number of Species in the Rotation.* Angewandte Wissenschaft No. 361, Schriftenreihe des Bundesministers für Ernahrung, Landwirtschaft und Forsten, A, German Federal Republic, 101 pp.

Brants, I. and Harms, H. (1998) Herbicide tolerant sugar beet. *Proceedings of the Institut International de Recherches Betteravières*, Vol. 61, pp. 195–204.

Braunack, M.V. (1997) The effect of soil physical properties on growth and yield of sugarcane. *Sugar Cane*, Vol. 2, pp. 4–12.

Braunack, M.V. and Hurney, A.P. (1996) Reducing soil physical constraints under sugarcane. In

Sugar 2000 Symposium: Sugarcane: Research Towards Efficient and Sustainable Production (eds J.R. Wilson, D.M. Hogarth, J.A. Campbell and A.L. Garside), pp. 217–219. CSIRO Division of Tropical Crops and Pasture, Brisbane.

Brejski, E. (1998) Influence of modernizing the heat economy on the coal consumption and the decrease of impurity emission to atmosphere, on the example of Brzesc Kujawski SA sugar factory. *Gazeta Cukrownicza*, Vol. 106, No. 10, pp. 192–194.

Brentrup, F., Kusters, J., Kuhlmann, H. and Lammel, J. (2001) Application of the Life Cycle Assessment methodology to agricultural production: an example of sugar beet production with different forms of nitrogen fertilizers. *European Journal of Agronomy*, Vol. 14, No. 3, pp. 221–233.

Brodie, J.E., Hicks, W.S., Richards, G.N. and Thomas, F.G. (1984) Residues related to agricultural chemicals in the ground waters of the Burdekin River Delta, North Queensland. *Environmental Pollution*, Vol. 8, pp. 187–215.

Brunotte, J. (1990) *Agricultural Engineering Measures to Protect and Preserve the Soil during Sugarbeet Growing*. No. 183, Forschungsbericht Agrartechnik des Arbeitskreises Forschung und Lehre der Max-Eyth-Gesellschaft (MEG), 205 pp.

Brunotte, J. and Sommer, C. (1993) Tramlines in sugar beet cultivation. *Landtechnik*, Vol. 48, No. 8–9, pp. 395, 468–470.

Brzesowsky, W.J. and van Vilsteren, A.E.M. (1988) An economic evaluation of sugar cane production under different water supply systems in Thailand. *Agricultural Water Management*, Vol. 13, No. 1, pp. 83–91.

Buchholz, K., Jordening, H.J. and Nahle, C. (1992) Installation of scrubbers for elimination of organic compounds from [pulp-] dryer vapours. *Zuckerindustrie*, Vol. 117, No. 12, pp. 991–992.

Buchs, W., Harenberg, A. and Zimmermann, J. (1997) The invertebrate ecology of farmland as a mirror of the intensity of the impact of man? An approach to interpreting results of field experiments carried out in different crop management intensities of a sugar beet and an oil seed rape rotation including set-aside. *Biological Agriculture and Horticulture*, Vol. 15, No. 1–4, pp. 83–107.

Bugaenko, I.F. (1996) Purification of wastewaters from sugar manufacture abroad. *Sakharnaya Promyshlennost'*, No. 5, pp. 27–29.

Buniak, W., Dmowski, Z. and Szyszkowski, P. (1996) The yield and quality of sugarbeet roots as affected by sprinkler irrigation and nitrogen fertilizer application. *Zeszyty Problemowe Postepow Nauk Rolniczych*, Vol. 438, pp. 267–272.

Bunn, S.E., Davies, P.M. and Kellaway, D.M. (1997) Contributions of sugar cane and invasive pasture grass to the aquatic food web of a tropical lowland stream. *Marine and Freshwater Research*, Vol. 48, No. 2, pp. 173–179.

Burcky, K. and Vierling, G. (2001) Experiences with BISZ – the interactive consulting and information package via the internet for the South German sugar beet cultivation. *Proceedings of the Institut International de Recherches Betteravières*, Vol. 64, pp. 89–92.

Burke, R.A., Molina, M., Cox, J.E., Osher, L.J. and Piccolo, M.C. (2003) Stable carbon isotope ratio and composition of microbial fatty acids in tropical soils. *Journal of Environmental Quality*, Vol. 32, No. 1, pp. 198–206.

Businelli, D., Tombesi, E. and Trevisan, M. (2001) Modelling herbicide treatment impact on groundwater quality in a central Italy area. *Agronomie*, Vol. 21, No. 3, pp. 267–276.

Buzzanell, P.J. (2000a) Organic sugar: short term fad or long term growth opportunity? In *Hot Issues for Sugar (Proceedings of the 9th ISO Seminar, London, 21–22 November 2000)*, pp. 24–43. International Sugar Organization, London.

Buzzanell, P.J. (2000b) The US organic sugar market. *Sugar y Azucar*, Vol. 95, pp. 18–30.

Byrnes, B.H. (1990) Environmental effects of N fertilizer use – an overview. *Fertilizer Research*, Vol. 26, pp. 209–215.

Calero, L.M., Lopez, C. and Cortes, E. (2001) Analysis of the dispersion of particulate material from fixed sources in a Colombian sugarcane mill. *Proceedings of the International Society of Sugar Cane Technologists*, Vol. 24, pp. 258–260.

Camp, C.R., Lamm, F.R., Evans, R.G. and Phene, C.J. (2000) Subsurface drip irrigation – past, present, and future. National irrigation symposium. In *Proceedings of the 4th Decennial Symposium, Phoenix, Arizona, USA, 14–16 November 2000*, pp. 363–372. American Society of Agricultural Engineers, St. Joseph.

Camposeo, S., Rubino, P. and Caliandro, A. (2001) Irrigation scheduling of sugar beet with the drip method in Southern Italy. *Agricoltura Mediterranea*, Vol. 131, No. 3/4, pp. 164–175.

Canadian Forest Service (1995) *Sugar Maple: Characteristics, Ecology and Management. L'érable à sucre: caractéristiques, écologie et aménagement*, xviii + 394 pp. Ministere de l'Agriculture, des Pecheries et d'Alimentation, Quebec.

CANEGROWERS (1998) *Code of Practice for Sustainable Cane Growing in Queensland*. CANEGROWERS, Brisbane.

Cantele, A. and Zanin, G. (1992) Influence of crop rotation, irrigation and fertilization on quali-quantitative composition of the potential flora. *Rivista di Agronomia*, Vol. 26, No. 4, pp. 470–481.

Capone, L.T., Izuno, F.T., Bottcher, A.B., Sanchez, C.A., Coale, F.J. and Jones, D.B. (1995) Nitrogen concentrations in agricultural drainage water in south Florida. *Transactions of the ASAE*, Vol. 38, No. 4, pp. 1089–1098.

Cardenas, G.J. (1993) Ethanol from bagasse as fuel. Contribution to lowering of atmospheric CO_2. *Ingenieria Quimica*, Vol. 25, No. 288, pp. 113–116.

Caron, J., Espindola, C.R. and Angers, D.A. (1996) Soil structural stability during rapid wetting – influence of land use on some aggregate properties. *Soil Science Society of America Journal*, Vol. 60, pp. 901–908.

Carruthers, I.D. (1987) Irrigation design for management – irrigation under threat: a warning brief for management. *Irrigation Design for Management. Asian Regional Symposium*, Vol. 2, pp. 33–44. Hydraulics Research, Wallingford.

Cavalli, A.C., Lombardi Neto, F., Garcia, G.J. and Zullo Junior, J. (2000) Use of AVHRR/NOAA-14 multi-temporal data to evaluate soil degradation. *Acta Scientiarum*, Vol. 22, No. 4, pp. 1037–1043.

Cavanagh, J.E., Burns, K.A., Brunskill, G.J. and Coventry, R.J. (1999) Organochlorine pesticide residues in soils and sediments of the Herbert and Burdekin river regions, North Queensland – implications for contamination of the Great Barrier Reef. *Marine Pollution Bulletin*, Vol. 39, No. 1/12, pp. 367–375.

Cavazza, L., Venturi, G. and Amaducci, M.T. (1976) Outlines on the state of irrigation of the sugar beet in the world. *Proceedings of the Institut International de Recherches Betteravières*, Vol. 39, pp. 211–264.

Ceccatelli, M. and Peruzzi, A. (1995) Development of flame weeding machinery. *Macchine e Motori Agricoli*, Vol. 53, No. 4, pp. 11–18.

Ceddia, M.B., dos Anjos, L.H.C., Lima, E., Ravelli Neto, A. and da Silva, L.A. (1999) Sugar cane harvesting systems and changes on physical properties of a Yellow Podzolic soil in Espirito Santo, Brazil. *Pesquisa Agropecuaria Brasileira*, Vol. 34, No. 8, pp. 1467–1473.

Centurion, J.F., Cardoso, J.P. and Natale, W. (2000) Effect of method of management on some physical and chemical properties of a red latosol in different agroecosystems. *Revista Brasileira de Engenharia Agricola e Ambiental*, Vol. 5, No. 2, pp. 254–258.

Ceotto, E. (2001) Enhancing plant utilization of livestock manure: a cropping system perspective. *Gron Viden, Husdyrbrug*, No. 21, pp. 134–143.

Cerri, C.C. and Andreux, F. (1990) Changes in organic carbon content in Oxisols cultivated with sugar cane and pasture, based on 13C natural abundance measurement. In *14th International Congress of Soil Science*, Kyoto, Vol. IV, pp. 98–103.

Chabanne, A., Boyer, J., Michellon, R. and Seguy, L. (2001) Impact of plant covers on *Pelargonium* × *asperum* yields and soil biology (macrofauna) in Reunion (France). In *Conservation Agriculture, a Worldwide Challenge. First World Congress on Conservation Agriculture, Madrid, Spain, 1–5 October, 2001* (eds L. Garcia-Torres, J. Benites and A. Martinez-Vilela), Vol. 2: *Offered Contributions*, pp. 389–393. XUL, Cordoba.

Chambers, B.J. and Smith, K. (1999) Utilising organic manures with minimum risk to sugar beet quality. *British Sugar Beet Review*, Vol. 67, pp. 25–28.

Chang, C.J. and Lee, J.T. (1989) A practical report on a sieve plate scrubber for bagasse boiler at sugar factories. *Taiwan Sugar*, Vol. 36, No. 5, pp. 9–11.

Chang, C.J. and Lee, J.T. (1991) A practical report on an effective exhaust scrubbing system for bagasse boiler. *Taiwan Sugar*, Vol. 38, No. 5, pp. 9–16.

Chang, S.R. and Li, S.W. (1989) The disposal of alcoholic slops on sugar cane fields. *Proceedings of the International Society of Sugar Cane Technologists*, Vol. 20, pp. 551–557.

Chapman, L.S. (1997) Sugarcane yield responses from furrow irrigation at Mackay. *Proceedings of the Australian Society of Sugar Cane Technologists*, Vol. 19, pp. 285–292.

Chapman, L.S. and Usher, J.F. (1996) Social, economic and environmental issues associated with the biodunder by-product from the Sarina ethanol distillery. In *Sugar 2000 Symposium: Sugarcane: Research Towards Efficient and Sustainable Production* (eds J.R. Wilson, D.M. Hogarth, J.A. Campbell and A.L. Garside), pp. 295–297. CSIRO Division of Tropical Crops and Pasture, Brisbane.

Chapman, L.S., Haysom, M.B.C. and Saffigna, P.G. (1994) The recovery of ^{15}N from labelled urea fertilizer in crop components of sugarcane and in soil profiles. *Australian Journal of Agricultural Research*, Vol. 45, pp. 1577–1585.

Chapman, L.S., Haysom, M.B.C. and Usher, J.F. (1995) Canegrowers accept biodunder as a plant nutrient source at Mackay. *Proceedings of the Australian Society of Sugar Cane Technologists*, Vol. 17, pp. 199–206.

Chapman, R.N. and Milford, B.J. (1997) Economic constraints and opportunities in the Australian sugar industry. In *Intensive Sugarcane Production: Meeting the Challenges Beyond 2000* (eds B.A. Keating and J.R. Wilson), pp. 55–70. CAB International, Wallingford, UK.

Charleston, D.S., Kfir, R., van Rensburg, N.J., Barnes, B.N., Hattingh, V., Conlong, D.E., Visser, D. and Prinsloo, G.J. (2003) Integrated pest management in South Africa. In *Integrated Pest Management in the Global Arena* (eds K.M. Maredia, D. Dakouo and D. Mota-Sanchez), pp. 169–195. CAB International, Wallingford, UK.

Chartres, C.J. (1981) Land resources assessment for sugar-cane cultivation in Papua New Guinea. *Applied Geography*, Vol. 1, pp. 259–271.

Chaudhry, F.H. and Leme, E.J.A. (1996) Optimal irrigation scheduling of sugar cane by net profit maximization through dynamic programming. *Evapotranspiration and Irrigation Scheduling. Proceedings of the International Conference, San Antonio, Texas, USA, 3–6 November 1996* (eds C.R. Camp, E.J. Sadler and R.E. Yoder), pp. 1122–1127. American Society of Agricultural Engineers, St Joseph.

Chauhan, H.S. (1995) Issues of standardisation and scope of drip irrigation in India. In *Microirrigation for a Changing World: Conserving Resources/Preserving the Environment. Proceedings of the Fifth International Microirrigation Congress, Orlando, Florida, USA, 2–6 April 1995* (ed. F.R. Lamm), pp. 446–451. American Society of Agricultural Engineers, St Joseph.

Chawla, A.S., Raghuvanshi, C.S. and Anil Kumar (1989) Access of small and marginal farmers to groundwater – a field study. *IAHS Publication* No. 188, pp. 29–40.

Cheeroo-Nayamuth, F.C., Robertson, M.J., Wegener, M.K. and Nayamuth, A.R.H. (2000) Using a simulation model to assess potential and attainable sugar cane yield in Mauritius. *Field Crops Research*, Vol. 66, No. 3, pp. 225–243.

Chen, M.H. (1984) Effects of sulfur dioxide on organelles of crops. *Journal of Agricultural Research of China*, Vol. 33, No. 4, pp. 387–394.

Chen Han, Li Yan and Tao GuiZi (1995) The optimization of crop distribution and rational exploitation of ground water resources. *Ziranziyuan*, No. 1, pp. 50–54.

Cherry, M. (1983) Preventing soil blow. *Big Farm Management*, February, pp. 25–27.

Cherry, R. (2003) The effect of harvesting and replanting on arthropod ground predators in Florida sugarcane. *Florida Entomologist*, Vol. 86, No. 1, pp. 49–52.

Chinloy, T. and Hogg, B.M. (1968) Rotation of pangola grass and sugarcane on a Jamaican sugar estate. *Proceedings of the International Society of Sugar Cane Technologists*, Vol. 13.

Chinnusamy, C. and Jayanthi, C. (2000) Computation of crop evapotranspiration for scheduling irrigation in sugarcane. *Madras Agricultural Journal*, Vol. 87, No. 1/3, pp. 44–47.

Chisci, G. (1986a) Influence of change in land use and management on the land degradation phenomenon in Appennines hilly areas. In *Soil Erosion in the European Community – Impact of Changing Agriculture* (eds G. Chisci and R.P.C. Morgan), pp. 3–16. A.A. Balkema, Rotterdam.

Chisci, G. (1986b) Time-lapse analysis for planning purposes of landslides and erosion hazard, in relation to changes in land use and management. In *Soil Erosion in the European Community – Impact of Changing Agriculture* (eds G. Chisci and R.P.C. Morgan), pp. 77–88. A.A. Balkema, Rotterdam.

Chisci, G. and Boschi, V. (1988) Runoff and erosion control with hill farming in the sub-coastal Apennines climate. *Soil and Tillage Research*, Vol. 12, No. 2, pp. 105–120.

Chisci, G. and Morgan, R.P.C (eds) (1986) *Soil Erosion in the European Community – Impact of Changing Agriculture.* A.A. Balkema, Rotterdam.

Cho, T. and Kuroda, M. (1987) Soil moisture management and evaluation of water-saving irrigation on farms. *Journal of Irrigation Engineering and Rural Planning*, No. 12, pp. 25–40.

Choluj, D., Podlaski, S., Karwowska, R. and Chrobak, Z. (2001) Effect of soil moisture on seed germination and growth of sugar beet. In: *Proceedings of the Institut International de Recherches Betteravières*, Vol. 64, pp. 365–370.

Choudry, B.A. and Corpuz, I.T. (1984) Profitability of fertilizers and dolomite application to sugar cane in the Philippines. *Sugar y Azucar*, Vol. 79, pp. 29–44.

Christensen, B.T. (2004) Tightening the nitrogen cycle. In *Managing Soil Quality: Challenges in Modern Agriculture* (eds P. Schjonning, S. Elmholt and B.T. Christensen), pp. 47–67. CAB International, Wallingford, UK.

Christenson, D.R., Brimhall, P.B., Hubbell, L. and Bricker, C.E. (2000) Yield of sugar beet, soybean, corn, field bean and wheat as affected by lime application on alkaline soils. *Communications in Soil Science and Plant Analysis*, Vol. 31, pp. 1145–1154.

Christiansen, I. (1999) The code of practice for sustainable cane growing in Queensland – its development and future directions. *Proceedings of the Australian Society of Sugar Cane Technologists*, Vol. 21, pp. 267–273.

Christiansen, I. and Hunt, R. (2000) Research, extension and industry – working together can achieve results. *Marine Pollution Bulletin*, Vol. 41, No. 7/12, pp. 310–318.

Ciavatta, C. and Cecchini, M. (2002) Effect of leather-based fertilizers and irrigation on sugarbeet. *Informatore Agrario*, Vol. 58, No. 50, pp. 33–34.

Clay, W.D. (1998) What does sugar mean for health? In *Conflictual Issues in World Sugar (7th ISO Seminar)*, unpaginated. International Sugar Organization, London.

Coale, F.J. (1993) Sugarcane cultivar response to dolomite amendment of an acidic sandy soil. *Journal of the American Society of Sugar Cane Technologists*, Vol. 13, pp. 87–94.

Coale, F.J. and Schueneman, T.J. (1993) Sugarcane response to limestone and gypsum application on acidic sandy soils. *Journal of the American Society of Sugarcane Technologists*, Vol. 13, pp. 94–97.

Cock, J.H., Torres, J.S., Villegas, F., Amaya, A., Luna, C.A., Rangel, H., Viveros, C., Carbonel, J., Cortes, E., Larrahondo, J., Cassalett, C., Cruz, R., Gomez, L. and Victoria, J. (1999) Development of a green cane agriculture system. *International Sugar Journal (Worldwide Sugar Edition)*, Vol. 101, No. 1201, pp. 28–35.

Cock, M.J.W. (ed.) (1985) *A Review of Biological Control of Pests in the Commonwealth Caribbean and Bermuda up to 1982*. Commonwealth Agricultural Bureaux, Slough.

Coghlan, A. (2003a) Farming 1, wildlife 0. *New Scientist*, Vol. 180, No. 2418, p. 21.

Coghlan, A. (2003b) GM crops benefit the environment. *New Scientist*, Vol. 180, No. 2424, p. 17.

Colwell, R.R. (1992) Environmental impact – workshop. In *The Release of Genetically Modified Microorganisms: REGEM 2: Proceedings of a Symposium Held under the Auspices of the Federation of European Microbiological Societies, 29–31 August 1991, in Nottingham, UK* (eds D.E.S. Stewart-Tull, and M. Sussman), pp. 123–127. Plenum Publishing Co. Ltd., London.

Combres, J.C., Kamieniarz, C. and Maraux, F. (1996) Control of sugarcane irrigation under network constraints using IRRICANE. In *Sustainability of Irrigated Agriculture: Crop–Water–Environment Models. Proceedings of Workshop at the 16th ICID Congress, Cairo, Egypt, 17 September 1996* (eds R. Ragab, D.E.-D. El-Quosy, B. van Den Broek and L.S. Pereira), pp. 255–266. International Commission on Irrigation and Drainage (ICID), New Delhi.

Conte, C.A. (1999) Colonial science and ecological change. Tanzania's Mlalo basin, 1888–1946. *Environmental History*, Vol. 4, No. 2, pp. 220–244.

Contreras Moya, A.M., Herrera Rodriguez, Z. and Rosa Dominguez, E. (1998) Technical/economic analysis of physico-chemical treatment schemes for refinery wastewaters. *Centro Azucar*, Vol. 25, No. 3, pp. 82–86.

Cooke, D.A. and Scott, R.K. (eds) (1993) *The Sugar Beet Crop – Science into Practice*. Chapman and Hall, London.

Cortez, L., Freire, W.J. and Rosillo-Calle, F. (1998) Biodigestion of vinasse in Brazil. *International Sugar Journal (Cane Sugar Edition)*, Vol. 100, No. 1196, pp. 403–404, 409–413.

Courtenay, P. (1980) *Plantation Agriculture*, 2nd edn. Bell and Hyman, London.

Covarelli, G. and Onofri, A. (1998) Effects of timing of weed removal and emergence in sugar beet. In *Comptes-rendus 6ème symposium méditerranéen EWRS, Montpellier, France, 13–15 mai 1998* (ed J. Maillet), pp. 65–72. ENSA, Montpellier.

Crossland, C.J., Done, T.J. and Brunskill, G.J. (1997) Potential impacts of sugarcane production on the marine environment. In *Intensive Sugarcane Production: Meeting the Challenges Beyond 2000* (eds B.A. Keating and J.R. Wilson), pp. 423–436. CAB International, Wallingford, UK.

Cruz, R.L., Righetto, A.M. and Nogueira, M.A. (1991) Experimental investigation of soil and groundwater impacts caused by vinasse disposal. *Water Science and Technology*, Vol. 24, pp. 77–85.

Cukaliev, O. and Iljovski, I. (1993) Micro sprinkler irrigation and its effect on sugarbeet yield and sugar content. In *Integrated Resource Management and Landscape Modification for Environmental Protection. Proceedings of the International Symposium, Chicago, Illinois, USA, 13–14 December 1993* (ed. J.K. Mitchell), pp. 248–253. American Society of Agricultural Engineers, St Joseph.

Culverwell, T.L., Proksch, L. and Swartz, C. (1999) An extension pilot project to promote sustained irrigation scheduling of sugarcane crops by Umfolozi growers. *Proceedings of the Annual Congress – South African Sugar Technologists' Association*, No. 73, pp. 3–8.

Curry, J.P. and Purvis, G. (1982) Studies on the influence of weeds and farmyard manure on the arthropod fauna of sugar beet. *Journal of Life Sciences, Royal Dublin Society*, Vol. 3, No. 2, pp. 397–408.

Daberkow, S. (1997) Adoption rates for selected crop management practices. *Choices. The Magazine of Food, Farm, and Resources Issues*, No. 3, pp. 26–30.

Dahlman, R.C., Strain, B.R. and Rogers, H.H. (1985) Research on the response of vegetation to evaluated atmospheric carbon dioxide. *Journal of Environmental Quality*, Vol. 14, No. 1, pp. 1–8.

Dalri, A.B. and Cruz, R.L. (2002) Effect of subsurface drip irrigation frequency on growth of sugarcane (*Saccharum* spp.). *IRRIGA*, Vol. 7, No. 1, pp. 29–34.

Damandeep Singh (1994) India uses TV advertisements to counter dam protests. *Panoscope (London)*, No. 38, pp. 22–23.

Darwall, W. (2003) Water and biodiversity: a wise investment. *World Conservation*, Vol. 1, p. 24.

Das, S.K., Ahmad, A., Tripathi, S.K. and Singh, R.P. (1994) Effect of soil water conservation practices and sulphur fertilization on seed yield and quality of sunflower on eroded typic Ustochrept. *Journal of the Indian Society of Soil Science*, Vol. 42, No. 3, pp. 491–493.

da Silva, I. de F., de Andrade, A.P., Campos Filho, O.R. and de Oliveira, F.A.P. (1986) Effect of different crop canopies and soil conservation practices on erosion control. *Revista Brasileira de Ciencia do Solo*, Vol. 10, No. 3, pp. 289–292.

Dasrat, B., Rajkumar, A., Richards-Haynes, C., Quashie-Williams, C. and Eastwood, D. (1997) Evaluation of *Allorhogas pyralophagus* Marsh (Hymenoptera: Braconidae) for the biological control of *Diatraea* spp. (Lepidoptera: Pyralidae) in sugar-cane in Guyana. *Crop Protection*, Vol. 16, pp. 723–726.

Dasrat, L. (2001) Preliminary investigation of biological control of *Castniomera licus* (Drury) (Lepidoptera; Castniidae) by entomopathogenic nematodes. *Proceedings of the West Indies Sugar Technologists'*, Vol. 27.

Davidoff, B. and Hanks, R.J. (1989) Sugar beet production as influenced by limited irrigation. *Irrigation Science*, Vol. 10, pp. 1–17.

Davis, H.B. (1997) Cane productivity changes in the Guyana sugar industry from 1970 to 1996. *Proceedings of the West Indies Sugar Technologists*, Vol. 26, pp. 9–18.

Dawson, M.W., Lukacs, G.P. and Sockhill, D.G. (1995) Establishment of a constructed wetland at Victoria Mill. *Proceedings of the Australian Society of Sugar Cane Technologists*, No. 17, pp. 270–278.

de Boer, H.G. (1997) Strip tillage in sugar cane agriculture: techniques and challenges. *Zuckerindustrie*, Vol. 5, pp. 371–375.

de Brabander, D.L. and Boucque, C.V. (1992) Feeding dairy cattle in a more environment friendly way. *Revue de l'Agriculture*, Vol. 45, No. 2, pp. 253–264.

Debruck, J. and Lewicki, W. (1997) Sugarbeet vinasse: fertilizes soil and plants. *Neue Landwirtschaft*, Vol. 6, pp. 54–57.

Declerck, S., Strullu, D.G. and Plenchette, C. (1998) Monoxenic culture of the intraradical forms of *Glomus* sp. isolated from a tropical ecosystem: a proposed methodology for germplasm collection. *Mycologia*, Vol. 90, No. 4, pp. 579–585.

Defra (Department for Environment Food and Rural Affairs) (2002) *Sugar Beet and the Environment in the UK*. Report by the United Kingdom in accordance with Article 47(3) of Council Regulation 1260/2001 on the environmental situation of agricultural production in the sugar sector. Department for Environment Food and Rural Affairs, London. *http://www.defra.gov.uk/corporate/consult/eisugar/report.pdf*

de Geus, J.G. (1973) *Fertilizer Guide for the Tropics and Subtropics*. Centre d'Etude de l'Azotec, Zurich; The State of Food and Agriculture – Agriculture Series No. 29, FAO, Rome.

Deil, U. and Sundermeier, A. (1992) The influence of agricultural systems on both sides of the Straits of Gibraltar. In *IXe Colloque international sur la biologie des mauvaises herbes, 16–18 September 1992, Dijon, France*, pp. 207–218. ANPP, Paris.

de la Rosa, D. (1989) Land evaluation/land information system to predict crop growth potentials in Andalucia, Spain. *Agricultural Engineering. Proceedings of the 11th International Congress on Agricultural Engineering, Dublin, Irish Republic, 4–8 September 1989* (eds V.A. Dodd and P.M. Grace), Vol. 4, pp. 2697–2703. A.A. Balkema, Rotterdam.

Demmers-Derks, H., Mitchell, R.A.C., Mitchell, V.J. and Lawlor, D.W. (1998) Response of sugar beet (*Beta vulgaris* L.) yield and biochemical composition to elevated CO_2 and temperature at two nitrogen applications. *Plant, Cell and Environment*, Vol. 21, No. 8, pp. 829–836.

Denmead, O.T., Mayocchi, C.L. and Dunin, F.X. (1997) Does green cane harvesting conserve soil water? *Proceedings of the Australian Society of Sugar Cane Technologists*, Vol. 19, pp. 139–146.

den Nijs, H.C.M., Bartsch, D. and Sweet, J. (eds) (2004) *Introgression from Genetically Modified Plants into Wild Relatives*. CAB International, Wallingford, UK.

Dent, D.R., Holderness, M. and Vos, J.G.M. (2003) Integrated pest management at CAB International. In *Integrated Pest Management in the Global Arena* (eds K.M. Maredia, D. Dakouo and D. Mota-Sanchez), pp. 493–499. CAB International, Wallingford, UK.

De Ploey, J. (1986) Soil erosion and possible conservation measures in loess loamy areas. In *Soil Erosion in the European Community – Impact of Changing Agriculture* (eds G. Chisci and R.P.C. Morgan), pp. 157–164. A.A. Balkema, Rotterdam.

Dersch, G. and Bohm, K. (2001) Effects of agronomic practices on the soil carbon storage potential in arable farming in Austria. *Nutrient Cycling in Agroecosystems*, Vol. 60, pp. 40–55.

de Snoo, G.R. (1997) Arable flora in sprayed and unsprayed crop edges. *Agriculture, Ecosystems and Environment*, Vol. 66, No. 3, pp. 223–230.

de Snoo, G.R. and van der Poll, R.J. (1999) Effect of herbicide drift on adjacent boundary vegetation. *Agriculture, Ecosystems and Environment*, Vol. 73, No. 1, pp. 1–6.

de Snoo, G.R., de Jong, F.M.W., van der Poll, R.J., Janzen, S.E., van der Veen, L.J. and Schuemie, M.P. (1997) Variation of pesticide use among farmers in Drenthe: a starting point for environmental protection. *Mededelingen – Faculteit Landbouwkundige en Toegepaste Biologische Wetenschappen, Universiteit Gent*, Vol. 62, No. 2a, pp. 199–212.

Destain, J.P., Guiot, J., Francois, E. and Meeus-Verdinne, K. (1991) The contribution of ^{15}N balances in the study of nitrogen fertilization of cereals and its environmental impact. *Revue de l'Agriculture*, Vol. 44, No. 1, pp. 89–101.

Deville, J. (1999) Organic sugar production – the Mauritian experience. In *Proceedings of the XXIII ISSCT Congress, New Delhi, India, 22–26 February 1999* (eds V. Singh and V. Kumar), Vol. 1, pp. P64–P69.

de Vries, G.E. (ed.) (1999) *Past, Present and Future Considerations in Risk Assessment when Using GMO's. Workshop Proceedings, Leeuwenhorst Congress Centre Noordwijkerhout, the Netherlands, 5–6 March 1998.* 160 pp. ProBio Partners VOF, Overschild.

Dewar, A.M., May, M.J., Woiwod, I.P., Haylock, L.A., Champion, G.T., Garner, B.H., Sands, R.J.N., Qi AiMing and Pidgeon, J.D. (2003) A novel approach to the use of genetically modified herbicide tolerant crops for environmental benefit. *Proceedings of the Royal Society of London. Series B, Biological Sciences*, Vol. 270, No. 1513, pp. 335–340.

Dey, A.D. and Davis, H.B. (1997) Fertilizer nitrogen uptake by sugarcane in Guyana. *Proceedings of the West Indies Sugar Technologists*, Vol. 26, pp. 238–244.

Dey, A.D., Davis, H.B. and Friday, N. (1997) Potential for reduced tillage systems and legume fallow for management of clay and silty clay soils in Guyana sugar estates. *Proceedings of the West Indies Sugar Technologists*, Vol. 26, pp. 271–277.

Dhillon, G.S. and Panshi, B.S. (1987) Changing crop patterns in Punjab and their impact on declining watertable. *IPRI Research Magazine, Irrigation and Power Research Institute, Amritsar*, Vol. 13, No. 4, pp. 42–55.

Diaw, O.T., Vassiliades, G., Seye, M. and Sarr, Y. (1991) Epidémiologie de la bilhariose intestinale a *Schistosoma mansoni* à Richard-Toll Delta du fleuve Senegal – étude malacologique. *Bulletin de la Société de Pathologie Exotique et de ses Filiales*, Vol. 84, pp. 174–183.

Diaz, M.J., Madejon, E., Ariza, J., Lopez, R. and Cabrera, F. (2002a) Co-composting of beet vinasse and grape marc in windrows and static pile systems. *Compost Science and Utilisation*, Vol. 10, pp. 258–269.

Diaz, M.J., Madejon, E., Lopez, F., Lopez, R. and Cabrera, F. (2002b) Composting of vinasse and cotton gin waste by using two different systems. *Resources, Conservation and Recycling*, Vol. 34, pp. 235–248.

Dick, R.G. and Hurney, A.P. (1986) Technology change in sugarcane agriculture. *Proceedings of the Australian Society of Sugar Cane Technologists*, Vol. 8, 69–75.

Diegues, A.S. (1991) The role of cultural diversity and communal participation in wetland management in Brazil. *Landscape and Urban Planning*, Vol. 20, No. 1/3, pp. 61–66.

Diekkruger, B. and Roske, H. (1995) Modeling the population-dynamics of *Isotoma notabilis* (Collembola) on sites of different agricultural usage. *Pedobiologia*, Vol. 39, No. 1, pp. 58–73.

Dilip Yewalekar (1998) Micro irrigation: a foot step of the 21st century. In *Proceedings of the International Agricultural Engineering Conference, Bangkok, Thailand, 7–10 December 1998* (eds V.M. Salokhe and Z. Jianxia), pp. 684–693. Asian Institute of Technology, Bangkok.

Dinter, A. and Paarmann, W. (1989) Studies of the population ecology of arthropods of the sludge disposal site for [washed off] sugarbeet soils. *Verhandlungen Gesellschaft für Okologie*, Vol. 18, pp. 769–776.

Dixon, J., Cobb, A.H. and Sanders, G.E. (1995) The influence of phenmedipham and ozone pollution on cation leakage from sugarbeet (*Beta vulgaris* L. cv Saxon). In *Brighton Crop Protection Conference: Weeds. Proceedings of an International Conference, Brighton, UK, 20–23 November 1995*, Vol. 1, pp. 381–386. British Crop Protection Council, Farnham.

Dixon, J., Hull, M.R., Cobb, A.H. and Sanders, G.E. (1996) Phenmedipham-ozone pollution interactions in sugarbeet (*Beta vulgaris* L. cv. Saxon): a physiological study. *Pesticide Science*, Vol. 46, No. 4, pp. 381–390.

Dixon, T.F. and Burbidge, D. (2000) The future for cogeneration in the sugar industry. *Proceedings of the Australian Society of Sugar Cane Technologists*, Vol. 22, pp. 38–45.

Dobereiner, J., Urquiaga, S. and Boddey, R.M. (1995) Alternatives for nitrogen nutrition of crops in tropical agriculture. *Fertilizer Research*, Vol. 42, No. 1/3, pp. 339–346.

Dominy, C.S., Haynes, R.J. and van Antwerpen, R. (2001) Long-term effects of sugarcane production on soil quality in the south coast and the midlands areas of KwaZulu-Natal. *Proceedings of the Annual Congress – South African Sugar Technologists' Association*, No. 75, pp. 222–227.

Donaldson, R. (1990) Ripeners used in the sugar industry. *South African Sugar Journal*, Vol. 4, No. 9, pp. 290–291.

Doney, D.L. (1998) *Beta* evaluation and sugar beet enhancement from wild sources. In *Report on the 4th International Beta Genetic Resources Workshop and World Beta Network Conference* (eds L. Frese, L. Panella, H.M. Srivastava and W. Lange), pp. 73–76. International Crop Network Series No. 12.

dos Santos, C.Y.M., Azevedo, D. de A. and de Aquino Neto, F.R. de (2002) Selected organic compounds from biomass burning found in the atmospheric particulate matter over sugarcane plantation areas. *Atmospheric Environment*, Vol. 36, No. 18, pp. 3009–3019.

Doty, R. (1933) A bibliography on the use of molasses as a fertilizer. *Proc. Hawaiian Sugar Tech.*, pp. 165–168.

Downing, C.M., Hobson, P.A., Kent, G.A. and Burbidge, D. (2002) Is investment in a bagasse dewatering mill economically justifiable for cogeneration? *Proceedings of the Australian Society of Sugar Cane Technologists*, Vol. 24, pp. 347–351.

Draycott, A.P. (1993) Nutrition. In *The Sugar Beet Crop – science into practice* (eds D.A. Cooke and R.K. Scott), pp. 238–277. Chapman and Hall, London.

Draycott, A.P. and Christenson, D.R. (2003) *Nutrients for Sugar Beet Production – Soil–Plant Relationships*. CAB International, Wallingford, UK.

Draycott, A.P., Allison, M.F. and Armstrong, M.J. (1997) Changes in fertilizer usage in sugar beet production. *Proceedings of the Institut International de Recherches, Betteravières*, Vol. 60, pp. 39–54.

Dua, S.K. (1995) The future of microirrigation. Microirrigation for a changing world: conserving resources/preserving the environment. In *Proceedings of the Fifth International Microirrigation Congress, Orlando, Florida, USA, 2–6 April 1995* (ed. F.R. Lamm), pp. 341–346. American Society of Agricultural Engineers, St Joseph.

Dugan, P. (2003) Freshwater fisheries: maintaining the flow. *World Conservation*, Vol. 1, p. 25.

Dunham, R.J. (1993) Water use and irrigation. In *The Sugar Beet Crop – Science into Practice* (eds D.A. Cooke and R.K. Scott), pp. 278–310. Chapman & Hall, London.

Durai, R., Chinnaswami, K.N. and Kumaraswamy, K. (1996) Conservation of moisture through amendments under varying irrigation regimes on sugarcane (*Saccharum officinarum*). *Indian Journal of Agronomy*, Vol. 41, No. 3, pp. 468–471.

Duval, R. (2000) Long term effects of catch crop management on nitrate leaching in chalk soil. *Institut International de Recherches Betteravières Info*, Vol. 7, pp. 14–16.

Dyakov, K. (1997a) Evapotranspiration, yielding capacity and time of irrigation of sugarbeet. *Rasteniev'dni Nauki*, Vol. 34, No. 7/8, pp. 72–74.

Dyakov, K. (1997b) Effect of a deficient irrigation regime on some technological traits of sugarbeet. *Rasteniev'dni Nauki*, Vol. 34, No. 7/8, pp. 75–76.

Dzhodzhov, K. and Georgiev, G.S. (1980) The protective action of a single shelterbelt against wind erosion. *Gorskostopanska Nauka*, Vol. 17, No. 1, pp. 60–64.

Eastwood, D., Persaud, L., Dasrat, L., Niles, D.O. and Taylor, I.W. (1997a) Some thoughts on agrochemical management by a sugarcane industry. *Proceedings of the West Indies Sugar Technologists*, Vol. 26, pp. 28–39.

Eastwood, D., Persaud, L., Davis, H.B., Niles, D.O., Dasrat, L. and Taylor, I.W. (GuySuCo Agrochemical Safety Committee) (1997b) *The Safe Use of Pesticides*, 2nd edn. Guysuco, Georgetown, Guyana.

Eavis, B.W. and Chase, K. (1973) Trash mulch and sugar cane yields in Barbados. *Proceedings of the West Indies Sugar Technologists*, Vol. 18, pp. 147–157.

Eck, H.V., Winter, S.R. and Smith, S.J. (1990) Sugar beet yield and quality in relation to residual beef feed-lot waste. *Agronomy Journal*, Vol. 82, pp. 250–254.

Eckhoff, J.L.A. and Bergman, J.W. (2001) Sugarbeet (*Beta vulgaris*) production under sprinkler and flood irrigation. In *Proceedings from the 31st Biennial Meeting (Agriculture) of the American Society of Sugar Beet Technologists, Vancouver, BC, Canada, 28 February–3 March 2001*, pp. 239–241. American Society of Sugar Beet Technologists, Denver.

Eeckhaut, R. (2001) Sugar beet production and its organization in Belgium. *Proceedings of the Institut International de Recherches Betteravières*, Vol. 64, pp. 1–10.

El Antri, M. (2001) Weed control in sugarbeet in Morocco. *Proceedings of the Institut International de Recherches Betteravières*, Vol. 64, pp. 479–486.

El-Debaby, A.S., El-Geddawi, I.H., Saad, A.M.M. and Azzazy, N.B. (1996) Effect of some irrigation systems and nitrogen fertilizer on growth characters of some sugar cane varieties. *Annals of Agricultural Science, Moshtohor*, Vol. 34, No. 3, pp. 811–821.

El-Geddawy, I.H., El-Debaby, A.S., Saad, A.M.M. and Azzazy, N.B. (1996) Irrigation systems and nitrogen fertilizer in relation to yield and quality of sugar cane varieties. *Annals of Agricultural Science, Moshtohor*, Vol. 34, No. 3, pp. 823–837.

El-Geddawy, I.H., El-Debaby, A.S., Saad, A.M.M. and Azzazy, N.B. (1997) Irrigation systems and nitrogen fertilizer in relation to yield and quality of sugar cane varieties. *Egyptian Journal of Agricultural Research*, Vol. 75, No. 4, pp. 1037–1053.

Elliott, L.F., Miller, D.E. and Richards, A.W. (1979) Phytotoxicity of incorporated cover crops to sugarbeet seedlings. *Plant Disease Reporter*, Vol. 63, No. 10, pp. 882–886.

Elliott, M.C. and Weston, G.D. (1993) Biology and physiology of the sugar-beet plant. In *The Sugar Beet Crop – Science into Practice* (eds D.A. Cooke and R.K. Scott), pp. 37–65. Chapman & Hall, London.

Elliott, M.C., Chen, D.F., Fowler, M.R., Kirby, M.J., Kubalakova, M., Scott, N.W. and Slater, A. (1996) Transgenesis – a scheme for improving sugar beet productivity. *Russian Journal of Plant Physiology*, Vol. 43, No. 4, pp. 544–551.

Ellis, R.D., Wilson, J.H. and Spies, P.M. (1985) Development of an irrigation policy to optimize sugar production during seasons of water shortage. *Proceedings of the South African Sugar Technologists' Association*, Vol. 59, pp. 142–147.

El-Maayar, M., Bhawan Singh, Andre, P., Bryant, C.R. and Thouez, J.P. (1997) The effects of climatic change and CO_2 fertilization on agriculture in Quebec. *Agricultural and Forest Meteorology*, Vol. 85, No. 3/4, pp. 193–208.

Elnaghy, M.A., Gehan, Shaban, H., El-Khodary, S.A. and Yaser, M.M. (1998) Effect of organic amendments on *Trichoderma* in different Egyptian soil types. *African Journal of Mycology and Biotechnology*, Vol. 6, No. 1, pp. 27–39.

Elsheemy, A.H. and Elsayed, M.A. (1993) An economic study of the effect of irrigation on sugar-cane productivity in Egypt. *Arab Universities Journal of Agricultural Sciences*, Vol. 1, No. 1, pp. 79–87.

El-Swaify, S.A. (1997) Factors affecting soil erosion hazards and conservation needs for tropical steeplands. *Soil Technology*, Vol. 11, pp. 3–16.

El-Swaify, S.A. and Cooley, K.R. (1980) *Sediment Losses from Small Agricultural Watersheds in Hawaii (1972–77)*. Science and Education Administration, Agricultural Reviews and Manuals, Western Series No. ARM-W-17, US Department of Agriculture.

El-Titi, A. and Landes, H. (1990) Integrated farming system of Lautenbach: a practical contribution toward sustainable agriculture in Europe. In *Sustainable Agricultural Systems* (eds C.A. Edwards, R. Lal, P. Madden, R.H. Miller and G. House), pp. 265–286. Soil and Water Conservation Society, Ankeny.

El Wali, A.M.O., Le Grand, F. and Gascho, G.J. (1980) Nitrogen leaching from soil and uptake by sugarcane from various urea-based fertilizers. *Soil Science Society of America Journal*, Vol. 44, pp. 119–122.

Epperlein, K. and Schmidt, H.W. (2001) Effects of pelleting sugarbeet seed with Gaucho(R) (imidacloprid) on associated fauna in the agricultural ecosystem. *Pflanzenschutz-Nachrichten Bayer*, Vol. 54, No. 3, pp. 369–398.

Eppink, L.A.A.J. (1986) Water erosion in the Netherlands: damage and farmers attitudes. In *Soil erosion in the European Community – Impact of Changing Agriculture* (eds G. Chisci and R.P.C. Morgan), pp. 173–182. A.A. Balkema, Rotterdam.

Esbjerg, P. (1998) Effects of reduced pesticide doses on flora and fauna of agricultural fields – a project linked to the Danish Pesticide Action Plan. *DJF Rapport, Markbrug*, No. 3, pp. 109–120.

Esteban Baselga, M.J.A. (2001) RECOM – interactive recommendations for sugar beet growers. *Proceedings of the Institut International de Recherches Betteravières*, Vol. 64, pp. 93–100.

Evans, H. (1962) Flood-fallowing – a consideration of its mechanisms, its advantages and disadvantages. *Sugar Journal*, November, pp. 27–30.

Evertz, E. (1991) Reduction of steam plant emissions in the sugar industry. *Zuckerindustrie*, Vol. 116, No. 2, pp. 122–128.

Evfimenko, V.F. (2001) A working state of ventilation systems at sugar factories. *Tsukor Ukraini*, No. 1/2, pp. 46–47.

Fawzy, G.M. and Amer, K.M. (2002) Utilization of conserved and augmented water in horizontal expansion in Egypt. In *Integration and Management of Irrigation, Drainage and Flood Control. 18th International Congress on Irrigation and Drainage, Montreal, Canada, 2002*, Vol. 1B, p. 17. International Commission on Irrigation and Drainage (ICID), New Delhi.

Fedorov, L.A. (1999) Persistent organic chemicals in the former Soviet Union. *Environmental Pollution*, Vol. 105, No. 3, pp. 283–287.

Feizi, M. (1998) Long term effect of using saline water on soil salinity and sodicity. In *The Use of*

Saline and Brackish Water for Irrigation. Implications for the Management of Irrigation, Drainage and Crops. Proceedings of the International Workshop at the Tenth ICID Afro-Asian Regional Conference on Irrigation and Drainage, Denpasas, Bali, Indonesia, 19–26 July 1998 (eds R. Ragab and G. Pearce), pp. 89–95. Indonesian National Committee on Irrigation and Drainage (INACID), Jakarta.

Ferenc, K., Gyozo, S. and Sandor, F. (1998) Carabids (Coleoptera: Carabidae) in the northwestern ecosystem of a Hungarian territory (Kisalfold). *Novenyvedelem*, Vol. 34, No. 1, pp. 3–10.

Fernandez Lavandera, O. and Pizarro Checa, A. (1983) Economic aspects of agriculture in the Costa del Sol. *Revista de Estudios Agro-Sociales*, No. 124, pp. 7–20.

Fernandez-Quintanilla, C., Garrido Valero, M. and Zaragoza Larios, C. (eds) (1999) *Integrated Control of Weeds: Good Farming Practices. Control integrado de las malas hierbas: buenas practicas agricolas*, 271 pp. Phytoma Espana, Valencia.

Ferrer, S.R.D. and Nieuwoudt, W.L. (1997) Factors affecting soil conservation decisions of Kwazulu-Natal commercial sugar cane farmers. *Agrekon*, Vol. 36, No. 4, pp. 453–460.

Ferrer, S.R.D. and Nieuwoudt, W.L. (1998) Choices of soil conservation methods on KwaZulu-Natal commercial sugarcane farms. *Agrekon*, Vol. 37, No. 4, pp. 463–471.

Fick, G.W., Loomis, R.S. and Williams, W.A. (1975) Sugar beet. In *Crop Physiology* (ed. L.T. Evans), pp. 259–295. Cambridge University Press, Cambridge.

Finegan, M. (1990) An Australian development in soil conservation. *Australian Journal of Soil and Water Conservation*, Vol. 3, No. 1, p. 18.

Fiorio, P.R., Dematte, J.A.M. and Sparovek, G. (2000) Chronology and environmental impact of land use on Ceveiro Microbasin in Piracicaba region, Brazil. *Pesquisa Agropecuaria Brasileira*, Vol. 35, No. 4, pp. 671–679.

Firake, N.N., Bangal, G.B., Kenghe, R.N. and More, G.M. (1990) Plastic tunnel and mulches for water conservation. *Agricultural Engineering Today*, Vol. 14, No. 3–4, pp. 35–39.

Fisher, S.J. and Kerr, S.P. (1998) Sugar beet varietal responses to irrigation and autumn growth. *Aspects of Applied Biology*, No. 52, pp. 173–178.

Floras, S.A. and Sgouras, I.D. (2002) Sustainable soil and crop management in the Larissa area, Greece, using remote sensing and GIS. In *Man and Soil at the Third Millennium. Proceedings International Congress of the European Society for Soil Conservation, Valencia, Spain, 28 March–1 April 2000* (eds J.L. Rubio, R.P.C. Morgan, S. Asins and V. Andreu), Vol. 2, pp. 1919–1932. Geoforma Edicions SL, Logrona.

Follett-Smith, R.R. (1934) The flood fallowing of cane fields in British Guiana. *Tropical Agriculture*, Vol. 10, pp. 91–95.

Fornalek, W. (1995) Sugar industry and the natural environment. *Gazeta Cukrownicza*, Vol. 103, No. 9, pp. 166–168.

Fornstrom, K.J. and Boehnke, R.D. (1976) A growing mulch tillage system to reduce wind erosion losses of sugar beets. *Journal of American Society of Sugar Beet Technologists*, Vol. 19, No. 1, pp. 64–73.

Frede, H.G. (1986) *Danger of Erosion in Agriculture.* KTBL-Arbeitspapier No. 104, 36 pp.

Freire, W.J. and Trevisan, G. (2002) Concentrated vinasse emulsifying process in order to reduce soil-vinasse water absorption. *Engenharia Agricola*, Vol. 22, pp. 288–295.

Freney, J.R., Denmead, O.T., Wood, A.W. and Saffigna, P.G. (1994) Ammonia loss following urea addition to sugarcane trash blankets. *Proceedings of the Australian Society of Sugarcane Technologists*, Vol. 16, pp. 114–121.

Friedemann, I. (1992) Proposal for a new concept for keeping air clean in the sugar industry. *Zuckerindustrie*, Vol. 117, No. 2, p. 127.

Frielingshaus, M., Roth, R., Cebulsky, I. and Clausnitzer, J. (1986) Evaluation of crops and cropping sequences in relation to the protection of soil from erosion by water. *Archiv für Acker- und Pflanzenbau und Bodenkunde*, Vol. 30, No. 6, pp. 337–345.

Frielingshaus, M., Roth, R. and Cebulsky, I. (1988) Criteria for evaluating crops for their suitability to protect soil against water erosion. *Archiv für Acker- und Pflanzenbau und Bodenkunde*, Vol. 32, No. 1, pp. 31–40.

Frizzone, J.A., Matioli, C.S., Rezende, R. and Goncalves, A.C.A. (2001) Economic viability of supplementary irrigation in sugar cane plantations, *Saccharum* spp., of northern state of São Paulo. *Acta Scientiarum*, Vol. 23, No. 5, pp. 1131–1137.

Fry, J. (1997) A global perspective of the sugar industry. In *Intensive Sugarcane Production: Meeting the Challenges Beyond 2000* (eds B.A. Keating and J.R. Wilson), pp. 1–16. CAB International, Wallingford, UK.

Fuchs, C., Jene, B., Murschel, B. and Zeddies, J. (1995) Balancing greenhouse gases CO_2 and N_2O and possibilities of reducing emissions in arable farming. *Agrarwirtschaft*, Vol. 44, No. 4/5, pp. 175–190.

Fuentes, O., Hatch, R.O., Emmons, J. and Grzemski, A.S. (2001) Benefits for a beet handling system by polymer assisted mud removal in a flume clarifier. In *Proceedings from the 31st Biennial Meeting (Operations) of the American Society of Sugar Beet Technologists, Vancouver, BC, Canada, 28 February–3 March 2001*, pp. 161–170. American Society of Sugar Beet Technologists, Denver.

Fuhr, F. (1986) Application of 14C-labeled herbicides in lysimeter studies. *Weed Science*, Vol. 33, Suppl. 2, pp. 11–17.

Fuhrer, J., Lehnherr, B. and Stadelmann, F.X. (1989) Air pollution and agricultural crops in Switzerland. Estimating crop yield losses with the help of a model. *Landwirtschaft Schweiz*, Vol. 11, No. 2, pp. 669–673.

Gabellini, E., Cecchini, M. and Biavati, E. (2001a) Management of the introduction of irrigation on sugar beet in northern Italy. *Industria Saccarifera Italiana*, Vol. 94, No. 2, pp. 39–43.

Gabellini, E., Esteban Baselga, J.A., Gomez de Silva, F. and Maslaris, N. (2001b) Breeding progress in south of Europe. *Proceedings of the Institut International de Recherches Betteravières*, Vol. 64, pp. 205–216.

Gajja, B.L., Sharma, V.P. and Joshi, P.K. (1994) Productivity variation and land irrigability class in Kakrapar canal command area in Gujarat State. *Indian Journal of Agricultural Economics*, Vol. 49, No. 4, pp. 609–616.

Gajja, B.L., Sharma, V.P. and Parshad, R. (1997) Impact of soil salinity and waterlogging on agricultural system in Ukai-Kakrapar canal command area. *Current Agriculture*, Vol. 21, No. 1/2, pp. 1–22.

Gajja, B.L., Parshad, R. and Purohit, M.L. (1998) Impact of environmental degradation on agricultural production, incomes and employment under Canal Command area. In *Ecological Agriculture and Sustainable Development. Proceedings of International Conference on Ecological Agriculture: Towards Sustainable Development, Chandigarh, India, 15–17 November 1997* (eds G.S. Dhaliwal, R. Arora, N.S. Randhawa and A.K. Dhawan), Vol. 2, pp. 499–511.

Gajja, B.L., Gupta, I.C. and Parshad, R. (2000) Impact of imperfect drainage in soils on agricultural production system under surface irrigation – a case study of irrigation projects in Gujarat State. *Role of Drainage and Challenges in 21st Century. Proceedings of the Eighth ICID International Drainage Workshop, New Delhi, India, 31 January–4 February 2000*, Vol. I, pp. 349–362. International Commission on Irrigation and Drainage, New Delhi.

Gajja, B.L., Khem Chand and Parshad, R. (2001) Consequences of land heterogeneity on agricultural productivity, income and employment in the Kakrapar Left Bank Canal Command Area, Gujarat State. *Current Agriculture*, Vol. 25, No. 1/2, pp. 27–36.

Galindo, M.C., Vece, M.B., Perondi, M.E., Monserrat Araoz, M., Garcia, G., del V. Hidalgo, M., Apella, M.C. and Blesa, M.A. (2001) Chemical behavior of the Sali River, Province of Tucuman, Argentina. *Environmental Geology*, Vol. 40, No. 7, pp. 847–852.

Gallardo-Lopez, F., Riestra-Diaz, D., Aluja-Schunemann, A. and Martinez-Davila, J.P. (2002) Factors determining agricultural diversity and production objectives for agroecosystems in the municipality Paso de Ovejas, Veracruz, Mexico. *Agrociencia (Montecillo)*, Vol. 36, No. 4, pp. 495–502.

Gardiner, P.J.B. and Cazalet, K. (1991) Bench terracing incorporation deep contour tillage on shallow coastal soils of Natal. *Proceedings of the Annual Congress – South African Sugar Technologists' Association*, No. 65, pp. 68–72.

Garo, R. (1994) Land usage in Albania and some improvement aspects and agricultural use of salty soils. In *International Conference on Land and Water Resources Management in the Mediterranean Region, Valenzano, Bari, Italy, 4–8 September 1994*, No. 2, pp. 573–581.

Garrity, D.P. (1993) Sustainable land-use systems for sloping uplands in southeast Asia. In *Technologies for Sustainable Agriculture in the Tropics: Proceedings of Two International Symposia held in San Antonio, TX and Denver, CO, 1990 and 1991 respectively* (eds J. Ragland and R. Lal), pp. 41–66. American Society of Agronomy, Madison.

Garside, A.L., Berthelsen, J.E., Richards, C.L. and Toovey, L.M. (1996) Fallow legumes on the wet tropical coast: some species and management options. *Proceedings of the Australian Society of Sugar Cane Technologists*, Vol. 18, pp. 202–208.

Garside, A.L., Bramley, R.G.V., Bristow, K.L., Holt, J.A., Magarey, R.C., Nable, R.O., Pankhurst, C.E. and Skjemstad, J.O. (1997a) Comparisons between paired old and new land sites for sugarcane growth and yield and soil chemical, physical, and biological properties. *Proceedings of the Australian Society of Sugar Cane Technologists*, Vol. 19, pp. 60–66.

Garside, A.L., Smith, M.A., Chapman, L.S., Hurney, A.P. and Magarey, R.C. (1997b) The yield plateau in the Australian sugar industry 1970–1990. In *Intensive Sugarcane Production: Meeting the Challenges Beyond 2000* (eds B.A. Keating and J.R. Wilson), pp. 103–124. CAB International, Wallingford, UK.

Garside, A.L., Bell, M.J. and Magarey, R.C. (2001) Monoculture yield decline – fact not fiction?

Proceedings of the International Society of Sugar Cane Technologists, Vol. 24, pp. 16–21.

Garside, A., Bell, M., Berthelsen, J. and Halpin, N. (2003) Intensity of the production system influences the impact of yield decline in sugarcane. *Sugar Cane International*, November, pp. 3–5.

Gasiorowska, B. (1997) The effect of irrigation and NPK fertilization on the chemical composition of sugar beet roots and leaves. *Biuletyn Instytutu Hodowli i Aklimatyzacji Roslin*, No. 203, pp. 181–186.

Gaudin, R. (1999) Impact of water supply on sugarcane cultivation. *Agriculture et Développement*, No. 24, pp. 4–8.

Gaus, C., Papke, O., Dennison, N., Haynes, D., Shaw, G.R., Connell, D.W. and Muller, J.F. (2001) Evidence for the presence of a widespread PCDD source in coastal sediments and soils from Queensland, Australia. *Chemosphere*, Vol. 43, No. 4/7, pp. 549–558.

Gauthier, R. (1996) Vertebrate pests, crop and soil: the case for an agroforestry approach to agriculture on recently deforested land in North Lampung. *Agrivita*, Vol. 19, No. 4, pp. 206–212.

Gawander, J.S. (1998a) The rise and fall of the vetiver hedge in the Fijian sugar industry. *Advances in Geoecology*, No. 31, pp. 959–964.

Gawander, J.S. (1998b) Fiji. In *Environmental Assessment for Agricultural Development in Asia and the Pacific: Report of an APO Study Meeting 25 June–5 July 1996, Tokyo, Japan*, pp. 176–187. Asian Productivity Organisation (APO), Tokyo.

Gawander, J.S. and Morrison, R.J. (1999) Long term study of changes in the properties of a Fijian Oxisol following sugarcane cultivation. *Proceedings of the International Society of Sugar Cane Technologists*, Vol. 24, pp. 61–69.

Geelen, P.M.T.M., Kwaad, F.J.P.M., van Mulligen, E.J., Wansink, A.G., Zijp, M. and van den Berg, W. (1995) The impact of soil tillage on crop yield, runoff and soil loss under various farming systems of maize and sugarbeet on loess soils. *Publikatie – Proefstation voor de Akkerbouw en de Groenteteelt in de Vollegrond, Lelystad*, No. 78A, pp. 146–151.

Gemtos, T.A. and Cavalaris, C.K. (2001) Soil tillage effect in the sugarbeet crop. In *Conservation Agriculture, a Worldwide Challenge. First World Congress on Conservation Agriculture, Madrid, Spain, 1–5 October 2001.* Vol. 2: *Offered Contributions* (eds L. Garcia-Torres, J. Benites and A. Martinez-Vilela), pp. 539–543. XUL, Cordoba.

Germeier, C.U. and Frese, L. (2001) The international database of *Beta* – an expert system for Beta Genetic Resources. *Proceedings of the Institut International de Recherches Betteravières*, Vol. 64, pp. 123–132.

Ghassemi, F., Jakeman, A.J. and Nix, H.A. (1995) *Salinization of Land and Water Resources – Human Causes, Extent, Management and Case Studies.* CAB International, Wallingford, UK.

Ghugare, R.V., Bankar, M.C., Mane, T.A. and Kenjale, S.T. (1994a) Effect of drip and surface methods of irrigation. (1) Yield and quality of Suru sugarcane (Co.7219). *Indian Sugar*, Vol. 44, No. 6, pp. 415–418.

Ghugare, R.V., Bankar, M.C., Mane, T.A. and Kenjale, S.T. (1994b) Effect of drip and surface methods of irrigation on (II) yield and quality of ratoon sugarcane (Co 7219). *Indian Sugar*, Vol. 44, No. 7, pp. 503–506.

Glanville, T.J., Titmarsh, G., Sallaway, M.M. and Mason, F. (1997) Soil erosion in cane land tillage systems. *Proceedings of the Australian Society of Sugar Cane Technologists*, Vol. 19, pp. 254–262.

Goldemberg, J., Monaco, L.C. and Macedo, I.C. (1993) The Brazilian fuel-alcohol program. In *Renewable Energy: Sources of Fuels and Electricity* (eds T.B. Johansson, H. Kelly, A.K.N. Reddy, R.H. Williams and L. Burnham), pp. 841–863. Earthscan Publications Ltd, London.

Gomes, M.A.F., Filizola, H.F. and Spadotto, C.A. (2002) Agricultural uses of 'new' areas in the Guarani aquifer region and implications for quality of groundwater. *Documentos – Embrapa Soja*, No. 180, pp. 36–42.

Gomez Toro, J.M. (1996) Effect of vinasse application on yield and quality of sugarcane. *Cana de Azucar*, Vol. 14, pp. 15–34.

Goncharov, V. (1994) The food industry in Russia: problems of development. *Mezhdunarodnyi Sel'skokhozyaistvennyi Zhurnal*, No. 5, pp. 40–43.

Goos, R.J., Johnson, B.E., Peterson, R.A. and Kobes, N. (2001) Effect of sugarbeet by-products on the solubility and availability of ferous sulfate in soil. *Journal of Sugar Beet Research*, Vol. 38, pp. 153–172.

Goudreddy, B.S., Chittapur, B.M., Shailendra, K. and Radder, G.D. (1994) Performance of sugarcane varieties on red lateritic soils of North Eastern Transitional Zone of Karnataka under well irrigation. *Karnataka Journal of Agricultural Sciences*, Vol. 7, No. 3, pp. 260–263.

Grabka, J. (1995) Directions of progress in sugar equipment. *Gazeta Cukrownicza*, Vol. 103, No. 9, pp. 164–165.

Graf, J.D., Borrelli, J., Fornstrom, K.J. and Jackson, G.D. (1983) Reduction of sediment loss by no-till cultivation. In *Advances in Irrigation and Drainage: Surviving External Pressures* (eds J. Borrelli, V.R. Hasfurther and R.D. Burman), pp. 368–376. American Society of Civil Engineers, New York.

Graham, M.H., Haynes, R.J. and Meyer, J.H. (1999) Green cane harvesting promotes accumulation of soil organic matter and an improvement in soil health. *Proceedings of the Annual Congress – South African Sugar Technologists' Association*, No. 73, pp. 53–57.

Graham, M.H., Haynes, R.J. and van Antwerpen, R. (2002) Size and activity of the soil microbial biomass in the row and inter-row of a sugarcane field under burning and green cane harvesting. *Proceedings of the Annual Congress – South African Sugar Technologists' Association*, No. 76, pp. 186–195.

Grange, I., Prammanee, P. and Prasertsak, P. (2002) Effect of sugarcane management systems on the soil physical and chemical properties in the central plains of Thailand. *Thai Journal of Agricultural Science*, Vol. 35, No. 2, pp. 175–186.

Greathead, D.J. (2003) Historical overview of biological control in Africa. In *Biological Control in IPM Systems in Africa* (eds P. Neuenschwander, C. Borgemeister and J. Langewald), pp. 1–26. CAB International, Wallingford, UK.

Greene, A.E. and Allison, R.F. (1994) Recombination between viral RNA and transgenic plant transcripts. *Science*, Vol. 263, pp. 1423–1425.

Grigorov, M. and Grigorov, S. (2001) Particularities of optimization of sprinkler irrigation in the zone of insufficient natural humidity. *Mezhdunarodnyi Sel'skokhozyaistvennyi Zhurnal*, No. 1, pp. 44–47.

Grosse, S. (1993) *Schistosomiasis and Water Resources Development: a Re-evaluation of an Important Environment–Health Linkage*. Technical Series No. 2, Environmental and Natural Resources Policy and Training Project (EPAT/MUCIA). pp. 1–32. Environmental and Natural Resources Policy and Training Project, Madison.

Groves, S.J. and Bailey, R.J. (1994) Strategies for the sub-optimal irrigation of sugar beet. *Aspects of Applied Biology*, No. 38, pp. 201–207.

Groves, S.J. and Bailey, R.J. (1997) The influence of sub-optimal irrigation and drought on crop yield, N uptake and risk of N leaching from sugarbeet. *Soil Use and Management*, Vol. 13, No. 4, pp. 190–195.

Grundwurmer, J. (1991) Experience with erosion protection in practical agriculture. *Berichte uber Landwirtschaft. Sonderheft*, Vol. 5, No. 3, pp. 117–133.

Gryllus, V. and Anyos, E. (1993) Impact of beet N-compounds upon technology and environment. *Cukoripar*, Vol. 46, No. 1, pp. 3–11.

Gudoshnikov, S. (2000) Organic sugar – a growth opportunity for producers? *International Sugar Journal*, Vol. 102, pp. 463–468.

Gudoshnikov, S. (2001) Organic sugar – niche commodity in the mainstream market. *F.O. Licht International Sugar and Sweetener Report*, Vol. 133, No. 22, 9 pp. (Text also made available by the International Sugar Organization.)

Gulati, J.M.L., Mishra, M.M., Paul, J.C. and Hati, N. (1995) Cane yield and water-use efficiency of autumn-planted sugarcane (*Saccharum officinarum*) under soil and intercropping stand at different levels of irrigation. *Indian Journal of Agronomy*, Vol. 40, No. 2, pp. 279–281.

Gunasekaran, N., Gurugovind, J. and Srockiasamy, S. (1999) Operation and maintenance of sugar mill effluent treatment plant. In *Proceedings of the 61st Annual Convention of the Sugar Technologists' Association of India, New Delhi, India, 7–9 September 1999*, pp. c58–c63. Sugar Technologists' Association of India, New Delhi.

Gupta, J.P. and Aggarwal, R.K. (1980) Effect of asphalt sub-surface moisture on water characteristics and productivity of sand soil. *Annals of Arid Zone*, Vol. 19, No. 4, pp. 445–450.

Gupta, V.K., Jain, C.K., Imran Ali, Chandra, S. and Agarwal, S. (2002) Removal of lindane and malathion from wastewater using bagasse fly ash – a sugar industry waste. *Water Research (Oxford)*, Vol. 36, No. 10, pp. 2483–2490.

Gura, T. (2001) The battlefields of Britain. *Nature (London)*, Vol. 412, No. 6849, pp. 760–763.

Gurav, B.S., Sadaphal, S.S., Chavai, A.M. and Khot, B.B. (2003) Constraints experienced by the sugarcane growing farmers in adoption of drip irrigation system and their suggestions for enhancing the rate of adoption of drip irrigation system. *Cooperative Sugar*, Vol. 34, No. 8, pp. 647–649.

Gurgel, P.V., Mancilha, I.M., Pecanha, R.P. and Siquera, J.F.M. (1995) Xylitol recovery from fermented sugarcane bagasse hydrolysate. *Bioresource Technology*, Vol. 52, pp. 219–223.

Guyot, R., Lefevre, M. and Swiderski, M. (2003) Beet yard and beet washer – latest developments. *Zuckerindustrie*, Vol. 128, No. 1, pp. 26–30.

Haaskma, J. and Vecchiettini, M. (1988) Beet vinasse – a protein source in animal nutrition. *Proceedings of the Institut International de Recherches Betteravières*, Vol. 51, pp. 315–332.

Hadlow, W. and Millard, E.W. (1978) Minimum tillage: a practical alternative to ploughing in the South African sugar industry. *Proceedings of the International Society of Sugar Cane Technologists*, Vol. 16, pp. 891–897.

Hadlow, W. and Millard, E.W. (1981) Minimum tillage. A practical alternative to ploughing in the South African sugar industry. *Sugar Journal*, Vol. 44, No. 2, pp. 15–17.

Hafez, S.L. and Sundararaj, P. (1999) Exploitation of nematicidal efficacy of trap crops for the management of *Heterodera schachtii* under sugarbeet ecosystem. *International Journal of Nematology*, Vol. 9, No. 1, pp. 27–33.

Hagen, R.J. (1974) Growers trials to combat soil blowing on mineral soils. *British Sugar Beet Review*, Vol. 42, No. 3, pp. 182–185.

Hallmark, W.B., Brown, L.P. and Hawkins, G.L. (1998) Effect of municipal fish and sugarmill wastes on sugarcane yields. *Louisiana Agriculture*, Vol. 41, pp. 9–10.

Hallmark, W.B., Brown, L.P. and Hawkins, G.L. (1999) Use of a nitrogen stabilization package to increase yields and reduce the nitrogen requirements of sugarcane. *Journal – American Society of Sugar Cane Technologists*, Vol. 19, pp. 25–34.

Ham, G.J., Cox, A.Z. and McMahon, G.G. (1997) A review of sodic soils research in the Queensland sugar industry. *Proceedings of the Australian Society of Sugar Cane Technologists*, Vol. 19, pp. 200–207.

Hamilton, D. and Haydon, G. (1996) *Pesticides and Fertilizers in the Queensland Sugar Industry – Estimates of Usage and Likely Environmental Fate*. Vol. 1 – *Pesticides*. Department of Primary Industries, Brisbane.

Hammad, E.A. and Dawelbeit, M.I. (2001) Effect of tillage and field condition on soil physical properties, cane and sugar yields in Vertisols of Kenana Sugar Estate, Sudan. *Soil and Tillage Research*, Vol. 62, No. 3/4, pp. 101–109.

Hance, T. and Gregoire-Wibo, C. (1987) Effect of agricultural practices on carabid populations. *Acta Phytopathologica et Entomologica Hungarica*, Vol. 22, No. 1–4, pp. 147–160.

Hanegraaf, M.C., Biewinga, E.E. and van der Bijl, G. (1998) Assessing the ecological and economic sustainability of energy crops. *Biomass and Bioenergy*, Vol. 15, No. 4/5, pp. 345–355.

Hansen, L. (1982) Problems of soil structure in monoculture of spring barley. In *Soil Degradation* (eds D. Boels, D.B. Davies and A.E. Johnston), pp. 87–94. Balkema, Rotterdam.

Hao, X., Chang, C. and Lindwall, C.W. (2001) Tillage and crop sequence effects on organic carbon and total nitrogen content in an irrigated Alberta soil. *Soil and Tillage Research*, Vol. 62, No. 3/4, pp. 167–169.

Hapase, D.G., Deshmukh, A.S., Gunjal, B.B. and Shinde, P.P. (1990) Evaluation of micro irrigation systems in sugarcane agriculture. In *Proceedings of the 11th International Congress on the Use of Plastics in Agriculture, New Delhi, India, 26 February–2 March 1990*, pp. B.41–B.47. A.A. Balkema, Rotterdam.

Hapase, D.G., Mankar, A.N. and Salunkhe, A.N. (1992) Techno-economic evaluation of drip irrigation for sugarcane crop. In *International Agricultural Engineering Conference. Proceedings of a Conference held in Bangkok, Thailand on 7–10 December 1992* (eds V.M. Salokhe, Gajendra Singh and S.G. Ilangantileke), Vol. III, pp. 897–904. Asian Institute of Technology, Bangkok.

Hari Om, Nepal Singh and Aryo, M.S. (1994) Combined effect of wastes of distillery and sugar mill on seed germination, seedling growth and biomass of okra (*Abelmoschus esculentus* (L) Moench). *Journal of Environmental Biology*, Vol. 15, No. 3, pp. 171–175.

Harland, J.I. (1993) By-products. In *The Sugar Beet Crop – Science into Practice* (eds D.A. Cooke and R.K. Scott), pp. 618–647. Chapman & Hall, London.

Harlander, S.K. (2002) Safety assessments and public concern for genetically modified food products: the American view. *Toxicologic Pathology*, Vol. 30, No. 1, pp. 132–134.

Hartemink, A.E. (1998) Soil chemical and physical properties as indicators of sustainable land management under sugar cane in Papua New Guinea. *Geoderma*, Vol. 85, No. 4, pp. 283–306.

Hartemink, A.E. (2003) *Soil Fertility Decline in the Tropics*. CAB International, Wallingford, UK.

Hartemink, A.E. and Kuniata, L.S. (1996) Some factors influencing yield trends of sugarcane in Papua New Guinea. *Outlook on Agriculture*, Vol. 25, pp. 227–234.

Hartmann, C., Blanchart, E., Albrecht, A., Bonneton, A., Parfait, F., Mahieu, M., Gaullier, C. and Ndandou, J.F. (1998) New soil preparation methods for market gardening in Vertisols of Martinique. Pedological and agro-economic impacts. *Agriculture et Développement*, No. 18, pp. 81–89.

Harveson, R.M. and Rush, C.M. (1998) Characterization of *Fusarium* root rot isolates from sugar beet by growth and virulence at different temperatures and irrigation regimes. *Plant Disease*, Vol. 82, No. 9, pp. 1039–1042.

Harveson, R.M. and Rush, C.M. (2002) The influence of irrigation frequency and cultivar blends on the severity of multiple root diseases in sugar beets. *Plant Disease*, Vol. 86, No. 8, pp. 901–908.

Harwell, M.A., Long, J.F., Bartuska, A.M., Gentile, J.H., Harwell, C.C., Myers, V. and Ogden, J.C. (1996) Ecosystem management to achieve ecological sustainability: the case of South Florida. *Environmental Management*, Vol. 20, No. 4, pp. 497–521.

Hassan, A.A., Yusufzai, S.U.K. and Sarkar, A.A. (1999) Response of sugarcane varieties to different levels of irrigation. *International Journal of Tropical Agriculture*, Vol. 17, No. 1/4, pp. 221–225.

Haynes, R.J. (1997) The concept of soil quality and its applicability to sugarcane production. *Proceedings of the Annual Congress – South African Sugar Technologists' Association*, No. 71, pp. 9–14.

Haynes, R.J. and Hamilton, C.S. (1999) Effects of sugarcane production on soil quality: a synthesis of world literature. *Proceedings of the Annual Congress – South African Sugar Technologists' Association*, No. 73, pp. 46–52.

Haynes, R.J., Dominy, C.S. and Graham, M.H. (2003) Effect of agricultural land use on soil organic matter status and the composition of earthworm communities in KwaZulu-Natal, South Africa. *Agriculture, Ecosystems and Environment*, Vol. 95, No. 1/3, pp. 453–464.

Hazell, D., Nott, R. and Shannon, M.F. (2003) *Review of the Project 'The Development of a Cane Toad Biological Control'*. Natural Heritage Trust, Environment Australia, Canberra.

Heijbroek, W. and van de Bund, C.F. (1982) The influence of some agricultural practices on soil organisms and plant establishment of sugar beet. *Netherlands Journal of Plant Pathology*, Vol. 88, No. 1, pp. 1–17.

Heisler, C. (1994) Effects of soil compaction on soil mesofauna: Collembola and Gamasina – a 3-year field trial. *Pedobiologia*, Vol. 38, No. 6, pp. 566–576.

Heissenhuber, A. and Schmidtlein, E.M. (1988) Ecological and economic problems of soil erosion. *Bayerisches Landwirtschaftliches Jahrbuch*, Vol. 65, No. 1, pp. 105–131.

Held, L.J., Jennings, J.W., Koch, D.W. and Gray, F.A. (1999) Trap crop radish: a sustainable alternative for nematicide in sugar beets. In *Proceedings Western Agricultural Economics Association, Selected Papers of the 1999 Annual Meeting, Fargo, North Dakota, USA, 11–14 July 1999*, 14 pp. Waite Library, Department of Agriculture and Applied Economics, University of Minnesota, St. Paul.

Helge, J.M. and Larson, D.O. (1993) Recent developments in beet transport water management. *Journal of Sugar Beet Research*, Vol. 30, No. 1–2, p. 100.

Hellin, J. and Larrea, S. (1998) Ecological and socio-economic reasons for adoption and adaptation of live barriers in Guinope, Honduras. *Advances in Geoecology*, No. 31, pp. 1383–1388.

Hellmann, D.B. (1977) The relative performance of six sugar cane varieties under three different [water] regimes. *Proceedings of the International Society of Sugar Cane Technologists*, Vol. 16, pp. 1363–1375.

Henne, L.J., Schneider, D.W. and Martinez L.M. (2002) Rapid assessment of organic pollution in a west-central Mexican river using a family-level biotic index. *Journal of Environmental Planning and Management*, Vol. 45, No. 5, pp. 613–632.

Henriksen, K., Berthelsen, L. and Matzen, R. (1998) Separation of liquid pig manure by flocculation and ion exchange. Part 2: Pilot-scale system. *Journal of Agricultural Engineering Research*, Vol. 69, No. 2, pp. 127–131.

Henriksson, L. and Hakansson, I. (1993) Soil management and crop establishment. In *The Sugar Beet Crop – Science into Practice* (eds D.A. Cooke and R.K. Scott), pp. 156–177. Chapman & Hall, London.

Henry, P.C. and Ellis, R.D. (1996) Soil as a factor in sugarcane ratoon yield decline on an irrigated estate in Swaziland. *Proceedings of the International Society of Sugar Cane Technologists*, Vol. 22, pp. 236–245.

Hermann, O., Wauters, A. and Dewar, A. (2001) Results of IIRB cooperative trials with insecticides in pelleted seed. *Proceedings of the Institut International de Recherches Betteravières*, Vol. 64, pp. 341–346.

Hernandez, I. and Babbar, L. (2001) Intensive animal production systems and the protection of the environment: current situation and opportunities. *Pastos y Forrajes*, Vol. 24, No. 4, pp. 281–289.

Herrera Coello, N. and Diaz Rodriguez, I. (1998) Alcohol industry residues as source of substrate for production of single-cell protein. *Centro Azucar*, Vol. 25, No. 3, pp. 16–19.

Herschy, R.W. (1998a) Irrigated land area: world. In *Encyclopaedia of Hydrology and Water Resources* (eds R.W. Herschy and R.W. Fairbridge), pp. 442–444. Kluwer Academic Publishers, Dordrecht.

Herschy, R.W. (1998b) Streamflow measurement. In *Encyclopaedia of Hydrology and Water Resources* (eds R.W. Herschy and R.W. Fairbridge), pp. 635–637. Kluwer Academic Publishers, Dordrecht.

Heszky, L., Bodis, L. and Kiss, E. (1999) Plant biodiversity of agro-ecosystems in Hungary. *Novenytermeles*, Vol. 48, No. 4, pp. 435–443.

Hidalgo-Gato, M.M., Rodriguez-Leon, R., Ricardo, N.E. and Ferras, H. (1999) Population dynamics of cicadellids (Homoptera: Cicadellidae) in a sugarcane ecosystem in Cuba. *Revista de Biologia Tropical*, Vol. 47, No. 3, pp. 503–512.

Hills, F.J., Broadbent, F.E. and Fried, M. (1978) Timing and rate of fertilizer nitrogen for sugarbeets related to nitrogen uptake and pollution potential. *Journal of Environmental Quality*, Vol. 7, No. 3, pp. 368–372.

Hindy, K.T. (1990) Deposition rate dependence upon particle size of heavy metals downwind of cane sugar industry in Middle Egypt. *Bulletin of the National Research Centre*, Vol. 15, No. 2, pp. 119–126.

Hodnett, M.G., Bell, J.P., Ah Koon, P.D., Soopramanien, G.C. and Batchelor, C.H. (1990) Control of drip irrigation of sugarcane using 'index' tensiometers: some comparisons with control by the water budget method. *Agricultural Water Management (Special Issue)*, Vol. 17, pp. 189–207.

Hoffmann, C. and Marlander, B. (2001) Breeding progress for yield and quality in sugar beet. *Proceedings of the Institut International de Recherches Betteravières*, Vol. 64, pp. 185–194.

Hoffmann-Campo, C.B., Oliveira, L.J., Moscardi, F., Gazzoni, D.L., Correa-Ferreira, B.S., Lorini, I.A., Borges, M., Panizzi, A.R., Sosa-Gomez, D.R. and Corso, I.C. (2003) Integrated pest management in Brazil. In *Integrated Pest Management in the Global Arena* (eds K.M. Maredia, D. Dakouo and D. Mota-Sanchez), pp. 285–299. CAB International, Wallingford, UK.

Hoi, L.W.S. and Shum, S.T.C. (1999) Plant materials as natural flocculant in cane juice clarification. *Proceedings of the International Society of Sugar Cane Technologists*, Vol. 23, pp. 7–16.

Holden, J., Hussey, B., McDougall, A., Mallon, K. and Shannon, E. (1998) Water check – improving irrigation efficiencies in the Queensland sugar industry. *Proceedings of the Australian Society of Sugar Cane Technologists*, Vol. 20, pp. 126–132.

Holman, K. and Miller, P. (1995) Precision techniques: right place, right dose. *British Sugar Beet Review*, Vol. 63, No. 3, pp. 36–37.

Holt, J.A. and Mayer, R.J. (1998) Change in microbial biomass and protease activities in soil associated with long-term sugar cane monoculture. *Biology and Fertility of Soils*, Vol. 27, pp. 127–131.

Holub, M. (1994) Vertical structure of the soil seed bank below wheat, sugar beet and lucerne. *Biologia (Bratislava)*, Vol. 49, No. 1, pp. 53–57.

Hopkins, S. (1992) Can sugar beet be green? *British Sugar Beet Review*, Vol. 60, No. 1, pp. 24–26.

Hossain, I.H. (1998) Development and management of land/water resources. The Everglades, agriculture, and south Florida. In *Proceedings of the Inter-American Sugar Cane Seminar, Crop Production and Mechanization, Miami, Florida, USA, 9–11 September 1998* (ed. A. Leyva), pp. 1–18, 263–281. Inter-American Sugar Cane Seminars, Miami.

Hossain, M.A., Yoshinaga, A., Zhao TingNing, Hossain, A.H.M.D. and Matin, M.A. (1999) Interaction of irrigation and interplant spacing on yield and sucrose content of sugarcane. *Forestry Studies in China*, Vol. 1, No. 2, pp. 54–63.

Hruska, M. (1991) Development trends for agricultural machinery in the 90s. *Mechanizace Zemedelstvi*, Vol. 41, No. 1, pp. 2–8.

Hunsigi, G. (1993) *Production of Sugarcane – Theory and Practice*. Springer-Verlag, Berlin.

Hunsigi, G. and Shankariah, C. (2001) Effect of biofertilizer on minimizing ground water pollution by nitrates in sugarcane soils. *Proceedings of the International Society of Sugar Cane Technologists*, Vol. 24, pp. 187–189.

Hunter, H.M., Walton, R.S. and Russell, D.J. (1996) Contempory water quality in the Johnstone River catchment. In *Downstream Impacts of Land Use* (eds H.H. Hunter, A.G. Eyles and G.E. Rayment), pp. 339–345. Department of Natural Resources, Brisbane.

Hussein, A.S.A. (1999) Grass ET estimates using Penman-type equations in central Sudan. *Journal of Irrigation and Drainage Engineering*, Vol. 125, No. 6, pp. 324–329.

Iguiniz, J. (1987) Peru: agriculture, crisis and macroeconomic policy. *CEPAL Review*, No. 33, pp. 157–170.

Ijumba, J.N., Mosha, F.W. and Lindsay, S.W. (2002) Malaria transmission risk variations derived from different agricultural practices in an irrigated area of northern Tanzania. *Medical and Veterinary Entomology*, Vol. 16, No. 1, pp. 28–38.

Inamdar, P.P., Pawar, J.R. and Sale, D.L. (1995) Economic efficiency of biwall drip irrigation in sugarcane production – a case study in Ankalkhop village in Sangli District of Maharashtra. *Bharatiya Sugar*, Vol. 22, No. 2, pp. 43–48.

Inamdar, P.P., Pawar, J.R. and Sale, D.L. (1996a) Economic efficiency of biwall drip irrigation in sugarcane production – a case study in Ankalkhop village in Sangli district of Maharashtra. *Bharatiya Sugar*, Vol. 22, No. 3, pp. 13–15, 17–18.

Inamdar, P.P., Pawar, J.R. and Sale, D.L. (1996b) Economic analysis of biwall drip irrigation unit for irrigating sugarcane crop in Maharashtra. *Bharatiya Sugar*, Vol. 22, No. 4, pp. 13–20.

Ingleby, K., Thuy, L.T.T., Phong, N.T. and Mason, P.A. (2000) Ectomycorrhizal inoculum potential of soils from forest restoration sites in South Vietnam. *Journal of Tropical Forest Science*, Vol. 12, No. 2, pp. 418–422.

Inglese, P. (1999) Evolutionary trends in fruit cultivation in Latin America. *Rivista di Frutticoltura e di Ortofloricoltura*, Vol. 61, No. 9, pp. 47–56.

Inklan, G. (1991) Assessment of the action on the environment of sugar factory wastewaters used for irrigation of agricultural crops. *Gigiena i Sanitariya*, No. 7, pp. 65–68.

Inman-Bamber, N.G., Culverwell, T.L. and McGlinchey, M.G. (1993) Predicting yield responses to irrigation of sugarcane from a growth model and field records. *Proceedings of the Annual Congress – South African Sugar Technologists' Association*, No. 67, pp. 66–72.

Inman-Bamber, N.G., Muchow, R.C., Holden, J.R., Robertson, M.J. and Ham, G.J. (1998) Soil water extraction by sugar cane beyond the readily available limit. *Proceedings of the Australian Society of Sugar Cane Technologists*, Vol. 20, pp. 112–117.

Inosako, K., Kuroda, M., Nakano, Y. and Yoshida, I. (1995) A model for irrigation scheduling utilizing soil moisture deficit. Application to a sugar cane field. *Bulletin of the Faculty of Agriculture, Tottori University*, Vol. 48, pp. 35–41.

Isermann, K. (1989) Vertical distribution of NO_3-N and denitrification parameters in the soil down to greater depths (about 10 m) under various soil management/nitrogen fertilization. In *Protection of Water Quality from Harmful Emissions with Special Regard to Nitrate and Heavy Metals. Proceedings of the 5th International Symposium of CIEC* (eds E. Welte and I. Szabolcs) pp. 131–139. CIEC, Budapest.

Ishikawa, S. and Takenaka, S. (2002) Weed community response to drilled sugar beet crop. *Proceedings of the Japanese Society of Sugar Beet Technologists*, No. 44, pp. 59–65.

Ismail, I. (1990) Critical land rehabilitation and conservation programmes in Sragen, Karanganyar and Sleman. *Berita – Pusat Penetitian Perkebunan Gula Indonesia*, No. 3, pp. 6–8.

ISO (International Sugar Organization) (2000) *Organic Sugar: Practices and Standards for Producing Organic Sugar, Demand Potential.* MECAS(00)10, ISO, London.

ISO (International Sugar Organization) (2001) *Sugar and the Environment: Survey of Environmental Legislation Affecting Sugar Crop Growers, Beet and Cane Processors/Millers*. MECAS(01)19, ISO, London.

Ivanoff, D.B., Reddy, K.R. and Robinson, S. (1998) Chemical fractionation of organic phosphorus in selected Histosols. *Soil Science*, Vol. 163, No. 1, pp. 36–45.

Izsaki, Z., Varkonyi, Z. and Csaba, L. (1993) New method of wastewater treatment and disposal at Sarkad sugar factory. *Cukoripar*, Vol. 46, No. 2, pp. 41–55.

Izuno, F.T., Bottcher, A.B., Coale, F.J., Sanchez, C.A. and Jones, D.B. (1995) Agricultural BMPs for phosphorus reduction in south Florida. *Transactions of the ASAE*, Vol. 38, No. 3, pp. 735–744.

Jacobs, L.W. (1995) *Utilization of Animal Manures for Crop Production.* Part II – *Manure Application to Cropland.* Extension Bulletin MM-2, Michigan State University, East Lansing.

Jadhav, G.S., Oza, S.R., Sondge, V.D. and Chavan, D.A. (1997) Response of pre-season sugarcane to irrigation and nitrogen levels. *Journal of Maharashtra Agricultural Universities*, Vol. 22, No. 3, pp. 313–315.

Jadhav, R.D., Pimpalkar, C.G., Thombare, M.T., Joshi, G.V., Tendulkar, S.R. and Khuspe, V.L. (1990) Syrup bleaching: by hydrogen peroxide for the production of white sugar. In *Thirty-eighth Annual Convention 1988: Deccan Sugar Technologists' Association (India)*, Part 1, pp. M1–M12. Deccan Sugar Technologists' Association, Pune.

Jafari-Darabjerdi, J. (1993) Development and structure of the irrigation system in Iran. Diskussionspapiere No. 13, Institut für Rurale Entwicklung, Universität Göttingen, 26 pp.

Jaggard, K.W., Dewar, A.M. and Pidgeon, J.D. (1998) The relative effects of drought stress and virus yellows on the yield of sugarbeet in the UK, 1980–95. *Journal of Agricultural Science*, Vol. 130, No. 3, pp. 337–343.

Jahns, G. and Kogl, H. (1993) Positioning and navigation of farm vehicles by means of satellite systems – a contribution to technology assessment. *Landbauforschung Volkenrode*, Vol. 43, No. 2/3, pp. 125–148.

Jambulingam, M., Murugesan, M., Patabi, S. and Srinivasan, T.R. (1999) Effect of irrigation frequencies on sugarcane juice quality and yield of early and midlate varieties. *Indian Sugar*, Vol. 49, No. 6, pp. 419–432.

Jambulingam, M., Murugesan, M., Patabi, S., Radhamani, R., Rakkiyappan, P. and Srinavasan, T.R. (2001) Effect of irrigation regimes on jaggery yield and quality of some sugarcane varieties. *Sugar Tech*, Vol. 3, No. 4, pp. 134–145.

James, G. and Hasibuan, M.A. (2002) The composting of sugarcane factory by-products. *Sugar Cane International*, May/June, pp. 24–30.

Jansen, R. and Burba, M. (2001) Progress in sugar beet quality by breeding. *Proceedings of the Institut International de Recherches Betteravières*, Vol. 64, pp. 195–204.

Jarkovsky, J. (1997) Evaluation of potential risks of soil salinization under influence of mineralized ground waters. *Vedecke Prace Vyskumneho Ustavu Zavlahoveho Hospodarstva v Bratislava*, No. 23, pp. 109–127.

Jarvis, P., Leeds, S. and Cormack, B. (2001) Organic sugar beet production in the UK. *British Sugar Beet Review*, Vol. 69, pp. 10–11.

Jayapaul, P., Uthayakumar, B., Pandian, B.J., Palchamy, A. and Swaminathan, V. (1995) Effect of irrigation schedule under varying land management and soil moisture conservation amendments on morpho-physiological characters of irrigated soybean (*Glycine max* L. Merr). *Crop Research (Hisar)*, Vol. 10, No. 1, pp. 45–50.

Jayapaul, P., Uthayakumar, B., Devasagayam, M.M., Pandian, B.J., Palchamy, A. and Balakrishnan, A. (1996) Effect of land configuration methods, irrigation regimes and soil moisture conservation amendments on soybean (*Glycine max* L. Merr.) yield and quality characters. *Crop Research (Hisar)*, Vol. 11, No. 3, pp. 253–257.

Jenicek, P., Zabranska, J. and Dohanyos, M. (1994) Anaerobic treatment of distillery slops in the circumstances of Central Europe. *Water Science and Technology*, Vol. 30, No. 3, pp. 157–160.

Jensen, A.S. (1995) Pulp drying with superheated steam under pressure – Part 1. *Zuckerindustrie*, Vol. 120, No. 10, pp. 855–861.

Jensen, A.S. (2002) Steam drying of beet pulp and bagasse. In *Advances in the Chemistry and Processing of Beet and Cane Sugar: Proceedings of the 2002 Sugar Processing Research Conference*, pp. 112–122. Sugar Processing Research Institute, New Orleans.

Jensen, A.S. (2003) Steam drying of beet pulp and bagasse. *International Sugar Journal*, Vol. 105, No. 1250, pp. 83–88.

Jhoty, I. and Autrey, J.C. (1999) Precision agriculture – perspectives for the Mauritian sugar industry. In *Proceedings of the Third Annual Meeting of Agricultural Scientists, Reduit, Mauritius, 17–18 November 1998* (eds J.A. Lalouette, D.Y. Bachraz and N. Sukerdeep), pp. 15–21. Food and Agricultural Research Council, Reduit.

Jhoty, I., Ramasamy, S. and Tulloo, P.K. (2001) A survey of irrigation practices for sugar cane in Mauritius. *Proceedings of the International Society of Sugar Cane Technologists*, Vol. 24, pp. 288–292.

Jobin, W.R. (1980) Sugar and snails: the ecology of bilharziasis related to agriculture in Puerto Rico. *American Journal of Tropical Medicine and Hygiene*, Vol. 29, No. 1, pp. 86–94.

Johansson, W. (1978) Water resources for irrigation in agriculture. 2. Water requirements of crops and irrigation. *Kungliga Skogs- och Lantbruksakademiens Tidskrift*, Vol. 117, No. 6, pp. 299–307.

Johnson, A.K.L., McDonald, G.T., Shrubsole, D.A. and Walker, D.H. (1997) Sharing the land – the sugar industry as part of the wider landscape. In *Intensive Sugarcane Production: Meeting the Challenges beyond 2000* (eds B.A. Keating and J.R. Wilson), pp. 361–380. CAB International, Wallingford.

Johnson, A.K.L., Ebert, S.P. and Murray, A.E. (2000) Land cover change and its environmental significance in the Herbert River catchment, north-east Queensland. *Australian Geographer*, Vol. 31, No. 1, pp. 75–86.

Johnston, A.E. (1989) Potable waters and the nitrate problem. *British Sugar Beet Review*, Vol. 57, pp. 22–23.

Johnston, M.A. and Wood, R.A. (1971) Soil compaction studies at Pongola. *Proceedings of the South African Sugar Technologists' Association*, Vol. 45, pp. 261–265.

Jolly, L. (2001) The commercial viability of fuel ethanol from sugar cane. *International Sugar Journal*, Vol. 103, No. 1227, pp. 117–125, 143.

Jolly, L. (2002) *Organic Sugar: a Viable Alternative?* MECAS(02)20, ISO (International Sugar Organization), London.

Joshi, P.K. and Tyagi, N.K. (1991) Sustainability of existing farming system in Punjab and Haryana – some issues on groundwater use. *Indian Journal of Agricultural Economics*, Vol. 46, No. 3, pp. 412–421.

Jossart, J.M., Vanhemelryck, J.L. and Ledent, J.F. (1995) Environmental impact of three energy crops, and energy balance. *Cahiers Agricultures*, Vol. 4, No. 5, pp. 377–382.

Joyce, K.N. and Dixon, T.F. (1999) Strategies for meeting stack emission regulations for the Australian sugar industry. *Proceedings of the Australian Society of Sugar Cane Technologists*, Vol. 21, pp. 438–442.

Kaczorowski, G. and Debek-Jankowska, A. (1997) Hoverflies (Syrphidae) in different types of thickets in central Poland. *Progress in Plant Protection*, Vol. 37, No. 2, pp. 85–87.

Kaffka, S., Dong DaXue and Peterson, G. (1999) Saline water can be reused to irrigate sugarbeets, but sugar may be low. *California Agriculture*, Vol. 53, No. 1, pp. 11–15.

Kaffka, S.R., Corwin, D.L. and Lesch, S.M. (2002) Sugar beet yield and quality variation at the field scale in response to soil salinity and correlated properties in California. *Proceedings of Institut International de Recherches Betteravières*, Vol. 65, pp. 159–170.

Kalaisudarson, S., Wahab, K. and Thirumurugan, M. (2002) Effect of irrigation, mulching and phosphobacteria on the growth and yield of sugarcane – Part I. *Plant Archives*, Vol. 2, No. 1, pp. 81–83.

Kallstrom, D.L., Carlson, J.L., Gruenuich, D.P. and Larson, D.O. (2001) Operational and economic considerations of using a belt press for dewatering soil washed from sugarbeets. In *Proceedings from the 31st Biennial Meeting (Operations) of the American Society of Sugar Beet Technologists, Vancouver, BC, Canada, 28 February–3 March 2001*, pp. 243–247. American Society of Sugar Beet Technologists, Denver.

Kammerbauer, H., Sanchez, M.D. and Dick, T. (1999) Plant response to motor exhaust emissions from gasoline and alcohol. *Revista Brasileira de Fisiologia Vegetal*, Vol. 11, No. 1, pp. 35–40.

Kammerling, B. (2001) LIZ – advisory systems on new path. *Proceedings of the Institut International de Recherches Betteravières*, Vol. 64, pp. 105–107.

Kandil, S.A., Abo El-Kheir, M.S.A. and Abou-Ellil, A.A. (2001) Physiological response of some sugar beet (*Beta vulgaris* L.) varieties to irrigation with different levels of chloride salinization. *Bulletin of the National Research Centre (Cairo)*, Vol. 26, No. 1, pp. 79–92.

Kannaiyan, S. (ed.) (2002) *Biotechnology of Biofertilizers*, xii + 375 pp. Kluwer Academic Publishers, Dordrecht.

Karczmarczyk, S., Koszanski, Z., Roy, M. and Sciazko, D. (1995) Effect of sprinkler irrigation and nitrogen fertilizer application on sugar and fodder beets cultivated on a good rye complex soil. II. Chemical composition of the crop. *Zeszyty Naukowe Akademii Rolniczej w Szczecinie, Rolnictwo*, Vol. 59, pp. 65–72.

Kareem, A.A. (1999) Micro-irrigation – need of the 21st century for conserving water resources. In *Irrigation under Conditions of Water Scarcity. 17th ICID International Congress on Irrigation and Drainage, Granada, Spain, 13–17 September 1999*, Vol. 1D, pp. 21–37. International Commission on Irrigation and Drainage, New Delhi.

Keating, B. and Wilson, J. (eds) (1997) *Intensive Sugarcane Production: Meeting the Challenge beyond 2000*. CAB International, Wallingford, UK.

Keating, B.A., Bauld, J., Hillier, J., Ellis, R., Weier, K.L., Sunners, F. and Connell, D. (1996) Leaching of nutrients and pesticides to Queensland groundwaters. In *Downstream Effects of Land Use* (eds H.M. Hunter, A.G. Eyles and G.E. Rayment), pp. 151–164. Department of Natural Resources, Brisbane.

Keating, B.A., Verburg, K., Huth, N.I. and Robertson, M.J. (1997) Nitrogen management in intensive agriculture: sugarcane in Australia. In *Intensive Sugarcane Production: Meeting the Challenges beyond 2000* (eds B.A. Keating and J.R. Wilson), pp. 221–242. CAB International, Wallingford, UK.

Keene, A.F., MacDonald, B.C.T., Melville, M.D. and Quirk, R.G. (2003) Using water quality and sediment characteristics to improve management of acid sulfate soils and drainage. *Proceedings of the Australian Society of Sugar Cane Technologists*, Vol. 25, p. 29.

Kent, R., Odum, H.T. and Scatena, F.N. (2000) Eutrophic overgrowth in the self-organization of tropical wetlands illustrated with a study of swine wastes in rainforest plots. *Ecological Engineering*, Vol. 16, No. 2, pp. 255–269.

Kenter, C. and Hoffmann, C. (2002) Yield and quality formation of sugarbeet as affected by temperature and water supply. *Zuckerindustrie*, Vol. 127, No. 9, pp. 690–698.

Keshavanath, P., Gangadhar, B., Ramesh, T.J., van Rooij, J.M., Beveridge, M.C.M., Baird, D.J., Verdegem, M.C.J. and van Dam, A.A. (2001) Use of artificial substrates to enhance production of freshwater herbivorous fish in pond culture. *Aquaculture Research*, Vol. 32, No. 3, pp. 189–197.

Khafagi, O.M.A. and El-Lawendy, W.I. (1997) Effect of different irrigation intervals on sugar beet growth, plant water relations and photosynthetic pigments. *Annals of Agricultural Science, Moshtohor*, Vol. 35, No. 1, pp. 305–319.

Khan, S.N., Said Rahman, Gulzar Ahmad, Buriro, U.A. and Jamro, G.H. (1998a) Predicting nitrogen requirements of sugarbeet based on different levels of irrigation. *Sarhad Journal of Agriculture*, Vol. 14, No. 4, pp. 277–280.

Khan, S.N., Said Rahman, Gulzar Ahmad, Buriro, U.A. and Jamro, G.H. (1998b) Root and sugar yield of sugarbeet as affected by nitrogen and irrigation levels. *Pakistan Sugar Journal*, Vol. 12, No. 1, pp. 17–20.

Kiewnick, S. and Jacobsen, B.J. (1998) Biological control of *Cercospora beticola* on sugar beet with phyllosphere bacteria. *Bulletin OILB/SROP*, Vol. 21, No. 9, pp. 279–282.

Kijne, J.W., Barker, R. and Molden, D. (eds) (2003) *Water Productivity in Agriculture: Limits and Opportunities for Improvement*. CAB International, Wallingford, UK.

King, N.J., Mungomery, R.W. and Hughes, C.G. (1953) *Manual of Cane Growing*. Australian Agricultural and Livestock Series, Angus and Robertson, Sydney.

King, R.B., Baillie, I.C., Abell, T.M.B., Dunsmore, J.R., Gray, D.A., Pratt, J.H., Versey, H.R., Wright, A.C.S. and Zisman, S.A. (1992) *Land Resource Assessment of Northern Belize*. Bulletin No. 43, 174 pp. Natural Resources Institute, Chatham.

Kingston, G. (1994) Benchmarking yield of sugarcane from estimates of crop water use. *Proceedings of the Australian Society of Sugar Cane Technologists*, Vol. 16, pp. 201–209.

Kingston, G. and Lawn, R.J. (2003) Managing natural resources used in sugar production systems: eight years on. *Proceedings of the Australian Society of Sugar Cane Technologists*, Vol. 25, p. 13.

Kingston, G., Hurney, A.P. and Aitken, R.L. (1996) Depression of CCS in sugarcane after use of liming products: economic significance and associations with soil chemistry and leaf nutrient composition. *Proceedings of Australian Society of Sugar Cane Technologists*, Vol. 18, pp. 164–173.

Kinnunen, H., Tiainen, J. and Tukia, H. (2001) Farmland carabid beetle communities at multiple levels of spatial scale. *Ecography*, Vol. 24, No. 2, pp. 189–197.

Kirchhoff, V.W.J.H., Marinho, E.V.A., Dias, P.L.S., Pereira, E.B., Calheiros, R., Andre, R. and Volpe, C. (1991) Enhancements of CO and O_3 from burnings in sugar cane fields. *Journal of Atmospheric Chemistry*, Vol. 12, No. 1, pp. 87–102.

Kirda, C., Kanber, R., Tulucu, K. and Gungor, H. (1999) Yield response of cotton, maize, soybean, sugar beet, sunflower and wheat to deficit irrigation. In *Crop Yield Response to Deficit Irrigation* (eds C. Kirda, P. Moutonnet, C. Hera and D.R. Nielsen), pp. 21–38. Kluwer Academic Publishers, Dordrecht.

Kittad, C.H., Jagtap, B.K., Ghugare, R.V. and Pharande, A.L. (1995) Effect of drip and surface irrigation methods on yield and quality of seasonal sugarcane (Co 7219). *Indian Sugar*, Vol. 45, No. 8, pp. 613–616.

Kleijer, G. (1993) Useful wild plants. *Naturopa*, No. 73, p. 11.

Klemes, J., Urbaniec, K., Vaccari, G., Mantovani, G., Bubnik, Z., Lentini, A., Kadlec, P. and Placek, I. (1999) Project SUCLEAN – research on minimization of energy and water use in sugar production by cooling crystallisation of concentrated raw juice. In *CITS Proceedings, 21st General Assembly, Antwerp, Belgium, 25–28 May 1999*, pp. 459–464. Verlag Dr. A. Bartens, Berlin.

Klimakhin, N.A., Nosoreva, E.V. and Ostroumov, V.B. (1998) Effects of effluents on the irrigation of agricultural crops. *Sakharnaya Promyshlennost'*, No. 4, pp. 16–17.

Klok, K.A., Charlesworth, P.B., Ham, G.J. and Bristow, K.L. (2003) Management of furrow irrigation to improve water use efficiency and sustain the groundwater resource – preliminary results from a case study in the Burdekin Delta. *Proceedings of the Australian Society of Sugar Cane Technologists*, Vol. 25, p. 38.

Kloos, H., Lo, C.T., Birrie, H., Ayele, T., Tedla, S. and Tsegay, F. (1988) Schistosomiasis in Ethiopia. *Social Science and Medicine*, Vol. 26, pp. 803–827.

Knappe, S. and Haferkorn, U. (2001) N leaching potential due to weather, soil and land use of typical pedohydrotopes of the Central German arid region (Parthe catchment area). In *9. Lysimetertagung zum Thema 'Gebietsbilanzen bei unterschiedlicher Landnutzung', 24–25 April 2001, HBLA Raumberg, Irdning, Austria*, pp. 83–87. Bundesanstalt fur alpenlandische Landwirtschaft Gumperstein, Irdning.

Knottnerus, D.J.C. (1976a) Stabilization of wind erodible land by an intermediate crop. *Mededelingen van de Faculteit Landbouwwetenschappen Rijksuniversiteit Gent*, Vol. 41, No. 1, pp. 73–77.

Knottnerus, D.J.C. (1976b) *Wind Erosion Research by Means of a Wind Tunnel. Measures to Control Wind Erosion of Soil and Other Materials for Reasons of Economy and Health*, 14 pp. Institut voor Bodemuruchtbaarheid, Haren.

Knox, J.W., Weatherhead, E.K. and Bradley, R.I. (1997) Mapping the total volumetric irrigation water requirements in England and Wales. *Agricultural Water Management*, Vol. 33, No. 1, pp. 1–18.

Kobayashi, H. (1996) Current approach to soil and water conservation for upland agriculture in Thailand. *JARQ, Japan Agricultural Research Quarterly*, Vol. 30, No. 1, pp. 43–48.

Kolenbrander, G.J. (1978) The nitrogen cycle and fertilizer requirements. *Proceedings of the Institut International de Recherches Betteravières Proceedings*, Vol. 41, pp. 1–11.

Kolomiets, A.P., Nedashkovskii, A.I. and Matsevetskaya, I.M. (1998) Growing sugarbeet on irrigated land. *Sakharnaya Svekla*, No. 12, pp. 12–13.

Korndorfer, G.H. and Anderson, D.L. (1997) Use and impact of sugar-alcohol residues vinasse and filter cake on sugarcane production in Brazil. *Sugar y Azucar*, Vol. 92, pp. 30–35.

Korytnik, V.N. and Malienko, A.M. (1994) Controlling weeds in good time. *Zemledelie*, No. 2, pp. 13–14.

Koszanski, Z., Karczmarczyk, S. and Roy, M. (1995) Effect of sprinkler irrigation and nitrogen fertilizer application on sugar and fodder beets cultivated on a good rye complex soil. I. Yields. *Zeszyty Naukowe Akademii Rolniczej w Szczecinie, Rolnictwo*, Vol. 59, pp. 57–64.

Koul, O., Dhaliwal, G.S. and Cuperus, G.W. (eds) (2003) *Integrated Pest Management: Potential, Constraints and Challenges*. CAB International, Wallingford, UK.

Kovac, K. and Zak, S. (2000) Ecological and technical-economic aspects of various sugarbeet cultivation methods. *Listy Cukrovarnicke a Reparske*, Vol. 116, No. 9/10, pp. 243–249.

Krebs, J.E., Tan, K.H. and Golley, F.B. (1974) A comparative study on chemical characteristics of tropical soils from volcanic material under forest and agriculture. *Communications in Soil Science and Plant Analysis*, Vol. 5, No. 6, pp. 579–596.

Kroes, S. and Dixon, T.F. (1998) NOx emissions from bagasse-fired boilers. *Proceedings of the Australian Society of Sugar Cane Technologists*, Vol. 20, pp. 453–457.

Kropff, M.J., Muchow, R.C., Rabbinge, R. and Aggarwal, P.K. (1997) Limits to intensive agricultural systems and opportunities for sustainable agricultural development. In *Intensive Sugarcane Production: Meeting the Challenges Beyond 2000* (eds B.A. Keating and J.R. Wilson), pp. 17–35. CAB International, Wallingford, UK.

Kumar, B.M.A., Murthy, K.N.K. and Chandrappa, H. (2000) Evaluation of drip irrigation systems adopted for sugarcane crop in Vertisols. *Environment and Ecology*, Vol. 18, No. 2, pp. 365–366.

Kumar, V. (1993) Effect of irrigation on yield, sucrose content and impurity levels of sugarbeet crop. *Journal of the Indian Society of Soil Science*, Vol. 41, No. 1, pp. 26–28.

Kumar Arindam (1999) Influence of carbonaceous sugar mill effluent on phytosociology (herbaceous). *Advances in Plant Sciences*, Vol. 12, No. 1, pp. 159–162.

Kumar Arindam (2001) Irrigational impact of industrial effluent on chemical constituents of soil and plant. *Advances in Plant Sciences*, Vol. 14, No. 2, pp. 351–358.

Kuniata, L.S. (1994) Importation and establishment of *Heteropsylla spinulosa* (Homoptera; Psyllidae) for the biological control of *Mimosa invasa* in Papua New Guinea. *International Journal of Pest Management*, Vol. 40, pp. 64–65.

Kuniata, L.S. (2000) Integrated management of sugarcane stemborers in Papua New Guinea. In *Sugarcane Pest Management Strategies in the New Millennium – Proceedings of the 4th ISSCT Sugarcane Entomology Workshop* (eds P.G. Allsopp and W. Suara-ard), pp. 37–50. International Society of Sugar Cane Technologists, Brisbane.

Kuniata, L.S. and Sweet, C.P.M. (1994) Management of *Sesamia grisescens* Walker (Lep.: Noctuidae), a sugar-cane borer in Papua New Guinea. *Crop Protection*, Vol. 13, pp. 488–493.

Kuniata, L.S. and Korowi, K.T. (2001) Biological control of the giant sensitive plant with *Heteropsylla spinulosa* (Hom.: Psyllidae) in Papua New Guinea. *Technical Bulletin* No. 288, pp. 145–151. Department of Primary Industries and Fisheries, Northern Territory of Australia, Darwin.

Kuniata, L.S., Chandler, K.J. and Korowi, K.T. (2001) Management of sugarcane pests at Ramu, Papua New Guinea. *Proceedings of the International Society of Sugar Cane Technologists*, Vol. 24, pp. 382–388.

Kunteova, L. (1996) The potential role of bioethanol. Trial in the Czech Republic on fermenting sugar beet thick juice. *International Sugar Journal (Beet Sugar Edition)*, Vol. 98, pp. 448–452.

Kunteova, L. (1997) What the beet brings – utilization of beet pulp and molasses slops. *Listy Cukrovarnicke a Reparske*, Vol. 113, pp. 259–262.

Kuzminski, C., Palencia, J.I., Fernandez-Polanco, Garcia, P.A. and Iza, J. (1991) Wastewater reclamation and reuse at the sugarbeet factory ACOR-1 (Valladolid, Spain). *Water Science and Technology*, Vol. 24, No. 9, pp. 303–312.

Kwon-Ndung, E.H., Busarii, L.D., Ishaq, M.N. and Misari, S.M. (2000) Sugarcane germplasm collecting expedition in northern Nigeria. *Sugar Cane International*, March, pp. 9–11.

Lacey, J. (1980) The microbiology of sugar cane bagasse. *Proceedings of the International Society of Sugar Cane Technologists*, Vol. 17, pp. 2442–2461.

Lal, R. (1990) *Soil Erosion in the Tropics – Principles and Management*. McGraw-Hill, New York.

Lanchote, V., Bonato, P.S., Cerdeira, A.L., Santos, N.A.G., de Carvallho, D. and Gomes, M.A. (2000) HPLC screening and GC-MS confirmation of tiazine herbicide residues in drinking water from a sugar cane area in Brazil. *Water, Air, and Soil Pollution*, Vol. 118, pp. 329–337.

Landrey, O.P. (1978a) Land use on steep slopes on an estate on the south coast of Natal. *South African Sugar Journal*, Vol. 62, No. 9, pp. 454–459.

Landrey, O.P. (1978b) Land use on steep slopes on an estate on the South Coast of Natal. *Proceedings, South African Sugar Technologists' Association*, No. 52, pp. 125–128.

Lanza, F. (1991) Yield variations in different cropping systems and environments: summary of six years of data from the Italian Ministry of Agriculture and Forests project on cropping systems (cereal, industrial and catch crops). *Annali dell'Istituto Sperimentale Agronomico*, Vol. 22, pp. 87–99.

Lara, L.B.L.S., Artaxo, P., Martinelli, L.A., Victoria, R.L., Camargo, P.B., Krusche, A., Ayers, G.P., Ferraz, E.S.B. and Ballester, M.V. (2001) Chemical composition of rainwater and anthropogenic influences in the Piracicaba River Basin, Southeast Brazil. *Atmospheric Environment*, Vol. 35, No. 29, pp. 4937–4945.

Larrahondo, J.E., Briceno, C.O., Rojas, M. and Palma, A.E. (2002) Postharvest sugar losses in sugar cane: varietal differences and effect of harvesting method. In *Advances in the Chemistry and Processing of Beet and Cane Sugar: Proceedings of the 2002 Sugar Processing Research Conference*, pp. 219–226. Sugar Processing Research Institute, New Orleans.

Larsen, J.L. (1982) *Vibrio anguillarum*: prevalence in three carbohydrate loaded marine recipients and a control. *Zentralblatt für Bakteriologie Mikrobiologie und Hygiene, I. Abteilung Originale, C*, Vol. 3, No. 4, pp. 519–530.

Larsen, J.L. (1985) *Vibrio anguillarum*: prevalence of typical and atypical strains in marine recipients with special reference to carbohydrate pollution. *Acta Veterinaria Scandinavica*, Vol. 26, No. 4, pp. 449–460.

Larsen, J.L. and Jensen, N.J. (1982) The ulcus-syndrome in cod (*Gadus morhua*). V. Prevalence in selected Danish marine recipients and a control site in the period 1976–1979. *Nordisk Veterinaermedicin*, Vol. 34, No. 7/9, pp. 303–312.

Larsen, J.L., Jensen, N.J. and Christensen, N.O. (1978) Water pollution and the ulcer syndrome in the cod (*Gadus morhua*). *Veterinary Science Communications*, Vol. 2, No. 3, pp. 207–216.

Last, P.J. and Draycott, A.P. (1975a) Growth and yield of sugar beet on contrasting soils in relation to nitrogen supply. I – Soil nitrogen analyses and yield. *Journal of Agricultural Science, Cambridge*, Vol. 85, pp. 19–26.

Last, P.J. and Draycott, A.P. (1975b) Growth and yield of sugar beet on contrasting soils in relation to nitrogen supply. II – Growth, uptake and leaching of nitrogen. *Journal of Agricultural Science, Cambridge*, Vol. 85, pp. 27–37.

Lau, L.S. (1979) Water reuse from sewage effluent by irrigation: a perspective for Hawaii. *Water Resources Bulletin*, Vol. 15, No. 3, pp. 740–752.

Lawrence, G. and Gray, I. (1997) Social factors shaping the Australian sugar industry. In *Intensive Sugarcane Production: Meeting the Challenges Beyond 2000* (eds B.A. Keating and J.R. Wilson), pp. 71–86. CAB International, Wallingford, UK.

Legrand, G. (2001) International beet library – 20,000 bibliographic references on sugar beet in 2001. *Proceedings of the Institut International de Recherches Betteravières*, Vol. 64, pp. 133–136.

Leitch, C. and Harbor, J. (1999) Impacts of land use change on freshwater runoff into the near-coastal zone, Holetown Watershed, Barbados: comparisons of long-term to single-storm effects. *Journal of Soil and Water Conservation (Ankeny)*, Vol. 54, No. 3, pp. 584–592.

Leng, R.A. and Preston, T.R. (1986) Constraints to the efficient utilization of sugar-cane and its by-products as diets for production of large ruminants. In *Ruminant Feeding Systems Utilizing Fibrous Agricultural Residues – 1985. Proceedings of the Fifth Annual Workshop of the Australian-Asian Fibrous Agricultural Residues Research Network held in Balai Penelitian Ternak, Ciawi, Bogor, 13–17 April 1985* (ed. R.M. Dixon), pp. 27–48. International Development Program of Australian Universities and Colleges Ltd, Canberra.

Lescure, J.P. and Affret, L. (1997) Agronomic exploitation downstream from the sugar factory. *Proceedings of the Institut International de Recherches Betteravières*, Vol. 60, pp. 307–311.

Lettinga, G. and van Haandel, A.C. (1993) Anaerobic digestion for energy production and environmental protection. In *Renewable Energy: Sources for Fuels and Electricity*, pp. 817–839. Earthscan, London.

Liao, M.C. (1972) Studies on soil conservation practices on sugar cane land. *Report of the Taiwan Sugar Experiment Station*, No. 55, pp. 75–89.

Liao, M.C. (1979) Improvement of hillside ditches in Taiwan. *Journal of Soil and Water Conservation*, Vol. 34, No. 2, pp. 102–104.

Lidon, B., Marlet, S. and Clopes, A. (1999) Evaluating the performance of sugarcane row irrigation. *Agriculture et Développement*, No. 24, pp. 114–125.

Liegeois, E. (1997) Discussion of genetically modified organisms. *Agricontact*, No. 291, pp. 29–32.

Lima, E., Boddey, R.M. and Dobereiner, J. (1987) Quantification of biological nitrogen fixation associated with sugar cane using a ^{15}N aided nitrogen balance. *Soil Biology and Biochemistry*, Vol. 19, pp. 165–170.

Lin, M.-S., Chen, M.-H. and Chang, J.-M. (1984) Research on SO_2 injury to important crops in Taiwan. *Journal of Agricultural Research of China*, Vol. 33, No. 2, pp. 153–158.

Lindau, C.W., Delaune, R.D. and Alford, P. (1997) Monitoring nitrogen pollution from sugarcane runoff using ^{15}N analysis. *Water, Air, and Soil Pollution*, Vol. 89, pp. 389–399.

Linedale, A.I. (1998) Monitoring of irrigation with urban waste water. *Proceedings of the Australian Society of Sugar Cane Technologists*, Vol. 20, pp. 56–62.

Lingle, S.E., Wiedenfeld, R.P. and Irvine, J.E. (2000) Sugarcane response to saline irrigation water. *Journal of Plant Nutrition*, Vol. 23, No. 4, pp. 469–486.

Lisson, S.N., Brennan, L.E., Bristow, K.L., Keating, B.A. and Hughes, D.A. (2003) DAM EA$Y –

software for assessing the costs and benefits of on-farm water storage based production systems. *Agricultural Systems*, Vol. 76, No. 1, pp. 19–38.

Liu, L.G. and Ho, F.J. (1989) Vacuum type sulphitation system. *Taiwan Sugar*, Vol. 36, No. 6, p. 16.

Liu, W.C., Hsieh, T.S., Chen, F. and Li, S.W. (1996a) Effect of swine lagoon effluent on soil heavy metal accumulation and sugarcane. *Report of the Taiwan Sugar Research Institute*, No. 152, pp. 1–17.

Liu, W.C., Hsieh, T.S., Chen, F. and Li, S.W. (1996b) Effects of swine lagoon effluents on soil heavy metal accumulation and sugarcane. *Taiwan Sugar*, Vol. 43, No. 6, pp. 9–18.

Liu, Z.C. (1985) Studies on the integrated control tactics of sugarcane insect pests. *Chinese Journal of Biological Control*, No. 3, pp. 10–12.

Lloyd, P. and Kuniata, L.S. (2000) The control of *Sesamia* at Ramu. *Sugar Cane International*, April, pp. 10–17.

Lopez-Lopez, E., Favari, L., Martinez-Tabche, L., Madrigal, M. and Soto, C. (2003) Hazard assessment of a mixture of pollutants from a sugar industry to three fish species of Western Mexico by the responses of enzymes and lipid peroxidation. *Bulletin of Environmental Contamination and Toxicology*, Vol. 70, No. 4, pp. 739–745.

Lora, E.S. and Jativa, I.H. (1999) Air pollution control in the sugar cane industry. *Proceedings of the International Society of Sugar Cane Technologists*, Vol. 23, pp. 321–330.

Lorenzoni, I., Jordan, A., Viner, D., Favis-Mortlock, D. and Hall, J. (2000) Sunflowers or sugar beet? Potential impacts of climate change on agriculture in Norfolk. *Ecos – British Association of Nature Conservationists*, Vol. 21, No. 3/4, pp. 47–53.

Lotz, L.A.P., Brussaard, L., Gillissen, L.J.W.J., Gorissen, A., Kempenaar, C., van Loon, J.J.A., Noordam, M.Y., Termorshuizen, A.J. and van Vliet, P.C.J. (2000) *Effects of the Large-scale Use of Transgenic Herbicide Resistant Cultivars: Development and Exploration of Scenarios*, 110 pp. Rapport No. 2, Plant Research International, Wageningen

Lowe, S., Browne, M., Boudjelas, S. and De Poorter, M. (2001) *100 of the World's Worst Invasive Alien Species – a selection from the global invasive species database.* Invasive Species Specialist Group (ISSG), Auckland.

Lu, C.K.H. and Chen, W.V.D. (1991) The recovery and utilization of sugar processing water. *Taiwan Sugar*, Vol. 38, No. 6, pp. 17–21.

Lu, X.Z. (1991) Soil problems in relation to tropical crop development and soil amendment measures in China. *Tropical Agriculture Research Series*, No. 24, pp. 30–39.

Lugo-Lopez, M.A., Abruna, F. and Perez-Escolar, R. (1981) *The Role of Crop and Industrial Residues in Erosion Control, Properties and Productivity of Some Major Soils of Puerto Rico.* Bulletin No. 266, Agricultural Experiment Station, University of Puerto Rico, 26 pp.

Lumkes, L.M. (1981) Control of wind erosion on beet fields with barley. *Landbouwmechanisatie*, Vol. 32, No. 3, pp. 223–227.

Lumkes, L.M. and te Velde, H.A. (1973) *Arable Cropping on Land Susceptible to Soil Blowing.* Publikatie No. 11, Proefstation voor de Akkerbouw, 36 pp.

Maassen, J. (2001) Beta(kwik), advisory system for the growing of sugar beet transferred to internet. *Proceedings of the Institut International de Recherches Betteravières*, Vol. 64, pp. 115–122.

McAughtrie, J. (2001) Sugar – much more than a bag of granulated. *British Sugar Beet Review*, Vol. 69, pp. 26–29.

McBurney, B. and McBurney, J.C. (1997) Designing bagasse fired boilers for low emissions. *International Sugar Journal (Worldwide Sugar Edition)*, Vol. 99, Nos. 1178, 1179, pp. 58–64, 133–134, 136, 138–139.

McCulloch, D.J. and Stranack, R.A. (1995) Some benefits and costs of implementing soil conservation practices on farms on the north coast of Kwazulu-Natal. *Proceedings of the Annual Congress – South African Sugar Technologists' Association*, No. 69, pp. 94–97.

Macedo, I. de C. (1992) The sugar cane agro-industry – its contribution to reducing CO_2 emissions in Brazil. *Biomass and Bioenergy*, Vol. 3, No. 2, pp. 77–80.

McGarry, D., Bigwood, R.D., Pillai-McGarry, U.P., Bray, S.G. and Moody, P.W. (1996) Effect of long term cane production on soil properties of a gleyed podzolic soil near Bundaberg, Queensland. In *Sugar 2000 Symposium: Sugarcane: Research Towards Efficient and Sustainable Production* (eds J.R. Wilson, D.M. Hogarth, J.A. Campbell and A.L. Garside), pp. 251–253. CSIRO Division of Tropical Crops and Pastures, Brisbane.

McGlinchey, M.G. (1998) Components of the water balance of an irrigated sugarcane crop in Swaziland. *Proceedings of the Annual Congress – South African Sugar Technologists' Association*, No. 72, pp. 114–116.

McGlinchey, M.G., Inman-Bamber, N., Culverwell, T.L. and Els, M. (1995) An irrigation scheduling method based on a crop model and an automatic weather station. *Proceedings of the Annual Congress – South African Sugar Technologists' Association*, No. 69, pp. 69–73.

Mach, J. and Svatos, M. (1998) The non-food use of sugarbeet especially for production of bioethanol. In *Sbornik praci z mezinarodni vedecke konference, Agrarni perspektivy VII. Evropska integrace a vyuzivani prirodnich zdroju, Dil I, Praha, Czech Republic, 17.-18. Zari 1998*, pp. 544–549. Provozne Ekonomicka Fakulta, Ceska Zemeldelska Universita v Praze, Prague.

McIlroy, W.A., Smith, R.J. and Ames, R.V. (1995) Environmental auditing: the CSR Sugar Mills Group experience. *Sugar y Azucar*, Vol. 90, No. 7, pp. 30–33.

McLean, F.C. (1982) Results of recent flood–fallow experimentation in Guyana. *Proceedings of the West Indies Sugar Technologists*, Vol. 21, pp. 243–255.

Maclean, N.R. (1975) Long-term effects of sugar cane production on some physical and chemical properties of soils in the Goondi Mill area. *Proceedings of the Queensland Society of Sugar Cane Technologists*, Vol. 42, pp. 123–126.

McMartin, D.W., Headley, J.V., Wood, B.P. and Gillies, J.A. (2003) Photolysis of atrazine and ametryne herbicides in Barbados sugar cane plantation soils and water. *Journal of Environmental Science and Health. Part B, Pesticides, Food Contaminants, and Agricultural Wastes*, Vol. 38, No. 3, pp. 293–303.

McNeely, J. (2003) Water and agriculture: food, water and biodiversity. *World Conservation*, Vol. 1, p. 10.

Madan, Y.P. (2001) Integrated pest management in sugarcane. *Indian Sugar*, Vol. 50, No. 12, pp. 867–871.

Madejon, E., Diaz, M.J., Lopez, R., Murillo, J.M. and Cabrera, F. (1995) Corn fertilization with three (sugarbeet) vinasse composts. *Fresenius Environmental Bulletin*, Vol. 4, pp. 232–237.

Madejon, E., Diaz, M.J., Lopez, R., Lozano, C. and Cabrera, F. (1996) Cotton fertilization with composts of (sugarbeet) vinasse and agricultural residues. *Fertilizer Research*, Vol. 43, pp. 179–182.

Madejon, E., Diaz, M.J., Lopez, R. and Cabrera, F. (2001a) A study on nitrogen leaching under crops fertilized with mineral and organic fertilizers in lysimeters. In *Control of Adverse Impacts of Fertilizers and Agrochemicals – Proceedings of an International Workshop of the ICID, Cape Town, South Africa, 22–27 October 2000* (eds A. Mermoud, A. Musy, L.S. Pereira and R. Ragab), pp. 21–30. International Commission on Irrigation and Drainage (ICID), Rome.

Madejon, E., Lopez, R., Murillo, J.M. and Cabrera, F. (2001b) Agricultural use of three (sugar-beet) vinasse composts: effect on crops and chemical properties of a Cambisol soil in the Guadalqivir river valley (SW Spain). *Agriculture, Ecosystems and Environment*, Vol. 84, pp. 55–65.

Madeley, J. (1981) Ethanol production runs into environmental difficulties. *Mazingira*, Vol. 5, No. 3, pp. 42–51.

Madsen, H.B., Hashalt, B. and Platou, S.W. (1986) The development of a computerized erodibility map covering Denmark. In *Soil Erosion in the European Community – Impact of Changing Agriculture* (eds G. Chisci and R.P.C. Morgan), pp. 143–154. A.A. Balkema, Rotterdam.

Madsen, K.H. and Sandoe, P. (2001) Herbicide resistant sugar beet – what is the problem? *Journal of Agricultural and Environmental Ethics*, Vol. 14, No. 2, pp. 161–168.

Madsen, K.H., Streibig, J.C. and Jensen, J.E. (1995) Herbicide resistant crops – impact on herbicide use and environment. *SP Rapport – Statens Planteavlsforsog*, No. 3, pp. 221–230.

Mafla Cifuentes, H. (1997) El Nino and its catastrophic consequences in the sugar industry of Ecuador, South America. *Sugar Journal*, Vol. 60, pp. 16, 20–21.

Magar, S.S. (1995) The status of microirrigation in Maharashtra, India. In *Microirrigation for a Changing World: Conserving Resources/Preserving the Environment. Proceedings of the Fifth International Microirrigation Congress, Orlando, Florida, USA, 2–6 April 1995* (ed. F.R. Lamm), pp. 452–456. American Society of Agricultural Engineers, St Joseph.

Magarey, R.C., Suma, S., Irawan, Kuniata, L.S. and Allsopp, P.G. (2002) Sik na binatang bilong suka – diseases and pests encountered during a survey of *Saccharum* germplast 'in the wild' in Papua New Guinea. *Proceedings of the Australian Society of Sugar Cane Technologists*, Vol. 24, pp. 219–227.

Magarey, R.C., Kuniata, L.S., Croft, B.J., Chandler, K.J., Irawan, Kristini, A., Spall, V.E., Samson, P.R. and Allsopp, P.G. (2003) International activities to minimize industry losses from exotic pests and diseases. *Proceedings of the Australian Society for Sugar Cane Technologists*, Vol. 25, p. 18.

Magwenzi, O.E. (2000) Evaluation of irrigation efficiency in the Swaziland sugar industry. *Proceedings of the Annual Congress – South African Sugar Technologists' Association*, No. 74, pp. 151–156.

Magwenzi, O.E. (2002) Development of a decision support tool for assessing economics of irrigation in sugarcane. *Proceedings of the Annual*

Congress – South African Sugar Technologists' Association, No. 76, pp. 135–146.
Mahmoody, B., Minassian, V. and Alizadeh, A. (1997) Life cycle of *Urophlyctis leproides*, the causal agent of sugar beet leaf and crown wart and the effect of irrigation on disease incidence in Khuzestan province. *Iranian Journal of Plant Pathology*, Vol. 33, No. 1/2, pp. 71–83 (Pe), 24–27 (En).
Maiti, D., Das, S., Mandal, B. and Chakraborty, P.B. (2000) Influence of irrigation on growth and disease incidence of sugarbeet (*Beta vulgaris* L.). *Indian Agriculturist*, Vol. 44, No. 1/2, pp. 63–69.
Malavia, D.D., Patel, K.P. and Ramani, V.B. (1992) Response of irrigation levels and trash mulch on yield and quality of sugarcane. *Indian Sugar*, Vol. 42, No. 2, pp. 97–100.
Malavolta, E. (1994) *Fertilizing for High Yield Sugarcane*. International Potash Institute, Basle.
Maliwal, G.L., Parmar, H.C., Kaswala, R.R. and Raman, S. (1999) Effect of irrigation, mulching and other agronomic manipulation on yield and water economy in sugarcane. *Annals of Agricultural Research*, Vol. 20, No. 1, pp. 118–119.
Mallawaarachchi, T. and Quiggin, J. (2001) Modelling socially optimal land allocations for sugar cane growing in North Queensland: a linked mathematical programming and choice modelling study. *Australian Journal of Agricultural and Resource Economics*, Vol. 45, No. 3, pp. 383–410.
Mallawaarachchi, T., Blamey, R.K., Morrison, M.D., Johnson, A.K.L. and Bennett, J.W. (2001) Community values for environmental protection in a cane farming catchment in Northern Australia: a choice modelling study. *Journal of Environmental Management*, Vol. 62, No. 3, pp. 301–316.
Malmary, G., Albet, J., Putranto, A., Hanine, H. and Molinier, J. (2000) Recovery of aconitic and lactic acids from simulated aqueous effluents of the sugar-cane industry through liquid–liquid extraction. *Journal of Chemical Technology and Biotechnology*, Vol. 75, No. 12, pp. 1169–1173.
Maly, V. (1986) Sensitivity of cultivars of some agricultural crops to SO_2. *Scientia Agriculturae Bohemoslovaca*, Vol. 18, No. 1, pp. 9–18.
Malzer, G.L. and Graff, T. (1995) Impact of turkey manure application on corn production and potential water quality concerns, Westport, MN, 1994. In *Field Research in Soil Science*, pp. 121–125. Miscellaneous Publication 88–1995, Agricultural Experimental Station, University of Minnesota, St Paul.
Mambelli, S., Benati, R., Amaducci, M.T. and Venturi, G. (1992) Effects of irrigation on the growth and qualitative characteristics of roots of sugarbeet (*Beta vulgaris* L.). *Rivista di Agronomia*, Vol. 26, No. 4, suppl., pp. 623–632.
Mangal Singh (1996) Process house upgradation – sugar mills in India. *Indian Sugar*, Vol. 46, No. 2, pp. 103–111.
Mangisoni, J.H. and Phiri, G.S. (1996) New perspectives on local conservation techniques: a case study from Malawi. In *Sustaining the Soil: Indigenous Soil and Water Conservation in Africa*, (eds C. Riej, I. Scoones and C. Toulmin), pp. 202–212. Earthscan Publications Ltd, London.
Manners, J.G. (1982) *Principles of Plant Pathology*. Cambridge University Press, Cambridge.
Manson, D.A. (1981) Bagasse – a sugar mill by-product useful for soil conservation. *Journal of the Soil Conservation Service of NSW*, Vol. 37, No. 3, pp. 171–172.
Mantovani, G. and Vaccari, G. (2001) Green power, green manufacture, green sugar. *International Sugar Journal*, Vol. 103, No. 1233, pp. 372, 374, 376.
Marciniak, K. (1989) The next stage in arranging the water-and-effluent economy of Melno sugar factory. *Gazeta Cukrownicza*, Vol. 97, No. 4, pp. 71–72.
Marciniak, K. (2000) Environmental protection in a typical sugarbeet factory on the threshold of the 21st century and the example of the Melno factory. *Gazeta Cukrownicza*, Vol. 108, No. 10, pp. 191–195.
Marcus, K.G. (1992) El Nino – its impact on 1992 world sugar production. *Sugar y Azucar*, Vol. 87, pp. 23, 26–28, 40, 42–44.
Maredia, K.M., Dakouo, D. and Mota-Sanchez, D. (eds) (2003) *Integrated Pest Management in the Global Arena*. CAB International, Wallingford, UK.
Margotti, A. (2000) Sugarbeet does not always require irrigation. *Irrigazione e Drenaggio*, Vol. 47, No. 3, pp. 8–11.
Marie-Jeanne, S. (1993) Dedusting of chimney gases. *Revue Agricole et Sucrière de l'Île Maurice*, Vol. 72, No. 3, pp. 25–32.
Marlander, B., Ackmann, I. and Deking, G. (1981) Protection against wind erosion on sandy soil. *DLG-Mitteilungen*, Vol. 96, No. 14, pp. 767–769.
Martin, L. (1979) Accelerated soil erosion from tractor wheelings: a case study in mid-Bedfordshire. In *Comptes-rendus, Colloque sur l'érosion agricole des sols en milieu tempéré non-Méditerranéen, Université Louis Pasteur, Strasbourg* (eds H. Vogt and T. Vogt), pp. 157–161. Université Louis Pasteur, Strasbourg.

Martin, T.G. and Catterall, C.P. (2001) Do fragmented coastal heathlands have habitat value to birds in eastern Australia? *Wildlife Research*, Vol. 28, No. 1, pp. 17–31.

Masilaca, A.S., Presad, R.A. and Morrison, R.J. (1986) The impact of sugar cane cultivation on three Oxisols from Vanua Levu, Fiji. *Tropical Agriculture*, Vol. 63, pp. 325–330.

Massoud, M.M. and Shalaby, E.M. (1998) Response of sugar beet (*Beta vulgaris*, L.) varieties to irrigation intervals in Upper Egypt. II. Water consumption and water use efficiency. *Assiut Journal of Agricultural Sciences*, Vol. 29, No. 5, pp. 23–30.

Massoud, M.M., Mohamed, K.A., Rayan, A.A. and Ally, M.K. (1999) Response of sugar beet (*Beta vulgaris*, L.) varieties to irrigation intervals in Upper Egypt. I. Yield and its quality. *Assiut Journal of Agricultural Sciences*, Vol. 30, No. 4, pp. 53–61.

Masundire, H. (2003) Water and ecosystems – high adventure: the ecosystem approach. *World Conservation*, Vol. 1, pp. 6–7.

Mathers, A.C. and Stewart, B.A. (1984) Manure effects of crop yields and soil properties. *Transactions of American Society of Agricultural Engineers*, Vol. 27, pp. 1022–1026.

Matthews, K. (1983) Beating the blow: available techniques. *British Sugar Beet Review*, Vol. 51, No. 1, pp. 65, 68–69.

Maud, R.R. (1959) The occurrence of two alkali soils in Zululand. *Proceedings of the South African Sugar Technologists' Association*, Vol. 33, pp. 138–144.

Maud, R.R. (1960) The compaction of sugar-belt soils at various moisture levels. *Proceedings of the South African Sugar Technologists' Association*, Vol. 34, pp. 154–156.

Mauremootoo, J. (2001) Case study 2.15 – Mauritius and La Reunion cooperate to prevent a sugar cane pest spreading. In *Invasive Alien Species: A Toolkit of Best Prevention and Management Practices* (eds R. Wittenberg and M.J.W. Cock), p. 42. CAB International (on behalf of the Global Invasive Species Programme, GISP), Wallingford, UK.

May, M.J. and Fisher, S.J. (2001) Information transfer for sugar beet production. In: *Proceedings of the Institut International de Recherches Betteravières*, Vol. 64, pp. 81–88.

Mbohwa, C. and Fukuda, S. (2003) Electricity from bagasse in Zimbabwe. *Biomass and Bioenergy*, Vol. 25, pp. 197–207.

Megias, M.D., Martinez-Teruel, A. and Hernandez, M.R. (1999) Potential environmental impact of effluents from the artichoke (*Cynara scolymus* L.) by-product ensiling process using additives. *Journal of Agricultural and Food Chemistry*, Vol. 47, No. 6, pp. 2455–2458.

Meier, E.A., Thorburn, P.J., Wegener, M.K. and Basford, K.E. (2002) A conceptual analysis of nitrogen from trash blankets in the wet tropics. *Sugar Cane International*, May/June, pp. 8–11.

Mekki, B.B. and El-Gazzar, M.M. (1999) Response of root yield and quality of sugar beet (*Beta vulgaris* L.) to irrigation with saline water and foliar potassium fertilization. *Annals of Agricultural Science (Cairo)*, Vol. 44, No. 1, pp. 213–225.

Merkes, R. (1983) Reducing erosion in sugarbeet growing. *DLG-Mitteilungen*, Vol. 98, No. 6, pp. 330–332.

Merry, R.E. (2003) 'Dripping with success': the challenges of an irrigation redevelopment project. *Irrigation and Drainage*, Vol. 52, pp. 71–83.

Meyer, E., Boevey, T.C. and Meyer, J.H. (1989) The incorporation of lime under minimum tillage conditions. *Proceedings of the South African Sugar Technologists' Association*, Vol. 63, pp. 142–146.

Meyer, G.M., Keeping, M.B., Watt, D.A., Cassim, T.N., Botha, F.C. and Huckett, B.I. (2000) A wild-type insecticidal gene from a bacterium is poorly expressed in sugarcane: an overview. *Proceedings of the South African Sugar Technologists' Association*, Vol. 74, pp. 184–185.

Meyer, J.H. and van Antwerpen, R. (1995) Trends in water quality of selected rivers in the South African sugar industry. *Proceedings of the Annual Congress – South African Sugar Technologists' Association*, No. 69, pp. 60–68.

Meyer, J.H. and van Antwerpen, R. (2001) Soil degradation as a factor in yield decline in the South African sugar industry. *Proceedings of the International Society of Sugar Cane Technologists*, Vol. 24, pp. 8–15.

Meyer, J.H., Turner, P.E.T. and Fey, M.V. (1991) Interim evaluation of phospho-gypsum as an ameliorant for soil acidity in sugarcane. *Proceedings of the South African Sugar Technologists' Association*, Vol. 65, pp. 41–46.

Meyer, J.H., Harding, R., Rampersad, A.L. and Wood, R.A. (1998) Monitoring long term soil fertility trends in the South African sugar industry using the FAS analytical database. *Proceedings of the South African Sugar Technologists' Association*, Vol. 72, pp. 61–67.

Meyer, W.S. (1997) The irrigation experience in Australia – lessons for the sugar industry. In *Intensive Sugarcane Production: Meeting the Challenges Beyond 2000* (eds B.A. Keating and J.R. Wilson), pp. 437–454. CAB International, Wallingford, UK.

Millette, E.D. (1991) Industrial wastewater monitoring at a cane sugar factory. *Industry and Environment*, Vol. 14, No. 1, pp. 30–34.

Milosevic, R., Manojlovic, S., Bogdanovic, D. and Ming, M.F. (1989) The dynamics of nitrate nitrogen as a function of soil type and sugar-beet fertilization. *Protection of Water Quality from Harmful Emissions with Special Regard to Nitrate and Heavy Metals. Proceedings of the 5th International Symposium of CIEC* (eds E. Welte and I. Szabolcs), pp. 141–144. CIEC, Budapest.

Mineyev, V.G., Rempe, E.Kh. and Solovei, I.N. (1993) Ecological assessment of the use of chemical means of plant protection in growing sugar beet on typical chernozem. *Moscow University Soil Science Bulletin*, Vol. 48, No. 2, pp. 26–31.

Minhas, Y.J., Khanzada, A.G. and Memon, M.I. (1992) Effect of delaying first irrigation after planting in sugar cane crop. *Pakistan Sugar Journal*, Vol. 6, No. 2, pp. 21–23.

Mishra, L.N., Pathak, R.A., Pandey, A.K. and Pratap, B. (2001) Effect of various mulches on the conservation of soil moisture in aonla + guava cropping system. *South Indian Horticulture*, Vol. 49, Special No., pp. 367–369.

Mitchell, A.W., Reghenzani, J.R., Hunter, H.M. and Bramley, R.G.V. (1996) Water quality and nutrient fluxes from river systems draining to the Great Barrier Reef Marine Park. In *Downstream Impacts of Land Use* (eds H.H. Hunter, A.G. Eyles and G.E. Rayment), pp. 23–33. Department of Natural Resources, Brisbane.

Mitchell, A.W., Reghenzani, J.R. and Furnas, M.J. (2001) Nitrogen levels in the Tully River – a long-term view. *Water Science and Technology*, Vol. 43, No. 9, pp. 99–105.

Mitra, A.K. (1989) Economic appraisal of irrigation in a water scarce region – an extended approach. *Irrigation Theory and Practice. Proceedings of the International Conference, Southampton, UK, September 1989* (eds J.R. Rydzewski and C.F. Ward), pp. 827–835. Pentech Press Ltd, London.

Miyagi, K. and Shimabuku, M. (1997) Effects of irrigation at different growth stages on yield and quality of sugarcane in Okinawa Miyako Island. *Report of the Kyushu Branch of the Crop Science Society of Japan*, No. 63, pp. 81–83.

Moenne-Loccoz, Y., Powell, J., Higgins, P., McCarthy, J. and O'Gara, F. (1998) An investigation of the impact of biocontrol *Pseudomonas fluorescens* F113 on the growth of sugarbeet and the performance of subsequent clover–*Rhizobium* symbiosis. *Applied Soil Ecology*, Vol. 7, No. 3, pp. 225–237.

Moens, L. (2002) Sugar cane as a renewable feedstock for the chemical industry: challenges and opportunities. In *Advances in the Chemistry and Processing of Beet and Cane Sugar: Proceedings of the 2002 Sugar Processing Research Conference*, pp. 26–41. Sugar Processing Research Institute, New Orleans.

Mohawed, S.M., Abdel-Hafez, S.I.I., El-Said, A.H.M. and Gherbawy, Y.A.M.H. (1999) Seasonal fluctuations of soil and root surface fungi of sugarcane (*Saccharum officinarum* L.) in Upper Egypt. *Egyptian Journal of Microbiology*, Vol. 34, No. 4, pp. 595–611.

Mohyuddin, A.I. (1992) Implementation of integrated pest management of sugarcane pests in Pakistan. In *Integrated Pest Management in the Asia–Pacific Region* (eds P.A.C. Ooi, G.S. Lim, T.H. Ho, P.L. Manalo and J.K. Waage), pp. 73–84. CAB International, Wallingford, UK.

Mohyuddin, A.I., Hamid, A. and Mohammad, A. (1994) *Integrated Pest Control of Sugar-cane Borers in the NWFP*. International Institute of Biological Control/Pakistan Agricultural Research Council, Islamabad.

Monnier, G. and Boiffin, J. (1986) Effect of the agricultural use of soils on water erosion – the case of cropping systems in western Europe. In *Soil Erosion in the European Community – Impact of Changing Agriculture* (eds G. Chisci and R.P.C. Morgan), pp. 17–32. A.A. Balkema, Rotterdam.

Moreira, D.R. and Cardoso, V.J.M. (1998) Effect of soil moisture content and the irrigation frequency on the sugarcane germination. *Pesquisa Agropecuaria Brasileira*, Vol. 33, No. 5, pp. 721–729.

Morgan, R.P.C. (1986) Soil degradation and soil erosion in the loamy belt of northern Europe. In *Soil Erosion in the European Community – Impact of Changing Agriculture* (eds G. Chisci and R.P.C. Morgan), pp. 165–172. A.A. Balkema, Rotterdam.

Morgan, R.P.C. (1995) *Soil Erosion and Conservation*. Longman, Harlow.

Moron, M.A., Deloya, C., Ramirez-Campos, A. and Hernandez-Rodriguez, S. (1998) Lamellicorne Coleoptera fauna of the region of Tepic, Nayarit, Mexico. *Acta Zoologica Mexicana*, No. 75, pp. 73–116.

Morris, J. (1994) The economics of irrigation in Britain. *Aspects of Applied Biology*, No. 38, pp. 321–331.

Morris, M. (1996) A green treatment for wastewater. *British Sugar Beet Review*, Vol. 64, No. 4, pp. 16–17.

Morris, M. and Herbert, R. (1997) The design and performance of a vertical flow reed bed for the treatment of high ammonia, low suspended solids organic effluents. *Water Science and Technology*, Vol. 35, No. 5, pp. 197–204.

Mosier, A.R. (1994) Nitrous oxide emissions from agricultural soils. *Fertilizer Research*, Vol. 37, pp. 191–200.

Moss, A.J. and Bennett, J. (1991) Focus on some north Queensland water quality issues. *Water*, December, pp. 16–19.

Moyart, M. and Sneessens, J.-F. (2001) Integral quality management project in the Belgian sugarbeet sector. *Proceedings of the Institut International de Recherches Betteravières*, Vol. 64, pp. 27–32.

Mrig, K.K. and Chaudhary, J.P. (1993) Comparative efficacy of different insecticides, applied with different irrigation levels for control of sugarcane shoot and root borers. *Cooperative Sugar*, Vol. 25, No. 1/2, pp. 33–37.

Mrig, K.K., Chaudhary, J.P. and Mahla, J.C. (1995) Effect of irrigation intervals on the incidence of shoot and root borers in sugarcane ratoon. *Journal of Insect Science*, Vol. 8, No. 2, pp. 188–191.

Mubien, B. (1996) Optimum sprinkler irrigation system for sugarcane plantation at Subang Sugar Factory. *Majalah Penelitian Gula*, Vol. 32, No. 3/4, pp. 24–28.

Muchembled, C. (1992) The technique of reduced doses. *Proceedings of the Institut International de Recherches Betteravières*, Vol. 55, pp. 315–319.

Muchow, R.C. and Keating, B.A. (1998) Assessing irrigation requirements in the Ord Sugar Industry using a simulation modelling approach. *Australian Journal of Experimental Agriculture*, Vol. 38, No. 4, pp. 345–354.

Muchow, R.C., Robertson, M.J., Wood, A.W. and Keating, B.A. (1996) Effect of nitrogen on the time-course of sucrose accumulation in sugarcane. *Field Crops Research*, Vol. 47, pp. 143–153.

Muchow, R.C., Robertson, M.J. and Keating, B.A. (1997) Limits to the Australian sugar industry: climatic and biological factors. In *Intensive Sugarcane Production: Meeting the Challenges Beyond 2000* (eds B.A. Keating and J.R. Wilson), pp. 37–54. CAB International, Wallingford, UK.

Muchow, R.C., Triglone, T., Plunkett, G.M., Sherrard, J.H., Wood, A.W. and Engelke, J.H. (2001) Field responses to irrigation in the Ord sugar industry. *Proceedings of the Australian Society of Sugar Cane Technologists*, Vol. 23, pp. 109–115.

Mukha, V. and Sviridov, V. (1999) The formation of a rational structure for the use of arable land: experience and problems. *Mezhdunarodnyi Sel'skokhozyaistvennyi Zhurnal*, No. 6, pp. 8–13.

Mulkins, L. (2000) From sugarcane monoculture to agro-ecological village. *LEISA*, Vol. 16, No. 4, pp. 22–23.

Munir Ahmad (2001) Agricultural productivity growth differential in Punjab, Pakistan: a district-level analysis. *Pakistan Development Review*, Vol. 40, No. 1, pp. 1–25.

Muraoka, T., Filho, J.O., Boaretto, A.E. and Ambrozano, E. (1995) Management of crop residues in sugarcane and cotton systems in Brazil. In *Soil Organic Matter Management for Sustainable Agriculture* (eds R.D.B. Lefro, G.J. Blair and E.T. Crasswell), pp. 27–31. Australian Centre for International Agricultural Research (ACIAR), Canberra.

Murathan, A.S. and Yildirim, E. (2001) Co-treatment of industrial and residential wastewater by the activated-sludge process for reduced organic loading on the Porsuk river, Turkey. *Fresenius Environmental Bulletin*, Vol. 10, No. 11, pp. 809–813.

Murombo, M., Takavarasha, E. and Wiseman, J. (1997) Green cane harvesting at Mkwasine estate, Zimbabwe. *Proceedings of the Annual Congress – South African Sugar Technologists' Association*, No. 71, pp. 30–37.

Murty, M.N. and Kumar, S. (2003) Win–win opportunities and environmental regulation: testing of porter hypothesis for Indian manufacturing industries. *Journal of Environmental Management*, Vol. 67, No. 2, pp. 139–144.

Murugesan, V. and Natarajan, S. (2000) Impact of micro-irrigation on sugarcane productivity – the Tamil Nadu experience. *Cooperative Sugar*, Vol. 31, No. 5, pp. 401–403.

Naf, W. (1991) Economic consequences of ozone pollution for Swiss agriculture. *Landwirtschaft Schweiz*, Vol. 4, No. 9, pp. 501–506.

Nair, N.V. and Somarajan, K.G. (2003) Diversity of *Saccharum* germplasm in Kerula, India. *Plant Genetic Resources Newsletter*, No. 135, pp. 40–43.

Narang, R.S., Nathu Singh and Mehra, S.P. (1992) Studies on the methods of irrigation and nitrogen application in sugarbeet. *Indian Sugar*, Vol. 41, No. 12, pp. 899–901.

Narayan, P.K. (1999) The causes and consequences of sugarcane burning in Fiji. *Development Bulletin*, July, pp. 110–112.

Nasir, N.M. and Quereshi, M.A. (1999a) Significance of ethanol stillage for soil and crop improvement. *Pakistan Sugar Journal*, Vol. 14, pp. 17–21.

Nasir, N.M. and Quereshi, M.A. (1999b) Response of sugarcane to biocompost prepared from filter cake and stillage. *Pakistan Journal of Soil Science*, Vol. 16, pp. 75–80.

Nass, H., Camino, J.M., Carrasco, A., Nino, M. and Vargas, A. (2001) Red leaf sheath spot on sugarcane. *Bioagro*, Vol. 13, No. 3, pp. 125–130.

Nastari, P.M. (1992) Biomass ethanol: strategy for economic development that preserves the environment. *Boletin GEPLACEA*, Vol. 9, No. 9.

Ndlovu, L.S. (2000) The performance of subsurface drip irrigation at Royal Swaziland Sugar Corporation during its first season. *Proceedings of the Annual Congress – South African Sugar Technologists' Association*, No. 74, pp. 157–161.

Ndlovu, L.S., Ellis, R.D. and Street, W.H. (2001) Conversion from sprinkler to subsurface drip irrigation in Swaziland. *Proceedings of the International Society of Sugar Cane Technologists*, Vol. 24, pp. 283–287.

Neeteson, J.J. and Ehlert, P.A.I. (1988) Environmental aspects of applying inorganic fertilizers to sugar beet. *Proceedings of the Institut International de Recherches Betteravières*, Vol. 52, pp. 79–91.

Nelson, P. and Ham, G. (1998) Soil sodicity: its influence on cane yield in the Burdekin. *Proceedings of the Australian Society of Sugar Cane Technologists*, Vol. 20, pp. 248–250.

Nelson, P.N., Lawer, A.T. and Ham, G.J. (2002) Evaluation of methods for field diagnosis of sodicity in soil and irrigation water in the sugarcane growing districts of Queensland, Australia. *Australian Journal of Soil Research*, Vol. 40, No. 8, pp. 1249–1265.

Neururer, H. (1984) Effect of bitumen mulch on weed emergence and on the availability of soil-acting herbicides to plants. *Bodenherbiziden. Zeitschrift fur Pflanzenkrankheiten und Pflanzenschutz*, No. 10, pp. 293–295.

Ng Cheong, L.R., Ah-Koon, D. and Muthy, K. (1996) The effects of irrigation interval on soil moisture and sugar cane yield in a furrow-irrigated field. *Revue Agricole et Sucrière de l'Île Maurice*, Vol. 75, No. 2, pp. 1–5.

Ng Kee Kwong, K.F. and Deville, J. (1984) Nitrogen leaching from soils cropped with sugarcane under the humid tropical climate of Mauritius, Indian Ocean. *Journal of Enviornmental Quality*, Vol. 13, pp. 471–474.

Ng Kee Kwong, K.F. and Deville, J. (1987) Residual fertilizer nitrogen as influenced by timing and nitrogen forms in a silty clay soil under sugarcane in Mauritius. *Fertilizer Research*, Vol. 14, pp. 219–226.

Ng Kee Kwong, K.F. and Deville, J. (1988) Filter cake as a nitrogen fertilizer for plant cane in Mauritius. *Tropical Agriculture*, Vol. 65, pp. 364–366.

Ng Kee Kwong, K.F. and Deville, J. (1994a) Application of ^{15}N-labelled urea to sugar cane through a drip-irrigation system in Mauritius. *Fertilizer Research*, Vol. 39, No. 3, pp. 223–228.

Ng Kee Kwong, K.F. and Deville, J. (1994b) The course of nitrogen fertilizer uptake by rainfed sugarcane in Mauritius. *Journal of Agricultural Science*, Vol. 122, pp. 385–391.

Ng Kee Kwong, K.F. and Paul, J.P. (1997) Vinasse as a potassium fertilizer for sugar cane. *Revue Agricole et Sucrière de l'Île Maurice*, Vol. 76, pp. 34–37.

Ng Kee Kwong, K.F., Deville, J., Cavalot, P.C. and Rivière, V. (1987) Value of cane trash in nitrogen nutrition of sugarcane. *Plant Soil*, Vol. 102, pp. 79–83.

Ng Kee Kwong, K.F., Umrit, G. and Julien, M.H.R. (1996) Impact of sugarcane cultivation on water quality in Mauritius. In *Sugar 2000 Symposium: Sugarcane: Research Towards Efficient and Sustainable Production* (eds J.R. Wilson, D.M. Hogarth, J.A. Campbell and A.L. Garside), pp. 264–265. CSIRO Division of Tropical Crops and Pastures, Brisbane.

Ng Kee Kwong, K.F., Umrit, G. and Nowbotsing, M. (1998) Fertilizer and herbicide usage in sugar cane cultivation in Mauritius. *Revue Agricole et Sucrière de l'Île Maurice*, Vol. 76, pp. 41–47.

Ng Kee Kwong, K.F., Bholah, A. and Veerapen, S. (1999a) Gaseous nitrogen losses from soils under sugar cane in Mauritius. *Proceedings of the International Society of Sugar Cane Technologists* (eds V. Singh and V. Kumar), Vol. 23, pp. 70–79.

Ng Kee Kwong, K.F., Paul, J.P. and Deville, J. (1999b) Drip-fertigation – a means for reducing fertilizer nitrogen to sugarcane. *Experimental Agriculture*, Vol. 35, pp. 31–37.

Ng Kee Kwong, K.F., Bholah, A., Volcy, L. and Pynee, K. (2002) Nitrogen and phosphorus transport by surface runoff from a silty clay loam soil under sugarcane in the humid tropical environment of Mauritius. *Agriculture, Ecosystems and Environment*, Vol. 91, No. 1/3, pp. 147–157.

Nibit, M.F., Khvoshchins'ka, A.M. and Golovan', L.G. (1994) New designs in plans for water supply and effluent disposal at sugar factories. *Tsukor Ukraini*, No. 4, pp. 29–31.

Nievergelt, J. (2002) Nitrate leaching in a crop rotation from 1981 to 2000. *Agrarforschung*, Vol. 9, No. 1, pp. 28–33.

Nixon, D.J. and Simmonds, L.P. (2004) The impact of fallowing and green manuring on soil conditions and the growth of sugarcane. *Experimental Agriculture*, Vol. 40, pp. 127–138.

Noble, A.D. and Hurney, A.P. (2000) Long-term effects of lime additions on sugarcane yield and soil chemical properties in north Queensland. *Experimental Agriculture*, Vol. 36, pp. 397–413.

Noble, A.D., Moody, P. and Berthelsen, S. (2003) Influence of changed management of sugarcane on some soil chemical properties in the humid wet tropics of north Queensland. *Australian Journal of Soil Research*, Vol. 41, pp. 1133–1144.

Nour, A.H., El-Wakeel, A.F. and Allam, A.I. (1989) Effect of salinity on sugarcane yield. *Proceedings of the International Society of Sugar Cane Technologists*, Vol. 20, pp. 532–535.

Nuttall, M. and Stevens, D.B. (1983) Beef cattle, the production of sugar beet top silage and its use in a finishing ration 1975–1980. In *75th Annual Report*, pp. 45–55. Norfolk Agricultural Station, Wymondham.

Nyati, C.T. (1996) Effect of irrigation regime on cane and sugar yields of variety NCo376 in the South-East lowveld of Zimbabwe. *Proceedings of the Annual Congress – South African Sugar Technologists' Association*, No. 70, pp. 59–62.

Olaoye, G. (2001) Estimate of ratooning ability in sugar cane under conditions of low available soil moisture in a savanna ecology of Nigeria. *Sugar Cane International*, March, pp. 8–13.

Olivier, F.C. and Singels, A. (2001) A database of crop water use coefficients for irrigation scheduling of sugarcane. *Proceedings of the Annual Congress – South African Sugar Technologists' Association*, No. 75, pp. 81–83.

Olsen, J.K., Gram, N.H. and Jorgensen, A.B. (2001) Genetic potential for breeding low tare. *Proceedings of the Institut International de Recherches Betteravières*, Vol. 64, pp. 529–532.

Olsson, R. and Nordstrom, T. (2002) Ways towards sustainable sugar beet production in Sweden. *Proceedings of the Institut International de Recherches Betteravières*, Vol. 65, pp. 83–89.

Olszyk, D.M., Thompson, C.R. and Poe, M.P. (1988) Crop loss assessment for California: modeling losses with different ozone standard scenarios. *Environmental Pollution*, Vol. 53, No. 1–4, pp. 303–311.

Ometo, J.P.H.B., Martinelli, L.A., Ballester, M.V., Gessner, A., Krusche, A.V., Victoria, R.L. and Williams, M. (2000) Effects of land use on water chemistry and macroinvertebrates in two streams of the Piracicaba river basin, south-east Brazil. *Freshwater Biology*, Vol. 44, No. 2, pp. 327–337.

Ooi, P.A.C., Lim, G.S., Ho, T.H., Manalo, P.L. and Waage, J.K. (eds) (1992) *Integrated Pest Management in the Asia–Pacific Region*. CAB International, Wallingford, UK.

Opanasenko, G.P. (1998) Energy-efficient technology for soil conservation. *Sakharnaya Svekla*, No. 10, p. 5.

Oriol, M., Rapanoelina, M. and Gaudin, R. (1995) Irrigation scheduling in sugarcane using tensiometers. *Agriculture et Developpement*, No. 6, pp. 39–47.

Otorowski, E. (1990) A beet sugar factory can also be energetically self-sufficient. *Gazeta Cukrownicza*, Vol. 98, No. 4, pp. 61–62.

Overholt, W.A., Conlong, D.E., Kfir, R., Schulthess, F. and Setamou, M. (2003) Biological control of gramineous lepidopteran stem borers in sub-Saharan Africa. In *Biological Control in IPM Systems in Africa* (eds P. Neuenschwander, C. Borgemeister and J. Langewald), pp. 131–144. CAB International, Wallingford, UK.

Oztas, T., Ozbek, A.K. and Turan, M. (2002) The cost of soil lost from fields due to removal on harvested sugarbeet: a case study in Turkey. *Soil Use and Management*, Vol. 18, No. 3, pp. 236–237.

Padiz, A.S. and Garcia, I.F. (1997) Taxonomic and ecological evaluation of edaphic Coleoptera in sugarcane. *Revista Biologia (Habana)*, Vol. 11, pp. 113–119.

Page, S., Siegert, F., O'Rieley, J.O., Boehm, H.V., Jaya, A. and Limin, S. (2002) The amount of carbon released from peat and forest fires in Indonesia during 1997. *Nature*, Vol. 420, pp. 61–65.

Pailles, C. and Moody, P.W. (1996) Phosphorus dynamics of sediments from the Johnstone Rivers. In *Downstream Impacts of Land Use* (eds H.H. Hunter, A.G. Eyles and G.E. Rayment), pp. 359–364. Department of Natural Resources, Brisbane.

Palacios Lazo, M., Lizarraga Travaglini, A., Velasquez Ochoa, R., Carranza Hernandez, E. and Segovia, I. (2003) Integrated pest management in Peru. In *Integrated Pest Management in the Global Arena* (eds K.M. Maredia, D. Dakouo and D. Mota-Sanchez), pp. 301–312. CAB International, Wallingford, UK.

Palaniswami, C. and Ramulu, U.S.S. (1994) Effects of continuous irrigation with paper factory effluent on soil properties. *Journal of the Indian Society of Soil Science*, Vol. 42, No. 1, pp. 139–140.

Palwarty, R.S., Eischeid, J. and Pulwarty, H. (2001) The impact of El Nino on rainfall and sugar production in Trinidad. *Proceedings of the West Indies Sugar Technologists Conference*, Vol. 27, unpaginated.

Pandey, K.P. (2002) Integrated pest management in sugarcane. In *Sugarcane Crop Management* (eds S.B. Singh, G.P. Rao and S. Easwaramoorthy), pp. 619–657. Sci Tech Publishing, Houston.

Pandian, B.J., Muthukrishnan, P. and Rajasekaran, S. (1992) Efficacy of different irrigation methods and regimes in sugarcane. *Indian Sugar*, Vol. 42, No. 4, pp. 215–219.

Panje, P., Matabwa, C.J. and Kumwenda, J.T.D. (1987) Soils of the drier areas of Malawi which are prone to drought, present agriculture on these soils and research which may increase the agricultural potential. In *Agronomic Adjustment to the Environment of the 400–600 mm Rainfall Zone of Southern Africa. Proceedings of an SADCC Workshop Held in Harare, Zimbabwe, September 1983*, pp. 47–61. Southern Africa Centre for Cooperation in Agricultural Research, Gaborone.

Pankhurst, C.E., Doube, B.M. and Gupta, V.V.S.R. (1997) Biological indicators of soil health: synthesis. In *Biological Indicators of Soil Health* (eds C.E. Pankhurst, B.M. Doube and V.V.S.R. Gupta), pp. 419–435. CAB International, Wallingford, UK.

Pankhurst, C.E., Hawke, B.G., Holt, J.A., Magarey, R.C. and Garside, A.L. (2000) Effect of rotation breaks on the diversity of bacteria in the rhizosphere of sugarcane and its potential impact on yield decline. *Proceedings of the Australian Society of Sugar Cane Technologists*, Vol. 22, pp. 77–83.

Pankhurst, C.E., Magarey, R.C., Stirling, G.R., Blair, B.L., Bell, M.J. and Garside, A.L. (2003) Management practices to improve soil health and reduce the effects of detrimental soil biota associated with yield decline of sugarcane in Queensland, Australia. *Soil and Tillage Research*, Vol. 72, pp. 125–137.

Papesch, J. and Steinert, K. (1997) Mulch and direct seeding in the Central German dry area. *Landtechnik*, Vol. 52, No. 3, pp. 128–129.

Parikh, M.M., Shrivastava, P.K., Savani, N.G. and Raman, S. (1992) Response of sugarcane crop to drip method of irrigation. *Cooperative Sugar*, Vol. 23, No. 10, pp. 673–677.

Park, S.E., Antony, G., Lisson, S.N. and Thorburn, P.J. (2003) A method for exploring the potential of agronomic practices to manage the greenhouse gas balance in sugarcane primary production. *Proceedings of the Australian Society of Sugar Cane Technologists*, Vol. 25, p. 21.

Parsana, G.J., Malavia, D.D. and Koshiya, D.J. (1994) Effect of drip irrigation on incidence of insect pests of sugarcane. *Gujarat Agricultural University Research Journal*, Vol. 20, No. 1, pp. 15–17.

Patrick, W.H., Gambrell, R.P., Parkpian, P. and Tan, F. (1994) Mercury in soils and plants in the Florida Everglades sugarcane area. In *Mercury Pollution – Integration and Synthesis* (eds C.J. Watras and J.W. Huckabee), pp. 609–614. Lewis Publishers, Boca Raton, Florida.

Paturau, J.M. (1989) *By-products of the Cane Sugar Industry – an Introduction to their Industrial Utilisation*, 3rd edn. Elsevier, Amsterdam.

Paulsen, H.M., Kucke, M. and Schnug, E. (1997) Irrigation of arable land with waste water of a sugar factory – legality and ecological consequences. *Zeitschrift für Kulturtechnik und Landentwicklung*, Vol. 38, No. 6, pp. 273–278.

Pawar, H.K. and More, V.D. (1993) Response of sugarcane to irrigation regimes and nitrogen. *Journal of Maharashtra Agricultural Universities*, Vol. 18, No. 2, pp. 222–226.

Pawar, H.K. and More, V.D. (1996) Irrigation and cane yield relationship in ratoon sugarcane and its economic analysis. *Journal of Maharashtra Agricultural Universities*, Vol. 21, No. 1, pp. 24–27.

Pawar, N.J., Pondhe, G.M. and Patil, S.F. (1998) Groundwater pollution due to sugar-mill effluent, at Sonai, Maharashtra, India. *Environmental Geology*, Vol. 34, No. 2/3, pp. 151–158.

Payne, J.H. (1991) *Cogeneration in the Cane Sugar Industry*. Elsevier, Amsterdam.

Paz, D., Octaviano, M.A. and Cardenas, G.J. (2001) Reduction of environmental impact of bagasse boilers by means of bagasse drying. *Revista Industrial y Agricola de Tucuman*, 2001, Vol. 78, No. 1/2, pp. 61–70.

Pearson, R.G. and Penridge, L.K. (1987) The effects of pollution by organic sugar mill effluent on the macro-invertebrates of a stream in tropical Queensland, Australia. *Journal of Environmental Management*, Vol. 24, pp. 205–215.

Peel, M.J.S. and Stalmans, M. (1999) The Systematic Reconnaissance Flight (SRF) as a tool in assessing the ecological impact of a rural development programme in an extensive area of the lowveld of South Africa. *African Journal of Ecology*, Vol. 37, No. 4, pp. 449–456.

Pene, C.B. (1999) Water supply diagnosis in sugarcane crops, and climatic risk management in Côte d'Ivoire. *Agriculture et Developpement*, No. 24, pp. 74–80.

Pene, C.B. and Dea, B.G. (2000) Interaction irrigation water – variety of sugarcane under conditions of water deficit in the north of Côte d'Ivoire. *Revue Agricole et Sucrière de l'Île Maurice*, Vol. 79, No. 2, pp. 1–21.

Pene, C.B.G. and Edi, G.K. (1999) Sugarcane yield response to deficit irrigation at two growth stages. In *Crop Yield Response to Deficit Irrigation* (eds C. Kirda, P. Moutonnet, C. Hera and D.R. Nielsen), pp. 136–147. Kluwer Academic Publishers, Dordrecht.

Pene, C.B., Chopart, J.-L. and Assa, A. (1997) Irrigation management on plots growing sugarcane (*Saccharum officinarum* L.) under a humid tropical climate: the northern and central regions of Côte d'Ivoire. *Sécheresse*, Vol. 8, No. 2, pp. 87–98.

Pene, C.B., Assa, D.A. and Dea, B.G. (2001) Interaction between irrigation water and sugarcane genotypes under water deficit conditions. *Cahiers Agricultures*, Vol. 10, No. 4, pp. 243–253.

Peng, S.Y. and Twu, L.T. (1980) *Studies on Chemical Weed Control for Sugarcane Intercropping in Taiwan*. Reports of the Taiwan Sugar Research Institute, No. 88, 18 pp.

Penin, M.E., Cuervo, E., Palacios, E., Valdes, E. and Alvarez, R. (1988) Environmental impact of bagasse storage liquor in the paper industry. How to eliminate by means of its utilization. *Revista ICIDCA sobre Los Derivados de la Cana de Azucar*, Vol. 22, No. 3, pp. 5–9.

Perry, T. (1995) *Vegetation Patterns of the Herbert River Floodplain*. Queensland Department of Primary Industries, Townsville.

Peruzzi, A., Ceccatelli, M., Silvestri, N. and Barberi, P. (1995) Flame weeding, prospects and agronomic aspects. *Macchine e Motori Agricoli*, Vol. 53, No. 5, pp. 11–17.

Petit, L.J., Petit, D.R., Christian, D.G. and Powell, H.D.W. (1999) Bird communities of natural and modified habitats in Panama. *Ecography*, Vol. 22, No. 3, pp. 292–304.

Pettitt, I.R., May, M.J. and Fisher, S.J. (2001) Sugar beet, a grower's guide – an encyclopaedic database in sugar beet agronomy. *Proceedings of the Institut International de Recherches Betteravières*, Vol. 64, pp. 109–114.

Phillips-Howard, K. (1996) The rapid evolution of small-basin irrigation on the Jos Plateau, Nigeria. In *Sustaining the Soil: Indigenous Soil and Water Conservation in Africa* (eds C. Riej, I. Scoones and C. Toulmin), pp. 213–218. Earthscan Publications Ltd, London.

Phipps, R.H. and Park, J.R. (2002) Environmental benefits of genetically modified crops: global and European perspectives on their ability to reduce pesticide use. *Journal of Animal and Feed Sciences*, Vol. 11, pp. 1–18.

Piccinni, G. and Rush, C.M. (2000) Determination of optimum irrigation regime and water use efficiency of sugar beet grown in pathogen-infested soil. *Plant Disease*, Vol. 84, No. 10, pp. 1067–1072.

Pickwell, G.C. (1974) Wind damage on sand land. *British Sugar Beet Review*, Vol. 42, No. 3, pp. 177–181.

Pidgeon, J.D. and Jaggard, K.W. (1998) Drought stress in sugar beet – the extent of the problem and future solutions. *Aspects of Applied Biology*, No. 52, pp. 65–70.

Pidgeon, J.D. and Soane, D.B. (1978) Soil structure and strength relations following tillage, zero tillage and wheel traffic in Scotland. In *Modification of Soil Structure* (eds W.W. Emerson, R.D. Bond and A.R. Dexter), pp. 371–378. Wiley, Chichester.

Pidgeon, J.D., Dewar, A.M. and May, M.J. (2001) Weed management for agricultural and environmental benefit in GMHT sugar beet. In *The BCPC Conference: Weeds, 2001, Volume 1 and Volume 2. Proceedings of an International Conference held at the Brighton Hilton Metropole Hotel, Brighton, UK, 12–15 November 2001*, pp. 373–380. British Crop Protection Council, Farnham.

Pigeon, O., Dardenne, P., de Vos, P., Mouteau, A., Bruyere, J., Jacobs, N., Verreck, A. and Tossens, A. (2001) Near infrared spectroscopy (NIR): a non-destructive, rapid and accurate method to determine imidacloprid on sugar beet pelleted seeds. *Proceedings of the Institut International de Recherches Betteravières*, Vol. 64, pp. 277–286.

Pinto, L.F.G. and Crestana, S. (2001) Soil and climate characteristics and socio-economic data in the diagnostic of agroecosystems of Sao Carlos region, SP, Brazil. *Pesquisa Agropecuaria Brasileira*, Vol. 36, No. 11, pp. 1325–1329.

Pinto, L.F.G. and de Santis Prada, L. (1999) Socio-environmental agricultural certification: a pilot initiative in sugarcane. *Informacoes Economicas – Instituto de Economia Agricola*, Vol. 29, No. 5, pp. 19–29.

Piotrowski, W., Smolinski, S., Maniewska, R., Gorlach, K. and Rzekanowski, C. (1996) The effect of irrigation on the number of micro-organisms in a very light soil under sugar beet cultivation. *Zeszyty Problemowe Postepow Nauk Rolniczych*, Vol. 438, pp. 281–289.

Platford, G.G. (1992) Land use planning in the South African sugarcane industry. In *Land Reclamation: Advances in Research and Technology* (eds T. Younos, P. Diplas and S. Mostaghimi), pp. 221–229. American Society of Agricultural Engineers, St Joseph.

Plichta, I. (2002) Thickening of sugar beet flume sludge. *Gazeta Cukrownicza*, Vol. 110, No. 4, pp. 104–105.

Plichta, I. and Nitzler, J. (1999) Waste water cleaning unit in the Chelmza sugarbeet factory after five years of utilization. *Gazeta Cukrownicza*, Vol. 107, No. 7, pp. 131–133.

Plunkett, G.M. and Muchow, R.C. (2003) Water extraction by sugarcane on soils of the Ord Irrigation Area. *Australian Journal of Experimental Agriculture*, Vol. 43, No. 5, pp. 487–495.

Podstawka-Chmielewska, E. and Ceglarek, S. (1995) Cultivation of sugarbeet and fodder beet on light soil under conditions of sprinkler irrigation and differentiated nitrogen fertilizer application. *Annales Universitatis*

Mariae Curie-Sklodowska. Sectio E, Agricultura, Vol. 50, pp. 21–27.

Podstawka-Chmielewska, E. and Malicki, L. (1997) The effect of cultivation intensity and sprinkle irrigation on the yield of sugarbeet on rendzina. *Biuletyn Instytutu Hodowli i Aklimatyzacji Roslin*, No. 202, pp. 173–177.

Poesen, J.W.A., Verstraeten, G., Soenens, R. and Seynaeve, L. (2001) Soil losses due to harvesting of chicory roots and sugar beet: an underrated geomorphic process? *Catena*, Vol. 43, No. 1, pp. 35–47.

Pohlan, J. and Borgman, J. (2000) Traditional methods of weed control in important crops of Central America: cause of soil fertility loss and erosion. *Zeitschrift für Pflanzenkrankheiten und Pflanzenschutz*, No. 17, pp. 761–768.

Polacik, S. (1992) Siting a factory with acceptance of environmental protection: example of the sugar factory at Nitra. *Listy Cukrovarnicke a Reparske*, Vol. 108, No. 1, pp. 12–21.

Polaszek, A. and Walker, A.K. (1991) The *Cotesia flavipes* species complex: parasitoids of cereal stemborers in the tropics. *Redia*, Vol. 74, pp. 335–341.

Polec, B. (2000) The state of waste water management in Polish sugar industry and European Union requirements. *Gazeta Cukrownicza*, Vol. 108, No. 8, pp. 150–156.

Polec, B. (2002) Impact of a new environmental protection law on sugarbeet factories. *Gazeta Cukrownicza*, Vol. 110, No. 11, pp. 325–330.

Polec, B. and Gozdek, K. (1994) Technical and technological progress in water economy and wastewater purification at [Polish] sugar factories in the years 1893–1993. *Gazeta Cukrownicza*, Vol. 102, No. 2, pp. 21–25.

Polec, B. and Kempnerska-Omielczenko, O. (1995) Environmental protection in the sugar industry: present state, hierarchy of requirements. *Gazeta Cukrownicza*, Vol. 103, No. 1, pp. 4–5.

Polec, B., Gozdek, K. and Zero, M. (1999) Effects of waste biodegradation in a biological sewer cleaning unit of the Koscian S.A. sugarbeet factory. *Gazeta Cukrownicza*, Vol. 107, No. 7, pp. 124–130.

Pollok, J.G. and Bosua, H. (1986) The design and operation of 560 hectares of drip irrigation at Simunye sugar estate (Swaziland). *Proceedings of the South African Sugar Technologists' Association*, Vol. 60, pp. 172–176.

Potyondi, L., Kimmel, J., Hamori, V. and Boros, J. (2001) Starting a new system of communication in the field of Hungarian sugarbeet production. *Proceedings of the Institut International de Recherches Betteravières*, Vol. 64, pp. 307–310.

Power, S. and de Araujo, L.M. (1993) Nature conservation with reference to the state of Alagoas, Brazil. *Environmentalist*, Vol. 13, No. 4, pp. 297–302.

Powlson, D.S., Jenkinson, D.S., Pruden, G. and Johnston, A.E. (1985) The effect of straw incorporation on the uptake of nitrogen by winter wheat. *Journal of Agricultural Science, Cambridge*, Vol. 36, pp. 29–30.

Preston, T.R. (1993) Sustainable systems of intensive livestock production for the humid tropics using local resources. *Occasional Publication* No. 16, pp. 101–105. British Society of Animal Production.

Proag, V. (1998) Effectiveness of irrigation in Mauritius. In *Proceedings of the Second Annual Meeting of Agricultural Scientists, Reduit, Mauritius, 12–13 August 1997* (eds J.A. Lalouette, D.Y. Bachraz, N. Sukurdeep and B.D. Seebaluck), pp. 61–72. Food and Agricultural Research Council, Reduit.

Prove, B.G., Truong, P.N.V. and Evans, D.S. (1986) Strategies for controlling caneland erosion in the wet tropical coast of Queensland. *Proceedings of the Australian Society of Sugarcane Technologists*, Vol. 8, pp. 77–84.

Prove, B.G., Doogan, V.J. and Truong, P.N.V. (1995) Nature and magnitude of soil erosion in sugarcane land on the wet tropical coast of northeastern Queensland. *Australian Journal of Experimental Agriculture*, Vol. 35, pp. 641–649.

Purvis, G. and Curry, J.P. (1984) The influence of weeds and farmyard manure on the activity of Carabidae and other ground-dwelling arthropods in a sugar beet crop. *Journal of Applied Ecology*, Vol. 21, No. 1, pp. 271–283.

Purvis, G., Fadl, A. and Bolger, T. (2001) A multivariate analysis of cropping effects on Irish ground beetle assemblages (Coleoptera: Carabidae) in mixed arable and grass farmland. *Annals of Applied Biology*, Vol. 139, No. 3, pp. 351–360.

Pussemier, L. (1991) The behaviour of pesticides in soil as an aid to research on decreased application rate. *Pédologie*, Vol. 41, No. 1, pp. 27–37.

Putthacharoen, S., Howeler, R.H., Jantawat, S. and Vichukit, V. (1998) Nutrient uptake and soil erosion losses in cassava and six other crops in a Psamment in eastern Thailand. *Field Crops Research*, Vol. 57, No. 1, pp. 113–126.

Qongqo, L.L. and van Antwerpen, R. (2000) Effect of long-term sugarcane production on physical and chemical properties of soils in KwaZulu-Natal. *Proceedings of the Annual Congress – South African Sugar Technologists' Association*, No. 74, pp. 114–121.

Quashie-Williams, C.A. (1991) Introduction of *Allorhogas pyralophagus* Marsh (Hym.: Brac.) for the biological control of the yellow headed borer, *Diatraea centrella* (Lep.: Pyral.) in Guyana. *Proceedings of the West Indies Sugar Technologists*, Vol. 24, pp. 80–87.

Quintana Perez, C., Gonzalez Gonzalez, R., Turino Hurtado, I.M. and Albernas Jimenez, A. (1998) Environmental impact due to combustion of cane biomass in the Cuban sugar industry. *Centro Azucar*, Vol. 25, No. 2, pp. 48–52.

Qureshi, M.E., Charlesworth, P.B., Bristow, K.L. and Wegener, M.K. (2002) Profitability of growing sugarcane under alternative irrigation systems in the Burdekin Delta. *Proceedings of the Australian Society of Sugar Cane Technologists*, Vol. 24, pp. 107–112.

Qureshi, S.A., Dodds, G.T. and Madramootoo, C.A. (2002a) Simulation modelling to optimize warabandi-type irrigation for sugarcane in Pakistan. *Sugar Cane International*, March/April, pp. 9–14.

Qureshi, S.A., Dodds, G.T. and Madramootoo, C.A. (2002b) Evaluation of irrigation schemes for sugarcane in Sindh, Pakistan, using SWAP93. *Agricultural Water Management*, Vol. 54, No. 1, pp. 37–48.

Ragen, A.K. (1992) Water survey in sugar factories. *Revue Agricole et Sucrière de l'Île Maurice*, Vol. 71, No. 2–3, pp. 350–357.

Raine, S. (1995) Increasing the efficiency of furrow irrigation for sugarcane production in the Burdekin. In *Proceedings of the 12th Irrigation Association of Australia Conference*, pp. 52–58. Tamworth.

Raine, S.R. and Bakker, D. (1996) Increased furrow irrigation efficiency through better design and management of cane fields. *Proceedings of the Australian Society of Sugar Cane Technologists*, Vol. 18, pp. 119–124.

Raine, S.R. and Shannon, E.L. (1996) Improving the efficiency and profitability of furrow irrigation for sugarcane production. In *Sugar 2000 Symposium: Sugarcane: Research towards Efficient and Sustainable Production* (eds J.R. Wilson, D.M. Hogarth, J.A. Campbell and A.L. Garside), pp. 211–212. CSIRO Division of Tropical Crops and Pastures, Brisbane.

Rajendra Gupta and Tripathi, S.K. (1998a) Scheduling of irrigation for sugarcane (*Saccharum officinarum*). *Indian Journal of Agricultural Sciences*, Vol. 68, No. 10, pp. 647–651.

Rajendra Gupta and Tripathi, S.K. (1998b) Scheduling of irrigation for sugarcane in India. *Cooperative Sugar*, Vol. 29, No. 10, pp. 693–700.

Rajendra Singh (2000) Pollution in Abu drainage – a preliminary report. *Advances in Plant Sciences*, Vol. 13, No. 1, pp. 43–45.

Ramdoyal, K., Badaloo, G.H. and Goburdhun, P. (2003) An international survey of sugar cane crossing programmes. *Sugar Cane International*, July/August, pp. 3–11.

Ramesh, P. (1998) Effect of paired row planting with skip furrow irrigation and trash mulch on yield, quality and water use of sugarcane (*Saccharum officinarum*). *Indian Journal of Agronomy*, Vol. 43, No. 2, pp. 338–341.

Ramesh, P., Kailasam, C. and Srinivasan, T.R. (1994) Performance of sugarcane (*Saccharum officinarum* L.) under surface drip, sub surface drip (Biwall) and furrow methods of irrigation. *Journal of Agronomy and Crop Science*, Vol. 172, No. 4, pp. 237–241.

Ramjeawon, T. and Baguant, J. (1995) Evaluation of critical BOD loadings from Mauritian sugar factories to streams and standards setting. *Journal of Environmental Management*, Vol. 45, No. 2, pp. 163–176.

Rao, P.N. (2000) Cane management under drought conditions or water stress. In *Proceedings of the 62nd Annual Convention of the Sugar Technologists' Association of India, Agra, India, 19–21 August 2000*, pp. A3–A8. Sugar Technologists' Association of India, New Delhi.

Rao, S.M. (1983) Nutrient value of distiller's waste and its effect on soil pH. *Proceedings of the Inter-American Sugar Cane Seminar*, Vol. 4, pp. 235–242.

Rao, S.R. and Narasimham, R.L. (1988) Effect of soil compaction in sugarcane. Part 1 – on soil moisture, porosity and nitrification. *International Journal of Tropical Agriculture*, Vol. 6, pp. 232–236.

Rao, U.S.V.P. and Rao, B.K. (1992) Pollution control measures in the K.C.P. Ltd., Vuyyuru Sugar unit. In *Proceedings of the 54th Annual Convention of the Sugar Technologists' Association of India, Bangalore, 17–19 January 1992*, pp. G23–G32. Sugar Technologists' Association of India, New Delhi.

Raskar, B.S. and Bhoi, P.G. (2001) Productivity of sugarcane as influenced by planting techniques and sources of fertigation under drip irrigation. *Indian Sugar*, Vol. 50, No. 11, pp. 801–810.

Rayment, G.E. (2002) Actual and potential pyritic acidity of NSW canelands. *Communications in Soil Science and Plant Analysis*, Vol. 33, pp. 3033–3044.

Rayment, G.E. (2003) Protecting the environment: from proposal to achievement. *Proceedings*

of the Australian Society of Sugar Cane Technologists, Vol. 25, p. 7.
Reeser, L.G., Acra, A.P.L. and Lee, T. (1994) *Renewable Energy's Role in Latin America's Future*. Paper No. 948010, American Society of Agricultural Engineers, St. Joseph. 13 pp.
Rhoades, J.D., Bingham, F.T., Letey, J., Pinter, P.J., Jr, Lemert, R.D., Alves, W.J., Hoffman, G.J., Replogle, J.A., Swain, R.V. and Pacheco, P.G. (1988) Reuse of drainage water for irrigation: results of Imperial Valley study. II. Soil salinity and water balance. *Hilgardia*, Vol. 56, No. 5, pp. 17–44.
Rice, R.W. and Izuno, F.T. (1998) Techniques for assessing BMP effectiveness in the absence of historical baseline data. *Applied Engineering in Agriculture*, Vol. 14, No. 4, pp. 381–389.
Rice, R.W., Izuno, F.T. and Garcia, R.M. (2002) Phosphorus load reductions under best management practices for sugarcane cropping systems in the Everglades Agricultural Area. *Agricultural Water Management*, Vol. 56, No. 1, pp. 17–39.
Richard-Molard, M. and Cariolle, M. (2001) Water and abiotic stress and genetic improvement. *Proceedings of the Institut International de Recherches Betteravières*, Vol. 64, pp. 153–158.
Rinaldi, I.M.P., Mendes, B. do P. and Cady, A.B. (2002) Distribution and importance of spiders inhabiting a Brazilian sugar cane plantation. *Revista Brasileira de Zoologia*, Vol. 19, Suppl. 1, pp. 271–279.
Rincones L., C. (1990) Response of eight sugarcane varieties to the application of two irrigation frequencies. *Cana de Azucar*, Vol. 8, No. 2, pp. 87–111.
Ripoli, T.C.C., Molina, W.F., Jr and Ripoli, M.L.C. (2000) Energy potential of sugar cane biomass in Brazil. *Scientia Agricola*, Vol. 57, No. 4, pp. 677–681.
Robertson, L.M. and Webster, D.E. (1995) Strategies for managing cane weevil borer. *Proceedings of the Australian Society of Sugarcane Technologists*, Vol. 17, pp. 83–87.
Robertson, M.J. and Muchow, R.C. (1994) Future research challenges for efficient crop water use in sugarcane production. *Proceedings of the Australian Society of Sugarcane Technologists*, Vol. 16, pp. 193–200.
Robertson, M.J., Inman-Bamber, N.G. and Muchow, R.C. (1997) Opportunities for improving the use of limited water by the sugarcane crop. In *Intensive Sugarcane Production: Meeting the Challenges beyond 2000* (eds B.A. Keating and J.R. Wilson), pp. 287–304. CAB International, Wallingford.
Robson, M.C., Fowler, S.M., Lampkin, N.H., Leifert, C., Leitch, M., Robinson, D., Watson, C.A. and Litterick, A.M. (2002) The agronomic and economic potential of break crops for ley/arable rotations in temperate organic agriculture. *Advances in Agronomy*, Vol. 77, pp. 369–427.
Rosenzweig, C. and Hillel, D. (1993) Agriculture in a greenhouse world. *Research and Exploration*, Vol. 9, No. 2, pp. 208–221.
Roth, D. and Rossler, U. (1998) The suitability of infrared thermometry for irrigation scheduling in humid regions. *Zeitschrift fur Bewasserungswirtschaft*, Vol. 33, No. 2, pp. 279–293.
Rover, A. and Buttner, G. (1999) Influence of drought stress on the internal quality of sugarbeet. *Proceedings of the Institut International de Recherches Betteravières*, Vol. 65, pp. 97–108.
Royal Society (2003) The farm scale evaluations of spring-sown genetically modified crops. *Philosophical Transactions of the Royal Society, Series B*, Vol. 358, Issue 1439, pp. 1775–1913.
Rozeff, N. (1997) A realistic view of sugarcane sustainability. *International Sugar Journal (Cane Sugar Edition)*, Vol. 99, No. 1185, pp. 469–474.
Rozeff, N. (1998a) Irrigation water salinity and macro yields of sugarcane in South Texas. *Sugar Cane*, No. 2, pp. 3–6.
Rozeff, N. (1998b) The USDA task force on agricultural air quality agricultural burning subcommittee. *Sugar Journal*, Vol. 61, No. 5, pp. 17–23.
Rudolph, R. and Klee, U. (1993) Ecological husbandry methods in row crops. *Landtechnik*, Vol. 48, pp. 164, 170–173.
Rufino, M., Romero, E.R., Scandaliaris, J., Perez Zamora, F. and Bulacio, A. (2001) Agronomic, industrial and economic impact of sugarcane chemical ripening. *Proceedings of the International Society of Sugar Cane Technologists*, Vol. 24, pp. 161–163.
Ruiz Traba, J. and Delgado Morejon, G. (1996) Economic effects of irrigation in sugarcane plantations. *ATAC, Revista de la Asociacion de Tecnicos Azucareros de Cuba*, No. 1, pp. 12–15.
Ruschel, A.P. and Vose, P.B. (1982) Nitrogen cycling in sugarcane. *Plant and Soil*, Vol. 67, pp. 139–146.
Ruschel, A.P., Tamas, F.R., Ahmad, N., Mendoza-Palacois, H.H. and Valdivia-Liga, S. (1982) Report of the work group on sugarcane. *Plant and Soil*, Vol. 67, pp. 395–397.
Russell, D.J. and Hales, P.W. (1996) Major fisheries resource and habitat issues in the Johnstone River catchment. In *Downstream Impacts of Land Use* (eds H.H. Hunter, A.G. Eyles and G.E. Rayment), pp. 353–358. Department of Natural Resources, Brisbane.

Ruzsanyi, L. (1996) Requirements of successful and effective irrigation in sugarbeet cultivation. *Cukoripar*, Vol. 49, No. 2, pp. 57–62.

Ruzsanyi, L. (2000) Water and nutrient supply of sugar-beet. *Cukoripar*, Vol. 53, No. 1, pp. 26–31.

Rzekanowski, C. and Rolbiecki, S. (1997) Actual sunshine hours and sugarbeet yield on a light soil under sprinkler irrigation. *Pamietnik Pulawski*, No. 110, pp. 121–127.

SAC (Sugar Association of the Caribbean) (2000) *Sugar Association of the Caribbean Handbook 1961–1999*. SAC Sugar Technologists Committee.

Sadeghian, S.Y. and Yavari, N. (2001) Germination response of drought tolerant sugar beets under mannitol induced osmotic stress. *Proceedings of the Institut International de Recherches Betteravières*, Vol. 64, pp. 371–376.

Saechu, M. (1995) Design of dry dust collectors from boilers with sugarcane bagasse as energy source. *Berita – Pusat Penelitian Perkebunan Gula Indonesia*, No. 12, pp. 11–14.

Sager, G. (1995) 'A burger and a coke' and other environmental problems: reflections on diet and the environmental crisis. *Sustainable Development (Bradford)*, Vol. 3, No. 3, pp. 149–157.

Sallaway, M.M. (1979) Soil erosion studies in the Mackay district. *Proceedings of the Australian Society of Sugar Technologists*, Vol. 1, pp. 125–132.

Sallaway, M.M. (1980) Effect on soil movement of continued cultivation into the wet season. *Proceedings of the Australian Society of Sugar Technologists*, Vol. 2, pp. 51–54.

Salman, A.G.A., Negm, A.A., Abo-Ghadir, M.F. and Temerak, S.A. (1978) The relative prevalence of arboreal and soil-inhabiting arthropods in sugarcane fields in Upper Egypt. *Proceedings of the Fourth Conference of Pest Control, 30 September–3 October 1978*, Part I, pp. 218–226. Academy of Scientific Research and Technology and National Research Centre, Cairo.

Saluchaite, A. (2000) Studies of beetles in farming lands with different agricultural technologies. *Transactions of the Estonian Agricultural University, Agronomy*, No. 209, pp. 178–179.

Salveter, R. (1998) Habitat use by adult syrphids (Diptera: Syrphidae) in a distinctly divided agricultural landscape. *Mitteilungen der Schweizerischen Entomologischen Gesellschaft*, Vol. 71, No. 1/2, pp. 49–71.

Samersov, V.F. and Skur'yat, A.F. (1985) Environmentally safe technology for plant protection. *Zashchita Rastenii*, No. 1, pp. 22–24.

Sanchez-Roman, R. (2000) Sugar cane drip irrigation. *Sugar Journal*, Vol. 63, No. 5, pp. 32–36.

Sarmento, C. and Robbins, T. (2001) Control of odors in the sugar beet processing industry. In *Proceedings from the 31st Biennial Meeting (Operations) of the American Society of Sugar Beet Technologists, Vancouver, BC, Canada, 28 February–3 March 2001*, pp. 219–228. American Society of Sugar Beet Technologists, Denver.

SASA (2002) *Manual of Standards and Guidelines for Conservation and Environmental Management in the South African Sugar Industry*. South African Sugar Association, Mount Edgecombe.

Sawarkar, V.B. (2000) A landscape approach to managing Terai ecosystems with reference to Uttar Pradesh (UP), India. In *Grassland Ecology and Management in Protected Areas of Nepal. Proceedings of a Workshop, Royal Bardia National Park, Thakurdwara, Bardia, Nepal, 15–19 March, 1999*. Vol. 2: *Terai protected areas* (eds C. Richard, K. Basnet, J.P. Sah and Y. Raut), pp. 3–15. International Centre for Integrated Mountain Development, Kathmandu.

Scaloppi, E.J., Leme, E.J.A., Saad, J.C.C. and Baptistella, J.R. (1989) Land disposal of sugarcane mill wastewater. In *Toxic Substances in Agricultural Water Supply and Drainage, 2nd Panamerican ICID Regional Conference, 1989*, pp. 195–207. US Committee, International Commission on Irrigation and Drainage, Denver.

Scandaliaris, J., Dantur, N., Perez Zamora, F. and Roncedo, M. (1995) Agricultural utilisation of residues from sugarcane agroindustry. Part 1: The use of filter cake. *Avance Agroindustrial*, Vol. 16, pp. 38–40.

Scandaliaris, J., Romero, E.R. and Perez Zamora, F. (1998) Impact of new harvesting technology on productivity in Argentina. *International Sugar Journal (Worldwide Sugar Edition)*, Vol. 100, No. 1193, pp. 257–262.

Scandaliaris, J., Perez Zamora, F., Rufino, M., Romero, E. and Morandini, M. (2002) Harvest in green as a strategy for reducing the environmental impact of sugarcane. *Avance Agroindustrial*, Vol. 23, No. 1, pp. 14–17.

Scanlon, J. (2003) Water and law: why we need robust legal frameworks. *World Conservation*, Vol. 1, p. 14.

Schjonning, P., Elmholt, S. and Christensen, B.T. (eds) (2004) *Managing Soil Quality: Challenges in Modern Agriculture*. CAB International, Wallingford, UK.

Schleiff, U. (1982) Salt dynamics in root-containing soil and its significance for the water supply of crops. *Zeitschrift für Kulturtechnik und Flurbereinigung*, Vol. 23, No. 1, pp. 38–49.

Schmidt, E.J. (1996) Irrigation of sugarcane: preliminary assessment of economic advantage at selected sites. *Proceedings of the Annual Congress – South African Sugar Technologists' Association*, No. 70, pp. 53–54.

Schmidt, E.J. (1997) Impacts of sugarcane production on water resources. *Proceedings of the Annual Congress – South African Sugar Technologists' Association*, No. 71, pp. 73–75.

Schmidt, E.J. (1998) The role of irrigation in the South African sugar industry. *Proceedings of the Annual Congress – South African Sugar Technologists' Association*, No. 72, pp. 108–113.

Schmidt, E.J. (2000) Improved water management for sugarcane production. In *6th International Micro-irrigation Congress – Micro-irrigation Technology for Developing Agriculture, Cape Town, South Africa, 22–27 October 2000*, pp. 1–7. International Commission on Irrigation and Drainage, Rome.

Schmidt, E.J. (2001) Decision support programmes for assessing the impact of irrigated sugarcane on water resources and profitability. *Proceedings of the Annual Congress – South African Sugar Technologists' Association*, No. 75, pp. 51–57.

Schmidtlein, E.M., Heissenhuber, A. and Steinhauser, H. (1987) Economic effects of the use of erosion control measures in silo maize and sugar beet culture. *Bayerisches Landwirtschaftliches Jahrbuch*, Vol. 64, No. 6, pp. 643–658.

Schmitz, H.J. and Fischer, P. (2003) Plant sourced fertilizers in substrates for organic vegetable production. *Gemuse (Munchen)*, Vol. 39, pp. 18–22.

Schroeder, B.L., Meyer, J.H., Wood, R.A. and Turner, P.E.T. (1993) Modifying lime requirements for sandy, to sandy clay loam soils in the Natal Midlands. *Proceedings of the South African Sugar Technologists' Association*, Vol. 67, pp. 49–52.

Schroeder, B.L., Robinson, J.B., Wallace, M. and Turner, P. (1994) Soil acidification: occurrence and effects in the South African sugar industry. *Proceedings of the South African Sugar Technologists Association*, Vol. 68, pp. 70–74.

Schrope, M. (2001) Save our swamp. *Nature*, Vol. 409, pp. 128–130.

Schulze, R.E., Horan, M.J.C. and Schmidt, E.J. (2000) Hydrological complexities in assigning rainfed sugarcane a 'Stream Flow Reduction Activity'. *Proceedings of the Annual Congress – South African Sugar Technologists' Association*, No. 74, pp. 140–150.

Schuster, R.P. (1997) Effect of crop on the diversity and the potential of fungal egg pathogens of cyst nematodes. In *Diagnosis and Identification of Plant Pathogens. Proceedings 4th International Symposium of the European Foundation for Plant Pathology, Bonn, Germany, 9–12 September 1996* (eds H.W. Dehne, G. Adam, M. Diekmann, J. Frahm, A. Mauler-Machnik and P. van Halteren), pp. 253–255. Kluwer Academic Publishers, Dordrecht.

Schwertmann, U. (1986) Soil erosion: extent, prediction and protection in Bavaria. In *Soil Erosion in the European Community – Impact of Changing Agriculture* (eds G. Chisci and R.P.C. Morgan), pp. 185–200. A.A. Balkema, Rotterdam.

Scigacz, J. (1995) Introduction of the 'cleaner production' procedure at Koscian sugar factory. *Gazeta Cukrownicza*, Vol. 103, No. 8, pp. 151–152.

Scott, R.K. and Jaggard, K.W. (1993) Crop physiology and agronomy. In *The Sugar Beet Crop – Science into Practice* (eds D.A. Cooke and R.K. Scott), pp. 178–237. Chapman & Hall, London.

Seeruttun, S. and Crossley, C.P. (1997) Use of digital terrain modelling for farm planning for mechanical harvest of sugar cane in Mauritius. *Computers and Electronics in Agriculture*, Vol. 18, No. 1, pp. 29–42.

Sehgal, J.L., Allam, A.I., Gupta, R.P. and Aziz, A. (1980) The suitability of the soils of the Mesopotamian plain for sugarcane cultivation. *Proceedings of the International Society of Sugar Cane Technologists*, Vol. 17, pp. 132–151.

Selman, M. (1976) Guard-crop systems becoming feasible. *Arable Farming*, Vol. 3, No. 4, pp. 47, 51.

Selvarajan, S. and Subramaniam, S.R. (1988) Water resource budgeting in Amaravathy River Basin. *Agricultural Situation in India*, Vol. 43, No. 4, pp. 295–299.

Sepaskhah, A.R. and Kamgar-Haghighi, A.A. (1997) Water use and yields of sugarbeet grown under every-other-furrow irrigation with different irrigation intervals. *Agricultural Water Management*, Vol. 34, No. 1, pp. 71–79.

Serase, J.I., Hall, D.O. and Stuckey, D.C. (1993) Bioethanol from grapes in the European Community. *Biomass and Bioenergy*, Vol. 5, No. 5, pp. 347–358.

Seru, V.B. (1998) Fiji. *Perspectives on Sustainable Farming Systems in Upland Areas: Proceedings of the APO-FFTC Study Meeting held in Tokyo and Takamatsu, Japan, 27 October–1 November 1997*, pp. 169–179. Asian Productivity Organization, Tokyo.

Setamou, M., Bernal, J.S., Legaspi, J.C. and Mirkov, T.E. (2002) Parasitism and location of sugarcane borer (Lepidoptera: Pyralidae) by *Cotesia flavipes* (Hymenoptera: Braconidae) on transgenic and conventional sugarcane. *Environmental Entomology*, Vol. 31, pp. 1219–1225.

Shafshak, S.E., El-Geddawy, I.H. and El-Shafai, A.M.A. (2001) Effect of irrigation regime and nitrogen fertilization on: 1 – growth characters and juice quality of sugarcane. *Pakistan Sugar Journal*, Vol. 16, No. 5, pp. 2–8.

Shaikh, A.A., Jadhav, A.S. and Bote, N.L. (1994a) Effects of planting methods, mulching and fertilizers on yield of rainfed pearl millet. *Journal of Maharashtra Agricultural Universities*, Vol. 19, No. 2, pp. 287–288.

Shaikh, A.A., Jadhav, A.S. and Bote, N.L. (1994b) Effect of moisture-conservation technique and fertilization on weed growth and yield of rainfed pearlmillet (*Pennisetum glaucum*). *Indian Journal of Agronomy*, Vol. 39, No. 2, pp. 307–308.

Shankaraiah, C., Nagaraju, M.S. and Gururaj Hunsigi (2001) Effect of associative nitrogen fixing bioagents on nitrogen nutrition of sugarcane and environmental quality. In *Proceedings of the 63rd Annual Convention of the Sugar Technologists' Association of India, Jaipur, India, 25–27 August 2001*, pp. A77–A87. Sugar Technologists' Association of India, New Delhi.

Shannon, E., McDougall, A., Kelsey, K. and Hussey, B. (1996) Watercheck – a co-ordinated extension program for improving irrigation efficiency on Australian cane farms. *Proceedings of the Australian Society of Sugar Cane Technologists*, Vol. 18, pp. 113–118.

Sharma, S.K. and Goel, D.K. (1997) Ethanol: an eco-friendly bio-fuel from sugar industry. In *Proceedings of the 59th Annual Convention of the Sugar Technologists' Association of India, Goa, India, 26 to 28 September 1997*, pp. C3–C23. Sugar Technologists' Association of India, New Delhi.

Sharma, R.K. and Verma, H.D. (1996) Effect of irrigation scheduling and other agronomic manipulations on yield and water economy in sugarcane. *Indian Journal of Agronomy*, Vol. 41, No. 1, pp. 122–126.

Sharmasarkar, F.C., Shankar Sharmasarkar, Held, L.J., Miller, S.D., Vance, G.F. and Zhang RenDuo (2001a) Agroeconomic analyses of drip irrigation for sugarbeet production. *Agronomy Journal*, Vol. 93, No. 3, pp. 517–523.

Sharmasarkar, F.C., Shankar Sharmasarkar, Held, L.J., Miller, S.D., Vance, G.F. and Zhang RenDuo (2001b) Assessment of drip and flood irrigation on water and fertilizer use efficiencies for sugarbeets. *Agricultural Water Management*, Vol. 46, No. 3, pp. 241–251.

Shaw, M.E.A. (1982) Aspects of the management of salinity on swelling clay soils in Jamaica. *Tropical Agriculture*, Vol. 59, No. 2, pp. 167–172.

Shehata, M.M. (1999) Effect of ground water depth and irrigation with saline water on sugar beet. II: Chemical constituents. *Egyptian Journal of Agricultural Research*, Vol. 77, No. 4, pp. 1675–1686.

Shinde, P.P. and Jadhav, S.B. (1998) Drip in sugarcane – an experience in India. In *Proceedings of the International Agricultural Engineering Conference, Bangkok, Thailand, 7–10 December 1998* (eds V.M. Salokhe and Zhang Jianxia), pp. 734–747. Asian Institute of Technology, Bangkok.

Shinde, P.P. and Jadhav, S.B. (2000) Water management with drip irrigation system for sugarcane. In *Proceedings of the 62nd Annual Convention of the Sugar Technologists' Association of India, Agra, India, 19–21 August 2000*, pp. A36–A41. Sugar Technologists' Association of India, New Delhi.

Shiva, V. and Bandyopadhyay, J. (1986) Desertification in India: trends and countertrends. *Desertification Control Bulletin*, No. 13, pp. 29–33.

Shrivastava, P.K., Parikh, M.M., Savani, N.G. and Raman, S. (1993) Response of sugarcane crop to sprinkler irrigation. *Indian Sugar*, Vol. 42, No. 10, pp. 761–763.

Shukla, G.L. (1995) Pollution control in the sugar industry: treatment and disposal of factory and distillery effluents. In *Sugarcane: Agro-industrial Alternatives* (eds G.B. Singh and S. Solomon), pp. 417–447. Oxford and IBH Publishing Co. Pvt. Ltd, New Delhi.

Sidiras, N., Kahnt, G. and Kubler, E. (1988) The effects of different soil tillage systems, crop rotations and short term cover crops on soil losses and runoff. *Journal of Agronomy and Crop Science*, Vol. 160, No. 1, pp. 22–28.

Sileika, A.S. (2000) Nutrient losses from agriculture in Lithuania. *Landbauforschung Volkenrode*, Vol. 50, No. 1/2, pp. 15–23.

Silva, M.A.L., Calasans, C.F., Ovalle, A.R.C. and Rezende, C.E. (2001) Dissolved nitrogen and phosphorus dynamics in the lower portion of the Paraiba do Sul river, Campos dos Goytacazes, RJ, Brazil. *Brazilian Archives of Biology and Technology*, Vol. 44, No. 4, pp. 365–371.

Silva Lora, E. and Olivares Gomez, E. (1995) Bagasse suspension burning and air pollution. *International Sugar Journal*, Vol. 97, No. 1164, pp. 683–688, 690, 693–694.

Simmons, S.R. and Dotzenko, A.D. (1975) Evaluation of conservation tillage effectiveness on sugarbeet fields in northeastern Colorado. *Journal of the American Society of Sugar Beet Technologists*, Vol. 18, No. 4, pp. 307–311.

Simonsen, P.E., Nega, A. and Furu, P. (1990) Intestinal schistosomiasis among children in a labour village of Wonji Sugar Estate, Ethiopia. *East African Medical Journal*, Vol. 67, No. 8, pp. 532–538.

Simpson, B.W., Blogg, D. and Ruddle, L.J. (2003) Simplifying environmental risk management for pesticides – a computer-based risk management tool for landholders and extension personnel. *Proceedings of the Australian Society of Sugar Cane Technologists*, Vol. 25, p. 36.

Singels, A. and Bezuidenhout, C.N. (1999) The relationship between ENSO and rainfall and yield in the South African sugar industry. *South African Journal of Plant and Soil*, Vol. 16, pp. 96–101.

Singels, A., Kennedy, A.J. and Bezuidenhout, C.N. (1998) Irricane: a simple computerised irrigation scheduling method for sugarcane. *Proceedings of the Annual Congress – South African Sugar Technologists' Association*, No. 72, pp. 117–122.

Singels, A., Bezuidenhout, C.N. and Schmidt, E.J. (1999) Evaluating strategies for scheduling supplementary irrigation of sugarcane in South Africa. *Proceedings of the Australian Society of Sugar Cane Technologists*, Vol. 21, pp. 219–226.

Singh, A., Singh, S. and Rao, S.N. (2003) Integrated pest management in India. In *Integrated Pest Management in the Global Arena* (eds K.M. Maredia, D. Dakouo and D. Mota-Sanchez), pp. 209–221. CAB International, Wallingford, UK.

Singh, B.B., Srivastava, G.K. and Gupta, R.P. (1996) Effect of industrial effluents on ground water quality, at Sardarnagar, Gorakhpur (U.P.). *Journal of Living World*, Vol. 3, No. 2, pp. 61–64.

Singh, D.P., Singh, P., Pannu, R.K. and Sharma, H.C. (1991) Carbon dioxide enrichment, climatic change and Indian agriculture: a preliminary analysis. In *Impact of Global Climatic Changes on Photosynthesis and Plant Productivity. Proceedings of the Indo-US Workshop held on 8–12 January 1991 at New Delhi, India* (eds Y.P. Abrol, Govindjee, P.N. Wattal, D.R. Ort, A. Gnanam and A.H. Teramura), pp. 279–296. Oxford and IBH Publishing Co. Pvt. Ltd, New Delhi.

Singh, H.P., Mishra, J.P. and Mahaver, L.R. (1998) Observation on certain physico-chemical characteristics of water in relation to pollution of river Ramganga. *Journal of Environmental Biology*, Vol. 19, No. 4, pp. 353–356.

Singh, I.P. and Sankhayan, P.L. (1991) Sustainability of water resources during the post-Green Revolution period in Punjab. *Indian Journal of Agricultural Economics*, Vol. 46, No. 3, pp. 433–439.

Singh, P.N. and Mohan, S.C. (1994) Water use and yield response of sugarcane under different irrigation schedules and nitrogen levels in a subtropical region. *Agricultural Water Management*, Vol. 26, No. 4, pp. 253–264.

Singh, S. (1994) Sustainable agriculture: management and policy issues for sugar cane economies. *Sugar Cane*, No. 2, pp. 18–21.

Sinha, S.K. (1993) Physico-chemical characteristics of effluent discharged from Lohat sugar factory in Bihar. *Environment and Ecology*, Vol. 11, No. 2, pp. 265–268.

Sitompul, S.M., Hairiah, K., Cadisch, G. and van Noordwijk, M. (2000) Dynamics of density fractions of macro-organic material after forest conversion to sugarcane and woodlots, accounted for in a modified Century model. *Netherlands Journal of Agricultural Science*, Vol. 48, pp. 61–73.

SKIL (Sugar Knowledge International) (2003) www.sucrose.com (accessed August–December 2003).

Skjemstad, J.O., Taylor, J.A., Janik, L.J. and Marvanek, S.P. (1999) Soil organic carbon dynamics under long-term sugarcane monoculture. *Australian Journal of Soil Research*, Vol. 37, pp. 151–164.

Sluyter, A. (1998) From archive to map to pastoral landscape: a spatial perspective on the livestock ecology of sixteenth-century New Spain. *Environmental History*, Vol. 3, No. 4, pp. 508–528.

Smedema, L.K. (1983) Evaluation of field drainage discharge from heavy land in the tropics. *Journal of Hydrology, Netherlands*, Vol. 67, No. 1/4, pp. 325–337.

Smid, J., Dekkers, W.A. and Dekking, A.J.G. (2001) Integrated arable farming: especially lower environmental impact. *PPO-Bulletin Akkerbouw*, No. 1, pp. 22–25.

Smith, F.W. (1997) Nutrient uptake and metabolism: prospects for molecular modification to improve efficiency. In *Intensive Sugarcane Production: Meeting the Challenges Beyond 2000* (eds B.A. Keating and J.R. Wilson), pp. 157–172. CAB International, Wallingford, UK.

Smith, J.C. (2001) Communicating with the public on environmental issues. In *Proceedings from the 31st Biennial Meeting (Operations) of the American Society of Sugar Beet Technologists, Vancouver, BC, Canada, 28 February–3 March 2001*, pp. 253–255. American Society of Sugar Beet Technologists, Denver.

Smith, K.A. and Arah, J.R.M. (1990) Measurement and modelling of nitrous oxide emissions from soils. *Ecological Bulletin*, Vol. 42, pp. 116–123.

Smith, R.E. (1995) Simulation of crop water balance with Opus. In *Crop-water-simulation Models in Practice. Selected Papers of the 2nd Workshop on Crop-Water-Models held on the Occasion of the 15th Congress of the International Commission on Irrigation and Drainage (ICID) at The Hague, The Netherlands in 1993* (eds L.S. Pereira, B.J.

van den Broek, P. Kabat and R.G. Allen), pp. 215–227. Wageningen Pers, Wageningen.

Snoad, J. (1995) Practical aspects of factory water usage reduction. *Proceedings of the Australian Society of Sugar Cane Technologists*, No. 17, pp. 257–264.

Snyder, G.H., Deren, C.W. and Glaz, B. (1999) Wetland crops versus wetland drainage. *HortScience*, Vol. 34, No. 1, pp. 46–49.

Soane, D.B., Dickson, J.W. and Campbell, D.J. (1982) Compaction by agricultural vehicles a review. III – Incidence and control of compaction in crop production. *Soil Tillage Research*, Vol. 2, pp. 3–36.

Sohrabi, T.M., Abbas Keshavarz and Abrahim Pazira (1998) An overview of land drainage in Iran. In *Drainage in the 21st Century: Food Production and the Environment. Proceedings of the Seventh International Drainage Symposium, Orlando, Florida, USA, 8–10 March 1998* (ed. L.C. Brown), pp. 70–81. American Society of Agricultural Engineers, St Joseph.

Soicher, A.J. and Peterson, F.L. (1997) Terrestrial nutrient and sediment fluxes to the coastal waters of West Maui, Hawai'i. *Pacific Science*, Vol. 51, No. 3, pp. 221–232.

Soil Survey Staff (1975) *Soil Taxonomy*. Agriculture Handbook 436. USDA Soil Conservation Service, Washington, DC.

Soil Survey Staff (1999) *Soil Taxonomy*, 2nd edn. USDA National Resources Conservation Services, Washington, DC.

Solomon, S., Ishwar Singh, Santa Deb, Joshi, P.C. and Sunil Kumar (2002) Sugarcane ripening response to dinitrosocifrol applied through irrigation water. In *Proceedings of the 64th Annual Convention of the Sugar Technologists' Association of India, Cochin (Kerala), India, 17–19 August 2002*, pp. A58–A61. Sugar Technologists' Association of India, New Delhi.

Soman, P. (2002) Impact of drip irrigation in sugarcane production. In *Proceedings of the 64th Annual Convention of the Sugar Technologists' Association of India, Cochin (Kerala), India, 17–19 August 2002*, pp. A173–A186. Sugar Technologists Association of India, New Delhi.

Sommer, C. (1989) Soil conservation by reduced tillage. In *EUR Report*, No. EUR 11258, pp. 17–28.

Sommer, C. and Lindstrom, M.J. (1987) Conservation tillage development in the Federal Republic of Germany. Optimum erosion control at least cost. In *Proceedings of the National Symposium on Conservation Systems, 14–15 December 1987, Chicago, Illinois, USA*, pp. 65–73. American Society of Agricultural Engineers, St Joseph.

Sommer, C. and Zach, M. (1984) Mulch sowing of sugar beet – a possibility of reducing soil erosion. *Mitteilungen der Deutschen Bodenkundlichen Gesellschaft*, Vol. 39, pp. 163–168.

Sommer, C., Durr, H.J. and Zach, M. (1995) Effects of soil compactions on plant yield and inferences for soil conservation. *KTBL-Schrift*, No. 362, pp. 176–188.

Sorella, P. (2000) Trickle irrigation: experiments and results with winter sugarbeet in southern Italy. *Industria Saccarifera Italiana*, Vol. 93, No. 1, pp. 6–12.

Sorensen, F.B. (2001) GrowCom – with day to day updating. *Proceedings of the Institut International de Recherches Betteravières*, Vol. 64, pp. 101–104.

Sotiriou, N. and Scheunert, I. (1994) Long-term experiments on the fate of ^{15}N urea in soil/plant systems under outdoor conditions. *Chemosphere*, Vol. 28, No. 2, pp. 333–340.

Southwick, L.M., Willis, G.H. and Selim, H.M. (1992) Leaching of atrazine from sugarcane in southern Louisiana. *Journal of Agricultural and Food Chemistry*, Vol. 40, pp. 1264–1268.

Southwick, L.M., Willis, G.H., Johnson, D.C. and Selim, H.M. (1995) Leaching of nitrate, atrazine, and meribuzin from sugarcane in southern Louisiana. *Journal of Environmental Quality*, Vol. 24, pp. 684–690.

Spain, A.V., Saffigna, P.G. and Wood, A.W. (1990) Tissue carbon sources for *Pontoscolex corethrurus* (Oligochaeta:Glossoscolecidae) in a sugarcane ecosystem. *Soil Biology and Biochemistry*, Vol. 22, No. 5, pp. 703–706.

Sparling, G.P. (1997) Soil microbial biomass activity and nutrient cycling as indicators of soil health. In *Biological Indicators of Soil Health* (eds C.E. Pankhurst, B.M. Doube and V.V.S.R. Gupta), pp. 97–119. CAB International, Wallingford, UK.

Sparovek, G. and Schnug, E. (2001a) Temporal erosion-induced soil degradation and yield loss. *Soil Science Society of America Journal*, Vol. 65, No. 5, pp. 1479–1486.

Sparovek, G. and Schnug, E. (2001b) Soil tillage and precision agriculture. A theoretical case study for soil erosion control in Brazilian sugar cane production. *Soil and Tillage Research*, Vol. 61, No. 1/2, pp. 47–54.

Sparovek, G., Weill, M.M., Ranieri, S.B.L., Schnug, E. and Silva, E.F. (1997) The life time concept as a tool for erosion tolerance definition. *Scientia Agricola*, Vol. 54, Special No., pp. 130–135.

Speare, R. (1990) A review of the diseases of the cane toad, *Bufo marinus*, with comments on biological control. *Australian Wildlife Research*, Vol. 17, pp. 387–410.

Sperlingson, C. (2003) What is the market value of precision and organically grown sugar? *Sveriges Utsadesforenings Tidskrift*, Vol. 113, pp. 29–33.

Spiess, E. and Diserens, E. (2001) *Sugarbeet: Harvesting Technique and Soil Conservation. Versatile Interactions between Technique, Harvesting Quality and Ecology.* FAT-Berichte, No. 567, Switzerland, 19 pp. Eidgenossische Forschungsanstalt fur Agrarwirtschaft und Landtechnik (FAT), Tanikon.

Srikanth, J., Easwaramoorthy, S., Kurup, N.K. and Santhalakshmi, G. (1997) Spider abundance in sugarcane: impact of cultural practices, irrigation and post-harvest trash burning. *Biological Agriculture and Horticulture*, Vol. 14, No. 4, pp. 343–356.

Srimannarayana, P. and Sudheer, K.J. (2000) Achieving zero pollution by totally recycling treated effluents as make-up water for cooling tower/spray pond. In *Proceedings of the 62nd Annual Convention of the Sugar Technologists' Association of India, Agra, India, 19–21 August, 2000*, pp. C39–C64. Sugar Technologists' Association of India, New Delhi.

Srivastava, A.K. (1984) Soil compaction: a problem in sugarcane culture. *Indian Sugar*, Vol. 34, pp. 503–504.

Srivastava, A.K. (1989) Pollution of the Faren stream by the effluents of sugar factory and distillery. *Advances in Plant Sciences*, Vol. 2, No. 2, pp. 308–310.

Srivastava, D.C. (1998) Integrated rodent management strategy in sugarcane agro-ecosystem. *Pest Management and Economic Zoology*, Vol. 6, No. 1, pp. 77–79.

Srivastava, R.C. and Upadhayaya, A. (1998) Study on feasibility of drip irrigation in shallow ground water zones of eastern India. *Agricultural Water Management*, Vol. 36, No. 1, pp. 71–83.

Srivastava, S.C. (1992) Sugarcane. In *IFA World Fertilizer Use Manual* (eds D.J. Halliday and M.E. Trenkel), pp. 257–266. International Fertilizer Industry Association, Paris.

Srivastava, T.K., Singh, G.B. and Srivastava, S.N. (1999) Efficacy of weed control measures and soil persistence of atrazine in sugarcane (*Saccharum officinarum*) as influenced by irrigation and nitrogen. *Indian Journal of Agronomy*, Vol. 44, No. 4, pp. 835–840.

Steiner, A. (2003) The water balance. *World Conservation*, Vol. 1, pp. 4–5.

Stekar, J.M.A. (1999) Silage effluent and the environment. *Sodobno Kmetijstvo*, Vol. 32, No. 2, pp. 73–77.

Stockfisch, N. and Marlander, B. (2002) Prospects for sustainable sugarbeet production. *Zuckerindustrie*, Vol. 127, No. 9, pp. 667–672.

Stoeven, K., Kohrs, K. and Schnug, E. (2002) Relation between N fertilization and N_2O release in different crop rotations. *Phyton (Horn)*, Vol. 42, No. 3, pp. 191–200.

Stout, B.A., Snyder, F.W. and Carleton, W.M. (1956) The effect of soil moisture and compaction on sugar beet emergence. *Journal of American Society of Sugar Beet Technologists*, Vol. 9, pp. 277–283.

Strohl, R. (1985) Jamaica must increase irrigated area to meet staple and export crop needs. *Agribusiness Worldwide*, July/August, pp. 3–7.

Strong, D.R., McCoy, E.D. and Rey, J.R. (1977) Time and number of herbivore species: the pests of sugarcane. *Ecology*, Vol. 58, pp. 167–175.

Stuck, J.D., Izuno, F.T., Pickering, N., Campbell, K.L. and Bottcher, A.B. (2001) Mathematical modeling of suspended solids and particulate phosphorus transport in farm conveyance systems of the Everglades Agricultural Area. *Transactions of the ASAE*, Vol. 44, No. 5, pp. 1117–1126.

Sturrock, R.F., Diaw, O.T., Talla, I., Niang, M., Piau, J.P. and Capron, A. (2001) Seasonality in the transmission of schistosomiasis and in populations of its snail intermediate hosts in and around a sugar irrigation scheme at Richard Toll, Senegal. *Parasitology*, Vol. 123, suppl., pp. S77–S89.

Styles, G.H. (1973) A technique for blowing soil in the West Midlands. *British Sugar Beet Review*, Vol. 41, No. 1, pp. 19–20.

Subagio, I. and Mumwandono (1992) Method of soil cultivation for replanting by deep tillage. *Berita – Pusat Penelitian Perkebunan Gula Indonesia*, No. 6, pp. 24–28.

Sullivan, D.J. and Sallaway, M.M. (1994) Development of soil conservation specifications in the coastal Burnett district. *Proceedings of the Australian Society of Sugar Cane Technologists*, Vol. 16, pp. 178–186.

Sumner, M.E. (1997) Opportunities for amelioration of soil physical and chemical constraints under intensive cropping. In *Intensive Sugarcane Production: Meeting the Challenges beyond 2000. Proceedings of the Sugar 2000 Symposium, Brisbane, Australia, 20–23 August 1996* (eds B.A. Keating and J.R. Wilson), pp. 305–326. CAB International, Wallingford.

Sundara, B. amd Reddy, B.S. (1994) Studies on sugarcane varietal performance and planting systems under saline water irrigation. *Cooperative Sugar*, Vol. 25, No. 11, pp. 457–461.

Sundaramurthy, V.T. (1998) Sugar distillery effluent activity against insect pests. *International*

Sugar Journal (Cane Sugar Edition), Vol. 100, pp. 566–567.

Sundarsingh, S.D., Rajagopal, A. and Ramiah, S. (1995) Irrigation management in sugarcane. *Madras Agricultural Journal*, Vol. 82, No. 1, pp. 41–44.

Sutton, M.R., Wood, A.W. and Saffigna, P.G. (1996) Long term effects of green cane trash management on Herbert River soils. In *Sugarcane: Research Towards Efficient and Sustainable Production* (eds J.R. Wilson, D.W. Hogarth, J.A. Campbell and A.L. Garside), pp. 178–180. CSIRO, Division of Tropical Crops and Pastures, Brisbane.

Svachula, V. (1993) Effect of irrigation on sugarbeet yields. *Rostlinna Vyroba*, Vol. 39, No. 6, pp. 543–550.

Svendsen, H., Hansen, S. and Jensen, H.E. (1995) Simulation of crop production, water and nitrogen balances in two German agro-ecosystems using the DAISY model. *Ecological Modelling*, Vol. 81, No. 1/3, pp. 197–212.

Swamy, K.R., Jamuna, P., Rao, T.K.V.V.M. and Raju, A.P. (1998) Effect of cane trash management on soil-moisture conservation, soil fertility and yield of sugarcane (*Saccharum officinarum*). *Indian Journal of Agricultural Sciences*, Vol. 68, No. 9, pp. 603–606.

Sweeney, C.A., Thorburn, P.J. and Bristow, K.L. (2001a) Water table contributions to sugarcane production: impacts on irrigation water use efficiency. *Proceedings of the Australian Society of Sugar Cane Technologists*, Vol. 23, pp. 116–121.

Sweeney, C.A., Thorburn, P.J. and Bristow, K.L. (2001b) Water table contributions to sugarcane production: impacts on irrigation water use efficiency. *Sugar Cane International*, August, pp. 16–19.

Swinford, J.M. and Boevey, T.M.C. (1984) The effects of soil compaction due to in-field transport on ratoon cane yields and soil physical characteristics. *Proceedings of the South African Sugar Technologists' Association*, Vol. 58, pp. 198–203.

Syers, J.K. (1998) Soil and plant potassium in agriculture. *Fertilizer Society Proceedings*, Vol. 411, pp. 1–32.

Syme, G.J., McDonald, G.T. and Butterworth, J. (1997) Doing things differently: the policy revolution. In *Intensive Sugarcane Production: Meeting the Challenges Beyond 2000* (eds B.A. Keating and J.R. Wilson), pp. 491–506. CAB International, Wallingford.

Szmrecsanyi, T. (1994) Technology and environmental degradation: the case of the sugarcane industry in the State of Sao Paulo. *Informacoes Economicas – Instituto de Economia Agricola*, Vol. 24, No. 10, pp. 73–81.

Tahir Sarwar, Kiyani, F.A. and Khan, M.J. (2001) On-farm irrigation evaluation of sugarcane fields at a civil canal in NWFP. *Sarhad Journal of Agriculture*, Vol. 17, No. 2, pp. 153–158.

Tai, P.Y.P. and Miller, J.D. (2001) A core collection of *Saccharum spontaneum* L. from the world collection of sugarcane. *Crop Science*, Vol. 41, pp. 879–885.

Tai, P.Y.P. and Miller, J.D. (2002) Germplasm diversity among four sugarcane species for sugar composition. *Crop Science*, Vol. 42, pp. 958–964.

Talla, I., Kongs, A., Verle, P., Belot, J., Sarr, S. and Coll, A.M. (1990) Outbreak of intestinal schistosomiasis in the Senegal River Basin. *Annales de la Société Belge de Médecine Tropicale*, Vol. 70, pp. 173–180.

Tarariko, A.G., Pirozhenko, G.S., Konchakov, A.V. and Litvin, A.G. (1990) A new concept of soil-conserving contour-meliorative agriculture and its effectiveness in Ukraine. *Vestnik Sel'skokhozyaistvennoi Nauki (Moskva)*, No. 11, pp. 113–118.

Tewari, D.D. and Archana Tripathi (2001) Effect of sugar effluent on seed germination of *Pisum sativum* (Var, Asauji and Arkel) at Balrampur. *Journal of Living World*, Vol. 8, No. 2, pp. 5–9.

Thanki, J.D., Patel, R.G., Holer, G.H., Patel, H.S. and Raman, S. (1999) Saving of irrigation water through mulch for sustainable production of sugarcane. *Cooperative Sugar*, Vol. 31, No. 3, pp. 207–209.

Thanki, J.D., Patel, R.G., Patel, H.S. and Raman, S. (2000a) Response of phasic irrigation in sugarcane crop under South Gujarat condition. *Indian Sugar*, Vol. 49, No. 10, pp. 831–834.

Thanki, J.D., Patel, R.G., Patel, H.S. and Raman, S. (2000b) Irrigation scheduling on the basis of climatological approach to sugarcane. *Indian Sugar*, Vol. 49, No. 11, pp. 905–907.

Tharme, R. (2003) Water and agriculture: true values. *World Conservation*, Vol. 1, p. 11.

Thielecke, K., Westphal, S. and Buchholz, K. (1991) Analysis of dryer exhaust gas components under actual technical conditions (parameters, equipment). In *Proceedings of the 19th General Assembly, Commission Internationale Technique de Sucrerie, Cambridge (UK), 24–28 June 1991*, pp. 407–440. CITS, Rain am Lech.

Thind, S.K. (1996) Effect of drought on leaf epidermal structure of sugarcane. *Sugar Cane*, No. 1, pp. 15–17.

Thomas, D.L., Perry, C.D., Evans, R.O., Izuno, F.T., Stone, K.C. and Gilliam, J.W. (1995)

Agricultural drainage effects on water quality in southeastern U.S. *Journal of Irrigation and Drainage Engineering*, Vol. 121, No. 4, pp. 277–282.

Thomas, J.R., Salinas, F.G. and Derther, G.F. (1981) Use of saline water for supplemental irrigation of sugarcane. *Agronomy Journal*, Vol. 73, No. 6, pp. 1011–1018.

Thompson, G.D. (1976) Water use by sugarcane. *South African Sugar Journal*, Vol. 60, pp. 593–600, 627–635.

Thompson, N., Barrie, I.A. and Ayles, M. (1981) *The Meteorological Office Rainfall and Evaporation Calculation System: MORECS (July 1981)*. Meteorological Office, Bracknell, UK, 69 pp.

Thomsen, I.K. and Christensen, B.T. (1996) Availability to subsequent crops and leaching of nitrogen ^{15}N-labelled sugar beet tops and oilseed rape residues. *Journal of Agricultural Science, Cambridge*, Vol. 126, pp. 191–199.

Thorburn, P.J., Sweeney, C.A. and Bristow, K.L. (1998) Production and environmental benefits of trickle irrigation for sugarcane: a review. *Proceedings of the Australian Society of Sugar Cane Technologists*, Vol. 20, pp. 118–125.

Thorburn, P.J., Dart, I.K. and Baillie, C.P. (2000) Nitrogen balances in trickle irrigated sugarcane. In *6th International Micro-irrigation Congress – Micro-irrigation Technology for Developing Agriculture, Cape Town, South Africa, 22–27 October 2000*. International Commission on Irrigation and Drainage, Rome.

Thorburn, P.J., Biggs, I.M., Weier, K.L. and Keating, B.A. (2003a) Nitrate in groundwaters of intensive agricultural areas in coastal Northeastern Australia. *Agriculture, Ecosystems and Environment*, Vol. 94, pp. 49–58.

Thorburn, P.J., Park, S.E. and Biggs, I.M. (2003b) Nitrogen fertilizer management in the Australian sugar industry: strategic opportunities for improved efficiency. *Proceedings of the Australian Society of Sugar Cane Technologists*, (ed. D.M. Hogarth), Vol. 25, p. 22.

Thuresson, M. (2001) *Pilot-scale Wetland Investigation in Treating Distillery Effluent with Three Different Emergent Macrophyte Species in Kenya*. Minor Field Studies No. 153, Swedish University of Agricultural Sciences (International Office), Uppsala, 41 pp.

Thuresson, S. (2001) *Treatment of Distillery Effluent in Vegetated Surface Flow Wetlands in Muhoroni, Kenya. Minor Field Studies* No. 144, Swedish University of Agricultural Sciences (International Office), Uppsala, 45 pp.

Tianco, A.P. (1983) Molasses as a fertilizer and soil amendment. *Proceedings of the Philippines Sugar Technologists Annual Convention*, Vol. 30, pp. 469–472.

Tinker, P.B. (1997) The environmental implications of intensified land use in developing countries. In *Land Resources: On the Edge of the Malthusian Precipice?* (eds D.J. Greenland, P.J. Gregory and P.H. Nye), pp. 1023–1033. The Royal Society, London.

Tiwari, T.N., Srivastava, R.P. and Singh, G.P. (1997) Salinity tolerance in sugarcane cultivars. *Sugar Cane*, Vol. 1, pp. 7–11.

Tobias, S., Hennes, M., Meier, E. and Schulin, R. (2001) Estimating soil resilience to compaction by measuring changes in surface and subsurface levels. *Soil Use and Management*, Vol. 17, No. 4, pp. 229–234.

Tognetti, R., Delfine, S., Sorella, P. and Alvino, A. (2002) Responses of sugarbeet to drip and low-pressure sprinkler irrigation systems: root yield and sucrose accumulation. *Agricoltura Mediterranea*, Vol. 132, No. 1, pp. 1–8.

Tomaszewska, A. and Polec, B. (1997) Evaluation of the state of purity of air in the vicinity of equipment connected with the wastewater economy of sugar factories. *Gazeta Cukrownicza*, Vol. 105, No. 10, pp. 186–189.

Tomov, B.W., Bernal, J.S. and Vinson, S.B. (2003) Impacts of transgenic sugarcane expressing GNA lectin on parasitism of Mexican rice borer by *Parallorhogas pyralophagus* (Marsh) (Hymenoptera: Braconidae). *Environmental Entomology*, Vol. 32, pp. 866–872.

Toor, S.S., Makhan Singh, Garcha, A.I.S. and Sarjit Singh (1999) Pre monsoon irrigation requirement of spring planted sugarcane under sub-tropical conditions. *Indian Sugar*, Vol. 49, No. 2, pp. 129–133.

Topham, M. and Beardsley, J.W. (1975) An influence of nectar source plants on the New Guinea sugarcane weevil parasite *Lixophaga sphenophori* (Villeneuve). *Proceedings of the Hawaiian Entomological Society*, Vol. 22, pp. 145–155.

Torres, J.S. (1998) A simple visual aid for sugarcane irrigation scheduling. *Agricultural Water Management*, Vol. 38, No. 1, pp. 77–83.

Torres, J.S. and Villegas, F. (1993) Differentiation of soil compaction and cane stool damage. *Sugar Cane*, Vol. 1, pp. 7–11.

Torres A., J.S., Cruz V., R. and Villegas T., F. (1996) *Technical Advances in Programming and Management of Irrigation in Sugarcane*. Serie Tecnica No. 19, CENICANA, Centro de Investigacion de la Cana de Azucar de Colombia, Cali, 53 pp.

Treadgold, B. (1998) *The Sugar Beet Industries of Emergent EU Countries: Poland, Slovak, Czech Republic and Hungary*. Nuffield Farming Scholarship Trust, Uckfield.

Tregaskis, K.L. and Prathapar, S.A. (1994) *Selected Papers Researching the Effects of Salinity and Waterlogging on Crops within the Murray Basin*. Technical Memorandum No. 94/11, Division of Water Resources, Institute of Natural Resources and Environment, CSIRO, Canberra. vi + 61 pp.

Trivelin, P.C.O., de Oliveira, M.W., Vitti, A.C., Gava, G.J. de C. and Bendassolli, J.A. (2002) Nitrogen losses of applied urea in the soil–plant system during two sugar cane cycles. *Pesquisa Agropecuaria Brasileira*, Vol. 37, pp. 193–201.

Trouse, A.C. and Humbert, R.P. (1961) Some effects of soil compaction on the development of sugar cane roots. *Soil Science*, Vol. 91, pp. 208–217.

Tudor-Owen, R.P.D. and Wyatt, J. (1991) A guide to the stabilization of water courses by planting indigenous trees. *Proceedings of the Annual Congress – South African Sugar Technologists' Association*, No. 65, pp. 73–76.

Tugnoli, V. (2001) Low-volume irrigation: new openings for Italian sugar beet. *Mondo Macchina*, Vol. 10, No. 11/12, pp. 14–16, 18–20.

Tugnoli, V. and Maini, M. (1992) Ecology and cost reduction: mechanization in modern agriculture. *Informatore Agrario*, Vol. 48, No. 31, pp. 27–30.

Tuitert, G. and Hofmeester, Y. (1994) Epidemiology of beet necrotic yellow vein virus in sugar beet at different initial inoculum levels in the presence or absence of irrigation: disease incidence, yield and quality. *European Journal of Plant Pathology*, Vol. 100, No. 1, pp. 19–53.

Turner, P.E., Meyer, J.H. and Wood, R.A. (1992) Some results of field experiments to test lime and phosphogypsum for aluminium amelioration. *Proceedings of the South African Sugar Technologists' Association*, Vol. 66, pp. 44–49.

Tzeng, J.S., Huang, C.M. and Wang, P.L. (2001) Recycling of TSC's wastewater. *Taiwan Sugar*, Vol. 48, No. 2, pp. 10–14.

Ullman, I. (1999) Subsurface drip irrigation (SDI) of sugarcane. *International Water and Irrigation Review*, Vol. 19, No. 4, pp. 39, 40, 42.

Umrit, G. and Ng Kee Kwong, K.F. (1999) Herbicide dissipation and runoff from soils under sugarcane in Mauritius. *Proceedings of the Annual Congress – South African Sugar Technologists' Association*, No. 73, pp. 24–29.

UNEP (1982) *Environmental Aspects of the Sugar Industry – an Overview*. Industry and Environment Office, United Nations Environment Programme.

Urbaniec, K. (1996) Energy conservation in the Polish sugar industry. *Hellenic Sugar Industry*, Vol. 1, pp. 29–41.

Urbaniec, K. and Suchecki, W. (1994) Methods of desulfurizing boiler flue gases. *Gazeta Cukrownicza*, Vol. 102, No. 4, pp. 65–67.

Urbano, P., Arroyo, J.M., Conde, J.R., Rojo, C. and Gonzalez, F. (1992) Sugarbeet irrigation. Trials in the Duero valley. *Agricultura, Revista Agropecuaria*, Vol. 61, No. 718, pp. 380–385.

Urbano, P., Rojo, J.M., Gonzalez, C. and Moya, M. (2000a) Influence of irrigation on the yield of sugarbeet cultivated in the South. *Agricultura, Revista Agropecuaria*, Vol. 69, No. 811, pp. 42–45.

Urbano, P., Rojo, J.M., Gonzalez, C. and Moya, M. (2000b) Influence of irrigation equipment on sugarbeet quality. *Agricultura, Revista Agropecuaria*, Vol. 69, No. 815, pp. 360–362.

Urquiaga, S., Cruz, K.H.S. and Boddey, R.M. (1992) Contribution of nitrogen fixation to sugar cane: nitrogen-15 and nitrogen-balance estimates. *Soil Science Society of America Journal*, Vol. 56, pp. 105–114.

Valdivia, S. (1982) Nitrogen gains and losses in sugarcane (*Saccharum* sp.) agro-ecosystems on the coast of Peru. *Plant and Soil*, Vol. 67, pp. 147–156.

Valli, F., Bucchi, R. and Tugnoli, V. (1996) Irrigation of sugarbeet at two localities in Northern Italy. *Informatore Agrario*, Vol. 52, No. 26, pp. 45–48.

Vallis, I., Parton, W.J., Keating, B.A. and Wood, A.W. (1996a) Simulation of the effects of trash and N fertilizer management on soil organic matter levels and yields of sugarcane. *Soil and Tillage Research*, Vol. 38, pp. 115–132.

Vallis, I., Catchpoole, V.R., Hughes, R.M., Myers, R.J.K., Ridge, D.R. and Weier, K.L. (1996b) Recovery in crops and soils of ^{15}N applied as sub-surface bands of urea to sugarcane. *Australian Journal of Agricultural Research*, Vol. 47, pp. 355–370.

van Antwerpen, R. and Meyer, J.H. (1996a) Evaluation of soil degradation under sugarcane cultivation in Northern Kwazulu-Natal [South Africa]. In *Sugar 2000 Symposium: Sugarcane: Research Towards Efficient and Sustainable Production 1996* (eds J.R. Wilson, D.M. Hogarth, J.A. Campbell and A.L. Garside), pp. 244–247. CSIRO Division of Tropical Crops and Pastures, Brisbane.

van Antwerpen, R. and Meyer, J.H. (1996b) Soil degradation under sugarcane cultivation in Northern Kwazulu-Natal. *Proceedings of the Annual Congress – South African Sugar Technologists' Association*, No. 70, pp. 29–33.

van Antwerpen, R. and Meyer, J.H. (1998) Soil degradation – II: effect of trash and inorganic fertilizer application on soil strength.

Proceedings of the Annual Congress – South African Sugar Technologists' Association, No. 72, pp. 152–158.

van Antwerpen, R., Haynes, R.J., Meyer, J.H. and Hlanze, D. (2003) Assessing organic amendments used by sugarcane growers for improving soil chemical and biological properties. *Proceedings of the South African Sugar Technologists' Association*, No. 77, pp. 293–304.

van Noordwijk, M., Cerri, C., Woomer, P.L., Nugroho, K. and Bernoux, M. (1997) Soil carbon dynamics in the humid tropical forest zone. *Geoderma*, Vol. 79, No. 1/4, pp. 187–225.

van Swaaij, A.C.P.M., Heijbroek, W. and Basting, J.L. (2001) Testing and improving seed vigour in sugar beet. *Proceedings of the Institut International de Recherches Betteravières*, Vol. 64, pp. 237–246.

Vanegas Chacon, E.A. and Vos, J. (1999) Land drainage in Guatemala. In *Annual Report 1998*, pp. 52–64. International Institute for Land Reclamation and Improvement.

Varis, A.L., Holopainen, J.K. and Koponen, M. (1984) Abundance and seasonal occurrence of adult Carabidae (Coleoptera) in cabbage, sugar beet and timothy fields in southern Finland. *Zeitschrift für Angewandte Entomologie*, Vol. 98, No. 1, pp. 62–73.

Ved Singh (2001) Effect of practice and schedule of irrigation on growth and yield of spring planted sugarcane (*Saccharum officinarum*). *Indian Journal of Agricultural Sciences*, Vol. 71, No. 4, pp. 277–278.

Vega, J., Alonso, A., Leyva, A. and Caballero, A. (1999) Diversity of agricultural crops in Jaruco municipality. *Cultivos Tropicales*, Vol. 20, No. 3, pp. 41–44.

Venegas V., R. and Aguilar G., C. (1992) Organic production of commercial crops. Transition in sugar beet (*Beta vulgaris*). In *Proceedings 9th International Scientific Conference IFOAM: Organic Agriculture, a Key to a Sound Development and a Sustainable Environment, 16 to 21 November 1992, São Paulo, Brazil* (eds U. Kopke and D.G. Schulz), pp. 181–188. IFOAM, Tholey-Theley.

Venturi, G. and Amaducci, M.T. (2002) Beet cultivation and the environment: the effects of nitrogen. *Industria Saccarifera Italiana*, Vol. 95, No. 1, pp. 3–15.

Verburg, K., Keating, B.A., Probert, M.E., Bristow, K.L. and Huth, N.I. (1998) Nitrate leaching under sugarcane: interactions between crop yield, soil type and management strategies. In *Agronomy – Growing Greener Future* (eds D.L. Michalk and J.E. Pratley), pp. 717–720. Australian Society of Agronomy, Wagga Wagga.

Ver Elst, P., Coppens, G., Vanongeval, L. and Geypens, M. (2001) Evaluation of N-fertilizer recommendations for sugar beets based in the N-INDEX expert system. *Proceedings of the Institut International de Recherches Betteravières*, Vol. 64, pp. 443–447.

Verschuur, G. and van Wijk, J. (1990) Ethanol. *Biotechnology and Development Monitor*, No. 2, pp. 9–11.

Vesely, D. (1975) Occurrence of diseases of emerging sugar beet in relation to abiotic factors. *Rostlinna Vyroba*, Vol. 21, No. 10, pp. 1045–1056.

Vesely, D. (1986) Relation between accompanying and plant pathogenic mycoflora of the rhizosphere of emerging sugar beet. *Sbornik UVTIZ, Ochrana Rostlin*, Vol. 22, No. 2, pp. 109–118.

Vigh, A. (1994) Water purification in the sugar industry. *Cukoripar*, Vol. 47, No. 1, pp. 9–17.

Villarias, L.J., Fombellida, A., Garcia Olano, O., Garcia Zumel, M., Garrido, F. and Garcia Tascon, J.R. (2001) Preliminary results of a study on the environmental impact of genetically modified varieties of sugarbeet resistant to glyphosate. In *Actas Congreso 2001 Sociedad Espanola de Malherbologia, Leon, Spain, 20, 21 y 22 de noviembre de 2001*, pp. 135–139. Spanish Weed Science Society, Madrid.

Visscher, P.T. and Duerr, E.O. (1991) Water quality and microbial dynamics in shrimp ponds receiving bagasse-based feed. *Journal of the World Aquaculture Society*, Vol. 22, No. 1, pp. 65–76.

Vitorello, V.A., Cerri, C.C., Andreux, F., Feller, C. and Victoria, R.L. (1989) Organic matter and natural carbon-13 distribution in forested and cultivated Oxisols. *Soil Science Society of America Journal*, Vol. 53, pp. 773–778.

Volgyi, L. (2002) Possibilities to improve the water management of the sugar factory. *Cukoripar*, Vol. 55, No. 1, pp. 7–14.

Volkov, S. (1986) Economic aspects of crop rotation. *Ekonomika Sel'skogo Khozyaistva*, No. 1, pp. 77–82.

von Bernhardt, K.G., Forstreuter, V. and Bruns, S. (1991) The influence of cultivation methods and of the seed bank on the composition of the accompanying vegetation of maize and beet fields taking the Osnabruck district as an example. *Bayerisches Landwirtschaftliches Jahrbuch*, Vol. 68, No. 2, pp. 211–223.

von der Meden, E.A. (1966) Note on salinity limits for sugarcane in Natal. *Proceedings of the South African Sugar Technologists' Association*, Vol. 40, pp. 273–275.

Vos, R., Velasco, M. and de Labastida, E. (1999) *Economic and Social Effects of 'El Nino' in*

Ecuador, 1997–8. Working Paper Series No. 292, Institute of Social Studies, The Hague, 51 pp.
Wagner, M., Rincones, C., Medina, G. and Mujica, M. (1995a) Effect of the chisel plough and of irrigation on development of sugarcane in a compacted soil. *Agronomia Tropical (Maracay)*, Vol. 45, No. 1, pp. 5–26.
Wagner, M., Rincones, C., Medina, G. and Mujica, M. (1995b) Evaluation of limiting factors of soils and water in sugarcane fields of the central region of Venezuela. *Cana de Azucar*, Vol. 13, No. 2, pp. 65–83.
Wakamura, Y. (2003) Utilization of bagasse energy in Thailand. *Mitigation and Adaptation Strategies for Global Change*, Vol. 8, pp. 253–260.
Wang, C.S., Chang, C.J. and Lee, J.T. (1990) Bagasse drying system by stack gases. *Taiwan Sugar*, Vol. 37, No. 3, pp. 15–19.
Wang, H.P., Davis, M. and Mauk, P. (1995) Effects of irrigation, planting depth, and fungicide seed treatment on sugarbeet stand establishment. *Journal of Sugar Beet Research*, Vol. 32, No. 1, pp. 1–7.
Wang, M.C., Tzeng, J.S., Chen, L.F. and Wang, P.L. (1999) Evaluation of soil treatment sites irrigated with wastewater from Nanching, Chishan, and Kaohsiung sugar mills. *Report of the Taiwan Sugar Research Institute*, No. 165, pp. 41–59.
Wang, P.L., Li, S.W., Chan, Y.Y., Tsai, C.S. and Chang, S.J. (1996) Nutrient recycling of the alcoholic slops by land furrow and sprinkle irrigation in sugar cane fields. *Proceedings of the International Society of Sugar Cane Technologists*, Vol. 22, pp. 56–62.
Waterhouse, D.F. (1994) *Biological Control of Weeds: Southeast Asian Prospects*. Australian Centre for International Agricultural Research, Canberra.
Watkinson, A.R., Freckleton, R.P., Robinson, R.A. and Sutherland, W.J. (2000) Predictions of biodiversity response to genetically modified herbicide-tolerant crops. *Science (Washington)*, Vol. 289, No. 5484, pp. 1554–1557.
Wegener, M.K., Muchow, R.C., Robertson, M.J. and Inman-Bamber, N.G. (2000) Economic evaluation of investment in research: the use of supplementary irrigation in Australian sugar cane production. *Proceedings of the Australian Society of Sugar Cane Technologists*, Vol. 22, pp. 178–185.
Weier, K.L. (1996) Trace gas emissions from a trash blanketed sugarcane field in tropical Australia. In *Sugarcane: Research Towards Efficient and Sustainable Production* (eds J.R. Wilson, D.M. Hogarth, J.A. Campbell and A.L. Garside), pp. 271–272. CSIRO Division of Tropical Crops and Pastures, Brisbane.
Weier, K.L. (1998) Sugarcane fields – sources or sinks for greenhouse gas emissions? *Australian Journal of Agricultural Research*, Vol. 49, pp. 1–9.
Weier, K.L., Keating, B.A. and Sunners, F. (1996) Nitrogen isotope ratios as indicator of the source of nitrate contamination in groundwater. In *Sugar 2000 Symposium: Sugarcane: Research Towards Efficient and Sustainable Production* (eds J.R. Wilson, D.M. Hogarth, J.A. Campbell and A.L. Garside), pp. 269–270. CSIRO Division of Tropical Crops and Pastures, Brisbane.
Weiergang, I., Pontoppidan, M.M., Bisgaard, S. and Pedersen, H.C. (2001) Enhanced seed quality and beet growth through interactions between microbes, seed and plant. *Proceedings of the Institut International de Recherches Betteravières*, Vol. 64, pp. 287–292.
Weigand, E. and Kirchgessner, M. (1980) Protein and energy value of vinasse for pigs. *Animal Feed Science and Technology*, Vol. 5, pp. 221–231.
Wei-Ping, K., Saffigna, P.G., Wood, A.W. and Freney, J.R. (1993) Evaluation of methods for control of ammonia volatilization from surface-applied nitrogen fertilizers to sugarcane trash in North Queensland. *Pedosphere*, Vol. 3, pp. 321–330.
Weiss, B., Buchs, W. and Harenberg, A. (1997) Development of the spider fauna after tillage of a six-year set-aside with natural succession. *Verhandlungen der Gesellschaft für Okologie*, Vol. 27, pp. 379–384.
Wevers, J.D.A. (1997) Reduced environmental contamination by new herbicide formulations. In *Proceedings of the 60th IIRB Congress, Cambridge, UK, 1–3 July 1997*, pp. 169–176.
Wevers, J.D.A. (2001) A scoring method to evaluate the environmental contamination of chemical crop protection treatments, applied to weed control in sugar beet. In *Proceedings from the 31st Biennial Meeting (Agriculture) of the American Society of Sugar Beet Technologists, Vancouver, BC, Canada, 28 February–3 March 2001*, pp. 116–119. American Society of Sugar Beet Technologists, Denver.
Whalen, S.A. (1989) Cane burning: environmental and health impacts. In *Creativity, Commitment, Cooperation: Tools for Advancement. Hawaiian Sugar Technologists 47th Annual Conference Reports, Honolulu, Hawaii, 14–16 November, 1988*, pp. A1–A5. Hawaiian Sugar Technologists, Aiea.
Whipps, J.M., de Leij, F.A.A.M., Lynch, J.M. and Bailey, M.J. (1998) Risk assessment associated with the release of a genetically modified *Pseudomonas fluorescens* in the field. *Bulletin OILB/SROP*, Vol. 21, No. 9, pp. 199–204.

Whitaker, D.A. (1978) The truth about bagassosis. In *The South African Sugar Year Book 1978/79*, p. 3.

White, L.H. (1996) Training for improved irrigation efficiency, a second sugar rehabilitation programme. *Proceedings of the West Indies Sugar Technologists*, Vol. 25, pp. 28–37.

Whittle, C.P., Vickers, R.A., Bellas, T.E., Kuniata, L.S. and Rumbo, E.R. (1993) Practical problems in the determination of a sex pheromone for a PNG cane borer. In *Pest Control and Sustainable Agriculture* (eds S.A. Corey, D.J. Dall and W.M. Milne), pp. 290–292. CSIRO Publications, Melbourne.

Whittle, C.P., Vickers, R.A., Kuniata, L.S., Bellas, T.E. and Rumbo, E.R. (1995) Identification of an attractant for the caneborer *Sesamia grisescens* Walker (Lepidoptera: Noctuidae). *Journal of Chemical Ecology*, Vol. 21, pp. 1409–1420.

Wiedenfeld, R.P. (1995) Effects of irrigation and N fertilizer application on sugarcane yield and quality. *Field Crops Research*, Vol. 43, No. 2/3, pp. 101–108.

Wijnands, F.G. and van Asperen, P. (1999) Reducing environmental pollution by choosing the correct pesticides. *PAV-Bulletin Akkerbouw*, Juni, pp. 28–37.

Wiklicky, L. (1981) A 'blueprint' for maximizing yields of sugar beet. In *Agricultural Yield Potentials in Continental Climates. Proceedings of the 16th Colloquium of the International Potash Institute*, pp. 215–226. International Potash Institute, Bern.

Willis, E.O. and Oniki, Y. (2002) Birds of a central São Paulo woodlot: 1. Censuses 1982–2000. *Brazilian Journal of Biology*, Vol. 62, No. 2, pp. 197–210.

Wilson, J.R., Hogarth, D.M., Campbell, J.A. and Garside, A.L. (eds) (1996) *Sugar 2000 Symposium: Sugarcane: Research Towards Efficient and Sustainable Production*, vi + 305 pp. CSIRO Division of Tropical Crops and Pastures, Brisbane.

Wilson, M. (ed.) (2003) *Optimizing Pesticide Use*. John Wiley & Sons, Chichester.

Wilson, T.E., Jr (1996) A comparison of raw sugar polarisation methods. High wavelength (NIR) versus ICUMSA method GS1/2/3–1 (1994). *International Sugar Journal (Cane Sugar Edition)*, Vol. 98, No. 1168, pp. 169–172, 174.

Winner, C. (1993) History of the crop. In *The Sugar Beet Crop – Science into Practice* (eds D.A. Cooke and R.K. Scott), pp. 1–36. Chapman and Hall, London.

Wischmeier, W.H. and Smith, D.D. (1965) *Predicting Rainfall-erosion Losses from Cropland East of the Rocky Mountains – Guide for Selection of Practices for Soil and Water Conservation*. Agricultural Handbook No. 282, USDA-ARS, Washington, DC.

Wischmeier, W.H. and Smith, D.D. (1978) *Predicting Rainfall Erosion Losses – a Guide to Conservation Planning*. Agricultural Handbook No. 537, USDA-ARS, Washington, DC.

Wischmeier, W.H., Johnson, C.B. and Cross, B.V. (1971) A soil erodibility nomograph for farmland and construction sites. *Journal of Soil and Water Conservation*, Vol. 26, pp. 189–193.

Witharama, W.R.G., Naylor, R.E.L. and Whytock, G.P. (1997) The seed bank and weed seedling emergence in sugarcane in Sri Lanka. In *1997 Brighton Crop Protection Conference: Weeds. Proceedings of an International Conference, Brighton, UK, 17–20 November 1997*, Vol. 2, pp. 663–664. British Crop Protection Council, Farnham.

Wittenberg, R. and Cock, M.J.W. (eds) (2001) *Invasive Alien Species: A Toolkit of Best Prevention and Management Practices*. CAB International (on behalf of the Global Invasive Species Programme, GISP), Wallingford, UK.

WMO (World Meteorological Organization) (1998) Irrigation and drainage. In *Encyclopaedia of Hydrology and Water Resources* (eds R.W. Herschy and R.W. Fairbridge), pp. 444–446. Kluwer Academic Publishers, Dordrecht.

Wolski, T. (1993) Effect of biogas emitted from an open fermentation chamber on atmospheric air quality. *Gazeta Cukrownicza*, Vol. 101, No. 12, pp. 253–255.

Wolski, T. (1995) Influence of biological effluent purifiers on the state of the atmosphere in a sugar factory region. *Gazeta Cukrownicza*, Vol. 103, No. 11, pp. 208–210.

Wong Sak Hoi, L., Ragen, A.K. and Julien, M.H.R. (1996) Sugar factory wastewater management strategies for environmental protection in Mauritius. In *Sugar 2000 Symposium: Sugarcane: Research Towards Efficient and Sustainable Production* (eds J.R. Wilson, D.M. Hogarth, J.A. Campbell and A.L. Garside), pp. 266–268. CSIRO Division of Tropical Crops and Pastures, Brisbane.

Wood, A.W. (1985) Soil degradation and management under intensive sugarcane cultivation in North Queensland. *Soil Use and Management*, Vol. 1, No. 4, pp. 120–124.

Wood, A.W. (1986) Green cane trash management in the Herbert Valley. *Proceedings of the Australian Society of Sugarcane Technologists*, Vol. 8, pp. 85–94.

Wood, A.W. (1991) Management of crop residues following green harvesting of sugarcane in north Queensland. *Soil and Tillage Research*, Vol. 20, pp. 69–85.

Wood, A.W., Muchow, R.C. and Robertson, M.J. (1996) Growth of sugarcane under high input conditions in tropical Australia. 3. Accumulation, partitioning and use of nitrogen. *Field Crops Research*, Vol. 48, pp. 223–233.

Wood, A.W., Bramley, R.G.V., Meyer, J.H. and Johnson, A.K.L. (1997) Opportunities for improving nutrient management in the Australian sugar industry. In *Intensive Sugarcane Production: Meeting the Challenges Beyond 2000* (eds B.A. Keating and J.R. Wilson), pp. 243–266. CAB International, Wallingford, UK.

Wood, A.W., Muchow, R.C., Sherrard, J., Triglone, T. and Vogelsang, H. (1998) Benchmarking irrigation practices in the Ord sugar industry. *Proceedings of the Australian Society of Sugar Cane Technologists*, Vol. 20, pp. 133–139.

Wood, D. and Lenne, J.M. (eds) (1999) *Agrobiodiversity: Characterisation, Utilization and Management*. CAB International, Wallingford.

Woods, E.J. (2000) Extension: tomorrow's innovation facilitators? *Proceedings of the Australian Society of Sugar Cane Technologists*, Vol. 22, pp. 16–22.

Woods, E.J., Cox, P. and Norrish, S. (1997) Doing things differently: the R, D and E revolution? In *Intensive Sugarcane Production: Meeting the Challenges Beyond 2000* (eds B.A. Keating and J.R. Wilson), pp. 469–490. CAB International, Wallingford, UK.

Workman, M., Scott, P.M. and Nixon, D.J. (1986) A review of the management and amelioration of saline/sodic soils at Mhlume (Swaziland) sugar company. *Proceedings of the South African Sugar Technologists' Association*, Vol. 60, pp. 162–167.

Wossink, G.A.A., de Buck, A. J., van Niejenhuis, J.H. and Haverkamp, H.C.M. (1997) Farmer perceptions of weed control techniques in sugarbeet. *Agricultural Systems*, Vol. 55, No. 3, pp. 409–423.

Wright, P.G. (1992) Reducing the usage of cooling water in the factory. *Proceedings of the Australian Society of Sugar Cane Technologists*, Vol. 14, pp. 223–230.

Wrigley, G. (1985) *Tropical Agriculture: the Development of Production*, 4th edn. English Language Books Society/Longman, Harlow.

Wu, C.Y. (1996) The recycling of pig wastes in TSC's farms. *Taiwan Sugar*, Vol. 43, No. 5, p. 21.

Wynne, A.T. and Meyer, J.H. (2002) An economic assessment of using molasses and condensed molasses solids as a fertilizer in the South African sugar industry. *Proceedings of the Annual Congress – South African Sugar Technologists' Association*, No. 76, pp. 71–78.

Yadav, R.L. (1986) Water management in sugarcane. *Indian Sugar Crops Journal*, Vol. 12, No. 2, pp. 1–4.

Yadav, R.L., Prasad, S.R., Ramphal Singh and Srivastava, V.K. (1994) Recycling sugarcane trash to conserve soil organic carbon for sustaining yields of successive ratoon crops in sugarcane. *Bioresource Technology*, Vol. 49, No. 3, pp. 231–235.

Yamamoto, Y. and Sukchan, S. (2002) Land suitability analysis concerning water resource and soil property. *JIRCAS Working Report*, No. 30, pp. 25–31.

Yang, P.Y., Chang, L.J. and Whalen, S.A. (1991) Anaerobic/aerobic pretreatment of sugarcane mill wastewater for application of drip irrigation. *Water Science and Technology*, Vol. 24, No. 9, pp. 243–250.

Yang LiTao and Li YangRui (1995) Studies on the effect of late irrigation on sugarcane yield, sucrose content and some physiological and biochemical characters of different sugarcane cultivars under drought conditions. *Acta Agronomica Sinica*, Vol. 21, No. 1, pp. 76–82.

Yonts, C.D., Palm, K.L. and Reichert, D.L. (1999) Late season irrigation management for optimum sugarbeet production. In *ASAE/CSAE-SCGR Annual International Meeting, Toronto, Ontario, Canada, 18–21 July 1999*, 10 pp. American Society of Agricultural Engineers, St. Joseph.

Young, G.R. and Kuniata, L.S. (1992) Life history and biology of *Sesamia grisescens* Walker (Lepidoptera: Noctuidae), a sugarcane borer in Papua New Guinea. *Journal of the Australian Entomological Society*, Vol. 31, pp. 199–204.

Zabaleta, J.M.T. (1997) Environmental concerns and self-regulation: the Philippine sugar industry experience. *Indian Sugar*, Vol. 47, No. 8, pp. 551–575.

Zadoks, J.C. and Schein, R.D. (1979) Southern corn leaf blight. In *Epidemiology and Plant Disease Management*, pp. 331–335. Oxford University Press, New York.

Zadrazil, H. (1990) Drag line irrigation: practical experiences with sugar cane. *Agricultural Water Management*, Vol. 17, No. 1–3, pp. 25–35.

Zajac, V. (2000) Environmental upgrading of a sugar beet factory in Opava. *Listy Cukrovarnicke a Reparske*, Vol. 116, No. 7/8, pp. 202–203.

Zak, S., Kovac, K., Lehocka, Z. and Babulicova, M. (2002) Agricultural and environmental effects of various sources of nitrogen in sugar beet cultivation. *Listy Cukrovarnicke a Reparske*, Vol. 118, No. 1, pp. 15–21.

Zaldivar, J., Nielsen, J. and Olsson, L. (2001) Fuel ethanol production from lignocellulose: a challenge for metabolic engineering and process integration. *Applied Microbiology and Biotechnology*, Vol. 56, No. 1/2, pp. 17–34.

Zamora, A., Elias, A. and Lara, A. (1996) Effect of the level of Gicabu in the performance of grazing dairy heifers. *Cuban Journal of Agricultural Science*, Vol. 30, pp. 143–147.

Zarrabi, M. and Ganbalini, G.N. (2001) Bionomics of *Bothynoderes obliquifasciatus* Menetries (Col.: Curculionidae) and its IPM strategy in sugar beet in Iran. *Proceedings of the Institut International de Recherches Betteravières*, Vol. 64, pp. 401–409.

Zaslawskij, M. (1982) Increasing yields on sloping sites and protection against erosion. *Internationale Zeitschrift der Landwirtschaft*, No. 5, pp. 470–472.

Zauner, E. (1988) Anaerobic treatment of solid wastes from agriculture and agroindustry. In *Agricultural Waste Management and Environmental Protection. Proceedings of the 4th International CIEC (International Scientific Centre of Fertilizers) Symposium held in Braunschweig, German Federal Republic, 11–14 May 1987*, Vol. 2, pp. 103–108. CIEC, Gottingen.

Zawislak, K. and Rychcik, B. (1997) Sugarbeet in crop rotations and monoculture: 20 years of static studies in north-eastern Poland. *Acta Academiae Agriculturae ac Technicae Olstenensis, Agricultura*, No. 64, pp. 227–236.

Zawislak, K. and Tyburski, J. (1992) The tolerance of root, industrial and fodder crops to continuous cultivation. *Acta Academiae Agriculturae ac Technicae Olstenensis, Agricultura*, No. 55, pp. 149–162.

Zeng, S.X. and Liu, Q.H. (1992) A study on optimum irrigation system of sugarbeet in Hetao irrigation area, China. *China Sugarbeet*, No. 3, pp. 19–22.

Zhan, P. and Hu, A. (1989) An investigation of microbes and evaluation of water quality in the zone polluted by sugar-refining wastewater in Harbin Reach of the Songhuajiang River during the freezing seasons. *Chinese Journal of Environmental Science/Huanjing Kexue*, Vol. 10, No. 4, pp. 27–30.

Zhang, Y.Q. (1989) Studies on structure and dynamics of spider communities in sugarcane fields. *Journal of Guangxi Agricultural College*, Vol. 8, No. 1, pp. 19–24.

Zia, M.H., Anwar-ul-Haq, M. and Iftikhar Ahmed (1999) Utilization of brackish groundwater for crops with EM technology. *Pakistan Journal of Agricultural Sciences*, Vol. 36, No. 3/4, pp. 186–188.

Zimmermann, B. and Zeddies, J. (2001) Evolution of sugar beet breeding and its contribution to value added. *Proceedings of the Institut International de Recherches Betteravières*, Vol. 64, pp. 217–230.

Zimmermann, H. (1974) Water consumption of sugar beet on diluvial soil. *Archiv für Acker- und Pflanzenbau und Bodenkunde*, Vol. 18, No. 11, pp. 825–840.

Ziska, L.H. and Bunce, J.A. (1997) Influence of increasing carbon dioxide concentration on the photosynthetic and growth stimulation of selected C4 crops and weeds. *Photosynthesis Research*, Vol. 54, No. 3, pp. 199–208.

Zund, P.R. and McDougall, A. (1997) Using soil survey data to estimate readily available water (raw) for sugarcane irrigation. *Proceedings of the Australian Society of Sugar Cane Technologists*, Vol. 19, pp. 310–314.

Appendices

Appendix 1. Recent examples of published research on the responses of sugarcane to management variables including irrigation.

Source	Locality	Examining	Varieties	Comments
Mubien (1996)	Subang sugar factory, Indonesia	Sprinkler irrigation	BZ 132	Optimum irrigation rate: 3.95–3.99 mm/day
Pene *et al.* (1997)	Northern and central regions of Côte d'Ivoire	Deficit irrigation	–	Yield losses due to deficit irrigation were greater at later (elongation and yield formation) growth stages than at earlier (tillering) stages
Singh and Mohan (1994)	Dehra Dun, Uttar Pradesh, India	Irrigation and N	COS 767 G.T. 54-9	Yield and irrigation efficiency was greatest at IW : CPE = 1. Up to 200 kg N/ha significantly increased yield. Irrigation and N had no significant effect on juice quality
Yang LiTao and Li YangRui (1995)	–	Late irrigation	Guitang 11; F 172	Various physiological and biochemical characteristics of either variety responded to (lack of) irrigation
Jadhav *et al.* (1997)	Marathwada Agricultural University, Parbhani	Irrigation and N	Co 7219	Irrigation at IW/CPE = 1 gave significantly higher cane yield than lower IW/CPE ratios. Maximum cane yield (150 t/ha) and WUE (661 kg/ha/cm) at 400 kg N/ha
Gulati *et al.* (1995)	Chiplima, Orissa	Levels of irrigation and intercropping with wheat or *Brassica juncea*	Co 6304	Cane yield increased with IW : CPE up to 1. Intercropping decreased cane yield by *c.* 8%. Net return and WUE were greatest with *B. juncea* intercrop
Wiedenfeld (1995)	Lower Rio Grande Valley, Texas	Irrigation and N	CP65-357	Various responses in plant and ratoon crops are reported

continued

Appendix 1. *Continued.*

Source	Locality	Examining	Varieties	Comments
Pene and Dea (2000)	CRNA Research Station, Ferkessedougou, northern Côte d'Ivoire	Irrigation and cane variety	Co 957; Co 997; R 570	Various responses in plant and first ratoon crops are reported
Shafshak *et al.* (2001)	Egypt	Irrigation and N	GT 54-9	Various responses in cane growth characters and juice quality are reported over two field seasons
Ruiz Traba and Delgado Morejon (1996)	Matanzas, Cuba	Irrigation and harvest timing	–	Irrigated yields exceeded unirrigated by 21.4 t/ha after 12 months, but after 24 months the difference had increased to only 27.8 t/ha
Pawar and More (1996)	Rahuri, Maharashtra	Irrigation and N	Co 7219	WUE highest with irrigation at 100 mm CPE. Yield increased with increasing N rate
Hossain *et al.* (1999)	Bangladesh	Irrigation and interplant spacing	Isd 16	Optimum treatment: flood irrigation and 45 cm spacing
Toor *et al.* (1999)	Punjab, India	Irrigation	CoJ 84	Greatest mean cane yield (65.3 t/ha) at IW : CPE = 1.5; not significantly different from yield (62.4 t) at IW : CPE = 0.75
Altaf-ur-Rehman and Said Rehman (1998)	Mardan, Pakistan	Irrigation and N	CP 65/357	Sugar yield increased with up to 150 kg N/ha; juice quality was greatest with no N. Sugar yield decreased with decreasing water availability, while juice quality increased
Bhale *et al.* (2002)	Parbhani, Maharashtra, India	Irrigation and intercropping with groundnut, cowpea or green gram		Sugar equivalent yield and gross monetary return (GMR) greatest in sugarcane/groundnut system. Cane yield, GMR and water expense component greatest at IW : CPE = 1.2; WUE greatest at IW : CPE = 1
Hassan *et al.* (1999)	–	Irrigation and cane variety	ISD 16; ISD 20; ISD 21	Cane yield averaged 64.0 t/ha without irrigation; no significant varietal differences. Greatest cane yield (171.8 t) with first irrigation (10 cm) 35–40 days after transplanting and then at Eo = 200 mm (IW : CPE = 1; four irrigations in total)
Pawar and More (1993)	Maharashtra	Irrigation and N	CO 7219	Cane yield response to water applied (50–125 mm cumulative Eo) and N application (187–312 kg N/ha) was linear

Source	Locality	Examining	Varieties	Comments
Jambulingam *et al.* (2001)	Sugarcane Breeding Institute, Coimbatore, Tamil Nadu, India	Irrigation and cane variety	Co 6806; Co 7201; CoC 671; Co 419; Co 6304; Co 1148	Various responses in relation to cane yield, juice quality and jaggery yield and quality are reported
Miyagi and Shimabuku (1997)	Okinawa Miyako Island	Irrigation at different growth stages	–	Yield and quality responses of sugarcane are reported
Pene *et al.* (2001)	CNRA Research Station, Ferkessedougou, northern Côte d'Ivoire	Irrigation and cane variety	R 570 and two others	Various responses in cane yield and quality in relation to irrigation and crop blooming are reported
Pene and Edi (1999)	Northern Côte d'Ivoire	Deficit irrigation at two growth stages	Co 449	Cane growth/yield much more sensitive to water stress at stem elongation than at tillering
Thanki *et al.* (2000b)	Gujarat Agricultural University, Navsari	Irrigation	–	Moisture stress during juvenile and main growth (but not mature) stages reduced cane yield considerably
Bhoi *et al.* (1999)	Maharashtra	Fertigation and planting (paired rows vs. four rows) under drip irrigation	CO 7219	Mean cane yield highest (171.4 t/ha) with four row planting + 20 splits of N (at 250 kg N/ha). Paired row planting produced similar yield (169.9 t) with 20 splits of N
Jambulingam *et al.* (1999)	Tamil Nadu	Irrigation and cane variety	CoC 671 and five others	Early maturing cv. CoC 671 produced highest sugar yield (17.30 t/ha) with irrigation at 25% ASM. Mid–late cvs produced higher cane yields, but early maturing cvs produced higher sugar yields
Minhas *et al.* (1992)	Tandojam, Pakistan	Delayed first irrigation	BL-4	Delaying irrigation decreased % germination, cane yield and % CSC
Rincones (1990)	–	Irrigation and cane variety	V68-78; PR62258; V64-10; V58-4; B6749; PR61632 and two others	PR61632, B6749, V64-10 and V58-4 showed good performance under irrigation and were recommended for use in areas of reduced water supply
Goudreddy *et al.* (1994)	Agricultural Research Station, Bidar (NE Transitional Zone, Karnataka)	Irrigation and cane variety	Co 7318; Co 7219; Co 740; Co 70A-645; Co 6907 and possibly others	Cane yield and quality responses of test varieties are reported

continued

Appendix 1. *Continued.*

Source	Locality	Examining	Varieties	Comments
Elsheemy and Elsayed (1993)	Egypt	Irrigation	–	Optimal trickle irrigation: 9000 m^3 and 1200 m^3 (first/second ratoons, respectively)
Ng Kee Kwong and Deville (1994a)	Belle Vue, Mauritius	Fertigation via drip irrigation	R 570	FUE was significantly higher with fertigation than with fertilizer incorporation
Dalri and Cruz (2002)	São Paulo, Brazil	Subsurface drip irrigation	–	Irrigation increased cane yield by over 45%, but irrigation frequency had no significant effect
Moreira and Cardoso (1998)	Araras, São Paulo, Brazil	Soil moisture content and irrigation	RB785148	Germination decreased with decreasing soil moisture, with no effect of irrigation interval
Nyati (1996)	South-east lowveld, Zimbabwe	Irrigation in years of low rainfall	NCo376	Irrigation increased yield but did not affect quality. 660 mm irrigation gave 69% of yield obtained under full irrigation (1520 mm). Irrigating at Et : Eo = 0.70 to 0.85 could save water without major yield reduction
Anwar *et al.* (2002)	Nawabshah, Sindh, Pakistan	Lodging and preharvest irrigation	–	Lodging and preharvest irrigation of mature cane adversely affect juice quality; irrigation to mature cane should halt a month before harvest
Chapman (1997)	Mackay, Australia	Irrigation	Q124; Q138 and four others	Irrigation increased cane yield, mainly due to increased stalk weight
Agrawal (2001)	Uttar Pradesh, India	Deficit irrigation, manure, cultural and cropping systems	Various	Effects on cane productivity and quality are highlighted
Ah-Koon *et al.* (2000a,b)	Mauritius	Deficit (drip) irrigation	R570	Cane yield declined with reduced irrigation, but could be optimized by applying reduced (0.50 ETc) irrigation over a large area versus full (1.0 ETc) irrigation over a small area
Muchow *et al.* (2001)	Ord, Australia	Irrigation scheduling	Q99	Less frequent irrigation is recommended, particularly during late growth
Ng Cheong *et al.* (1996)	Mauritius	Irrigation scheduling	–	Furrow irrigation every 20 days required more water than 10-day intervals; yield was greater with 10- than with 20-day intervals for the same volume of water; a larger area can be irrigated using the 10-day cycle

Source	Locality	Examining	Varieties	Comments
Others				
Oriol *et al.* (1995)	Morondava valley, Madagascar	Irrigation scheduling		The use of tensiometers for irrigation scheduling is discussed

See footnotes to Appendix 4 for abbreviations.

Appendix 2. Recent examples of published research on the responses of sugar beet to management variables including irrigation.

Source	Locality	Examining	Varieties	Comments
Azzazy (1998)	Shandaweel Research Station, Sohag, Egypt	Sowing date, irrigation interval and N	–	Root yield greater with 7- than with 14-day irrigation intervals in one of two growing seasons. Increasing N increased root yield, but decreased % sucrose in one of two growing seasons
Ciavatta and Cecchini (2002)	Emilia Romagna, Italy	Fertilizers and irrigation	–	Various impacts on yield are reported
Bonari *et al.* (1992)	North-central Italy	Drip irrigation	–	Root yield increased with increasing irrigation rate, but sugar content decreased; no significant difference between sugar yields at 60, 80 or 100% ET
Mambelli *et al.* (1992)	Cadriano, Bologna, Italy	Drip irrigation	Monofort; Kawegiga mono	Irrigation generally increased root yield; various impacts on quality parameters are reported over two growing seasons
Asad *et al.* (2000)	Badjgah and Kooshkak, Fars, Iran	Irrigation levels, N and time of N	–	Longer irrigation intervals decreased root yield and increased WUE, but had no effect on sugar content. Increasing N increased root yield but decreased sugar content at one of two sites. Timing of N application had no predictable effect
Borowczak and Grzes (2002a,b)	Poland	Irrigation, foliar fertilization and N	–	Irrigation increased yield of root, leaves and seed. Effects of fertilizer application in relation to irrigation are reported
Podstawka-Chmielewska and Malicki (1997)	Chelm, Poland	Cultivation intensity and sprinkler irrigation	–	Simplification of cultivation decreased leaf mass and increased protrusion of root tops above soil surface, but did not affect root or sugar yield if full cultivation was applied to previous crop residues. Sprinkler irrigation, independent of cultivation system, increased root, leaf and sugar yield (most greatly in drought years)
Khafagi and El-Lawendy (1997)	–	Irrigation interval	Sucropoly	Decreasing irrigation frequency led to decreased fresh and dry weight of leaves and roots, leaf number, leaf area, root length and water content of leaf and root, but increased chlorophyll and carotene concentration in leaves

Source	Locality	Examining	Varieties	Comments
Podstawka-Chmielewska and Ceglarek (1995)	Chelm district	Sprinkler irrigation and N	–	Irrigation increased root yield and sugar content, but increases varied considerably between years. Root yield increased with increasing N, but sugar content decreased
Svachula (1993)	Near Prague	Irrigation (long-term trials: 1967–1987)	Dobrovicka A	Irrigation increased root yield, but increases varied considerably between years. Highest average sugar yields in moist years with irrigation
Khan *et al.* (1998a,b)	Sugar Crops Research Institute, Mardan, Pakistan	Irrigation interval and N	KaweTerma; KawePak 294	Root and sugar yields were highest with the greatest application of N, but this reduced juice quality. Yields were higher under 7- or 14 (versus 21)-day irrigation intervals. KawePak 294 was higher yielding than KaweTerma
Gasiorowska (1997)	–	Irrigation and NPK	PN Mono 4	Various response in chemical composition of roots and leaves are reported
Rzekanowski and Rolbiecki (1997)	–	Sprinkler irrigation, fertilizer and sunshine hours	PN Mono 1	Irrigation affected the relationship between root and leaf yield parameters and sunshine hours
Dyakov (1997a,b)	Ruse, Bulgaria	Irrigation	–	Greatest WUE obtained with irrigation to maintain 60% of maximum ET; data for root yield and sugar content are given for a range of irrigation levels
Camposeo *et al.* (2001)	Southern Italy	Drip irrigation scheduling	–	Greatest root yield and best dense juice purity obtained under irrigation at 70% of total available water, with irrigation intervals of 11 days, irrigation volumes of 80 mm and only four irrigation events
Bazza and Tayaa (1999)	Doukkala, Morocco	Irrigation	–	Greatest profit with irrigation at 40 kPa of soil water potential or 60 mm/m of soil water depletion; data for root and sugar yields and WUE are given for various irrigation levels
Kirda *et al.* (1999)	Turkey	Deficit irrigation	–	Withholding irrigation during the ripening stage saved nearly 22% water with no significant yield decrease
Sorella (2000)	Torremaggiore, southern Italy	Trickle irrigation and time of harvest	–	Root and sugar yields decreased with decreasing irrigation rate, and were greatest with latest harvesting date

continued

Appendix 2. *Continued.*

Source	Locality	Examining	Varieties	Comments
Borowczak (1996)	–	Sprinkler irrigation and N (long-term trials: 1985–1992)	PN Mono 1	Irrigation and N enhanced seed yield. Irrigation increased germination by *c.* 2% for the whole period and *c.* 4% in dry years
Caliando *et al.* (1997)	Torremaggiore and San Severo, Italy	Irrigation	–	Root and sugar yields increased with irrigation rate
Buniak *et al.* (1996)	–	Sprinkler irrigation and N	–	Irrigation enhanced root yield; optimum N rate was 120 kg/ha. Increasing N rates reduced sugar and dry matter content. Various quality responses are reported
Massoud and Shalaby (1998); Massoud *et al.* (1999)	Sohag, Egypt	Irrigation interval	Recolta; Marina	Root and sugar yields were not significantly affected by irrigation treatment; WUE increased as irrigation interval increased. Sugar yield was greater in Recolta than in Marina
Urbano *et al.* (2000a,b)	Southern Spain	Irrigation	–	Best yields and quality were obtained with irrigation at 45 cbar
Wang *et al.* (1995)	Davis, California, USA	Irrigation, planting depth and fungicide seed treatment	–	Stand establishment was improved by irrigating before (versus immediately after) sowing, and by shallow (2 versus 4 cm) sowing of fungicide-treated seeds
Urbano *et al.* (1992)	Duero valley, Spain	Trickle irrigation and date of harvest	–	At four of five sites, root and sugar yields were significantly higher with irrigation of 30 and 45 cbar. Harvest date had no effect on root yield but sugar content and yield increased with lateness of harvest
Akinerdem (1992)	Eskisehir, Ankara and Konya, Turkey	Irrigation	–	Plant mortality 20–35% without irrigation, but decreased with increasing irrigation frequency. Five irrigation events gave best root and sugar yields (sugar yield doubled versus unirrigated)
Karczmarczyk *et al.* (1995); Koszanski *et al.* (1995)	Lipki	Sprinkler irrigation and N	PN Mono 4	Root yields increased with irrigation and N. Irrigation and increasing N reduced root sugar content. Other yield and chemical composition data are reported
Kumar (1993)	Sri Ganganagar, Rajasthan	Irrigation	Ramonskaya-06	Generally, sugar content and yield increased with increasing irrigation frequency, whilst impurity levels fell

See footnotes to Appendix 4 for abbreviations.

Appendix 3. Recent examples of published research on the responses of sugarcane to different irrigation methods and other management variables.

Source	Locality	Examining	Varieties	Comments
Ramesh (1998)	Coimbatore	PRP with AFI and trash mulch in unirrigated furrows	Co 8014	PRP with trash mulch in unirrigated furrows gave 7.5% higher cane yield and 11.0% lower water requirement compared with URP with EFI
Bakker *et al.* (1997)	Burdekin River irrigation area, Australia	AFI	–	AFI reduced yield compared to EFI (for the same irrigation frequency), but not when AFI was applied more frequently in response to the crop ET. AFI used less water than EFI, and improved WUE
Hapase *et al.* (1990, 1992)	Three different locations in India	Surface/ subsurface drip with daily/ alternate-day irrigation and PRP	Co 7219	Compared to furrow irrigation, drip systems showed a water saving of 50–55%, yield increase of 12–37%, increase in sugar recovery and a 2.7 times increase in WUE. Surface drip with PC mechanism, PRP and daily irrigation gave highest yield/ha
Azzazy *et al.* (1999)	El-Mattana Agricultural Research Station, Qena, Egypt	Drip irrigation	GT 54-9; F 153; G74-97	Sucrose and purity % were greater under drip than surface irrigation
Murugesan and Natarajan (2000)	Sevathur village, Vellore, Tamil Nadu, India	Turbulent PC Rain Tape irrigation	–	Rain Tape increased yield by 37% and reduced water requirements by 40% compared with conventional practice
Shinde and Jadhav (2000)	Vasantdada Sugar Institute, Pune	PC, Non-PC and in-line drip irrigation	–	PC and in-line drip irrigation used 50% less water, increased cane yield by 17–20% and increased WUE by 2.46 times compared with furrow irrigation
Ahluwalia *et al.* (1998)	Punjab Agricultural University, Ludhiana, India	Drip irrigation	–	At optimum irrigation levels, drip used 38% less water and increased WUE by 61% compared to surface irrigation. Drip induced earlier maturity and gave higher sugar yield, but lower juice extraction %
Kittad *et al.* (1995)	Rahuri, Maharashtra	Drip irrigation, AFI and PRP	Co 7219	Drip with pit planting or PRP increased cane yield by 21 or 14%, respectively, compared with furrow irrigation. Cane yield was lowest with AFI. Juice quality was greatest with drip irrigation/PRP

continued

Appendix 3. *Continued.*

Source	Locality	Examining	Varieties	Comments
Parikh *et al.* (1992)	Navsari, Gujarat	Drip irrigation	Co 6304	Cane yields were higher with drip, which also consumed 8–50% less water than furrow irrigation
Ramesh *et al.* (1994)	Coimbatore, Tamil Nadu	Surface/ subsurface (biwall) drip with daily/alternate-day irrigation	Co 6304; Co 8021	Subsurface drip increased cane yield parameters relative to surface drip or furrow irrigation in plant and ratoon crops
Pandian *et al.* (1992)	Cuddalore, Tamil Nadu	AFI, mulch and hydrophilic polymer (Jalshakti)	CoC 85061	Trash mulch in all rows with AFI gave highest cane yield; sugar yield was highest with EFI. WUE increased by *c.* 43–66% under AFI versus EFI, and was highest with mulch in all rows. Jalshakti amendment/EFI gave the lowest WUE
Sharma and Verma (1996)	Sehore, Madhya Pradesh	AFI, mulch and hydrophilic polymer (Jalshakti)	Co 7318	Jalshakti amendment with EFI at IW : CPE = 0.8 gave the highest mean cane yield and the highest sucrose content
Shrivastava *et al.* (1993)	Navsari, Gujarat	Sprinkler irrigation	Co 63304	Sprinkler irrigation at IW : CPE = 0.45 gave the highest yield and used 37.5% less water than furrow irrigation
Raskar and Bhoi (2001)	Rahuri, Maharashtra, India	Drip irrigation, fertigation, PRP and four row planting	–	Drip increased juice quality, cane yield by 20–30%, and reduced total water use by 42–52% versus surface irrigation. Juice quality increased with increasing levels and number of splits of fertigation
Kalaisudarson *et al.* (2002)	Annamalai University, Annamalainagar, Tamil Nadu, India	Irrigation level, mulching and phosphobacteria	CoG 93076	Irrigation at IW : CPE = 1 with 10 cm thick trash mulch increased cane yield; irrigation at IW : CPE = 0.50 reduced yield. Application of phosphobacteria with 100% recommended phosphorus favourably influenced cane yield
Durai *et al.* (1996)	Cuddalore, Tamil Nadu	Irrigation scheduling and soil amendments (coir waste; farmyard manure; pressmud (filter cake); hydrophilic polymer (Jalshakti))	CoC 671	Irrespective of irrigation regime, incorporation of 25 t/ha coir waste gave higher cane yield than other amendments. Jalshakti gave the next highest yield. Irrigation by recommended practice (every 10 days) gave the highest cane yield, but water consumption was higher and WUE lower than when irrigation was scheduled by IW : CPE or soil water depletion

Source	Locality	Examining	Varieties	Comments
Ved Singh (2001)	Rajasthan, India	Irrigation type and scheduling	CO 66-17	Cane yields greatest under irrigation in flat plot or AFI in rotation (odd/even) at IW : CPE = 0.9 or 1.2. Greatest WUE (and water saving of 30.3%) recorded under AFI in rotation (odd/even) compared to irrigation in flat plot and EFI
El-Debaby *et al.* (1996)	El-Matana, Egypt	Trickle irrigation and N	GT 54-9; F153; G74-96	Some cane growth parameters were greater under trickle than under furrow irrigation, others unaffected. Cane growth increased with increasing N, and was generally greatest in in cv. G.T. 54-9
El-Geddawy *et al.* (1996)	El-Matana, Egypt	Trickle irrigation and N	GT 54-9; F153; G75-96	Cane and sugar yields were highest in cv. GT 54-9, with trickle irrigation and maximum N rate. Juice quality was unaffected by irrigation system and cultivar, but declined with increasing N
Shinde and Jadhav (1998)	Pune, Maharashtra	Surface and subsurface irrigation, fertigation and mulching	–	Automatically controlled drip used up to 56% less water, increased yield by up to 52% and increased WUE by about 2.5–3-fold compared with conventional methods. Fertilizer rates could be reduced by drip fertigation. Mulch further decreased water use by 16%
Ghugare *et al.* (1994a,b)	Rahuri, Maharashtra	Drip irrigation, AFI and PRP	Co 7219	Plant crop juice quality was best with AFI, followed by drip irrigation. Ratoon cane yield, CSC and juice quality were greater with drip than with furrow irrigation
Sundarsingh *et al.* (1995)	Madurai, Tamil Nadu	AFI and coir pith amendment	CoC 671	Cane yield higher at IW : CPE = 0.90 than 0.75, higher under EFI than under AFI, and highest with 30 t/ha coir pith
El-Geddawy *et al.* (1997)	El-Mattana Research Station, Qena, Egypt	Drip irrigation, cane variety and N	GT54-9; F153; G74-96	Cane and sugar yield greater under drip than furrow irrigation, and increased with increasing N
Maliwal *et al.* (1999)	Khandha, Gujarat	AFI and mulch	CO 6304	Cane yield and net income greatest with EFI at IW : CPE = 0.9 and with trash mulch
Malavia *et al.* (1992)	Junagadh, Gujarat	Irrigation scheduling and mulch	Co 6304; CoC 671	Cane yield greater with trash mulch; sugar content unaffected by treatments

continued

Appendix 3. *Continued.*

Source	Locality	Examining	Varieties	Comments
Thanki *et al.* (2000a)	Gujarat	Irrigation scheduling, mulch and hydrophilic polymer (Jalshakti)	Co 6304	Cane yield significantly higher with irrigation at IW : CPE = 0.8 than 0.5; irrigation + Jalshakti produced the highest cane yield
Thanki *et al.* (1999)	Gujarat Agricultural University, Navsari, India	AFI and mulch	–	AFI with mulch reduced irrigation water use by 36% and gave yields similar to those under EFI

See footnotes to Appendix 4 for abbreviations.

Appendix 4. Recent examples of published research on the responses of sugar beet to different irrigation methods and other management variables.

Source	Locality	Examining	Varieties	Comments
Narang *et al.* (1992)	Kheri, Indian Punjab	Flood irrigation, EFI, AFI and N	Ramonsakaya-06	Root yields greatest under flood irrigation, then EFI, then AFI, and increased with increasing N
Yonts *et al.* (1999)	Nebraska, USA	Reduced late-season sprinkler and furrow irrigation	–	Yield parameters not influenced by water stress due to reduced or no irrigation late in the growing season
Arroyo *et al.* (1999)	Spain	Drip and sprinkler irrigation	Oryx	Crop yield did not vary significantly between irrigation systems at 90% Eo minus precipitation (P), but drip gave higher yields at 70 and 50% Eo – P. Highest sugar content at 90% Epan – P for drip and at 70% for sprinkler irrigation
Sepaskhah and Kamgar-Haghighi (1997)	Iran	AFI	–	AFI at 6-day intervals produced similar root yield to EFI at 10-day intervals, but used 23% less water. WUE was 43% higher for more frequent AFI versus less frequent EFI
Cukaliev and Iljovski (1993)	–	Micro-sprinkler and sprinkler irrigation	Al-omona	Highest yield and sugar content obtained with micro-sprinkler at 50–70% relative humidity (RH) versus sprinkler and unirrigated treatments
Tognetti *et al.* (2002)	Molise, southern Italy	Drip and low-pressure sprinkler irrigation and date of harvest	–	Increasing irrigation up to 100% of estimated ET gave improved root yield and sucrose accumulation. Drip was adjudged better than low-pressure sprinklers for sugar beet production and sugar quality in semi-arid environments. Growers should harvest in advance of traditional dates
Eckhoff and Bergman (2001)	Montana, USA	Low-pressure sprinkler irrigation	–	Higher sucrose content, root and sucrose yields, lower impurities and greater extraction under furrow-flood irrigation than under sprinkler irrigation

continued

Appendix 4. *Continued.*

Source	Locality	Examining	Varieties	Comments
Sharmasarkar *et al.* (2001b)	Wyoming, USA	Drip and flood irrigation	–	Greater root yields and sugar content under drip than flood irrigation; increasing trend in yields with increasing N. Drip used less water and resulted in higher WUE and FUE than flood irrigation. Beet production could be sustained with lower water and fertilizer use by using drip irrigation
Ruzsanyi (1996)	Hungary	Furrow and reel irrigation	Gisella; Emma; Astro; Magda; Hilma	Root yield greatest with 7 irrigations of 25 mm water under furrow (versus 5 or 3 applications giving a total of 180 mm via reel irrigation). Reducing number of irrigations or total amount of water applied reduced yields. Gisella and Emma gave the highest yield responses to irrigation

AFI, alternate furrow irrigation, skip furrow irrigation; ASM, available soil moisture; CSC, commercial sugar content; EFI, every furrow irrigation (conventional practice); Eo, evaporation from an open Class A pan; FUE, fertilizer use efficiency; IW : CPE (or IW/CPE), irrigation water to cumulative pan evaporation ratio; PRP, paired row planting; PC, pressure compensation; URP, uniform row planting (conventional practice); WUE, water use efficiency.

Index